Atmospheric Science and Meteorology

Atmospheric Science and Meteorology

Edited by Reina Phillips

SYRAWOOD
PUBLISHING HOUSE

New York

Published by Syrawood Publishing House,
750 Third Avenue, 9th Floor,
New York, NY 10017, USA
www.syrawoodpublishinghouse.com

Atmospheric Science and Meteorology
Edited by Reina Phillips

International Standard Book Number: 978-1-64740-404-8 (Hardback)

Cataloging-in-publication Data

Atmospheric science and meteorology / edited by Reina Phillips.
    p. cm.
Includes bibliographical references and index.
ISBN 978-1-64740-404-8
1. Meteorology. 2. Atmosphere. 3. Atmospheric physics. I. Phillips, Reina.
QC861.3 .M48 2023
551.5--dc23

TABLE OF CONTENTS

PREFACE

Meteorology is the science concerned with the atmosphere and its associated phenomena, including weather and climate. Atmosphere is a protective layer of gases that shelters all life forms present on the Earth. It consists of five layers including exosphere, thermosphere, mesosphere, stratosphere, and troposphere. At a broad level, atmosphere can be divided into two parts, namely, lower atmosphere and upper atmosphere. Lower atmosphere consists of the troposphere and lower stratosphere, where most of the weather phenomena occur. The lower atmospheric meteorology studies weather, climate and other aspects of the atmosphere. The role of trace gases in stratosphere and their influence on the photochemistry of ozone, and the variations of winds in the thermosphere are studied under upper atmospheric meteorology. This book provides comprehensive insights into atmospheric science and meteorology. The various studies that are constantly contributing towards advancing technologies and evolution of this field are examined in detail. Scientists and students actively engaged in this field will find it full of crucial and unexplored concepts.

The information contained in this book is the result of intensive hard work done by researchers in this field. All due efforts have been made to make this book serve as a complete guiding source for students and researchers. The topics in this book have been comprehensively explained to help readers understand the growing trends in the field.

I would like to thank the entire group of writers who made sincere efforts in this book and my family who supported me in my efforts of working on this book. I take this opportunity to thank all those who have been a guiding force throughout my life.

Editor

The Impact of Ensemble Meteorology on Inverse Modeling Estimates of Volcano Emissions and Ash Dispersion Forecasts: Grímsvötn 2011

Natalie J. Harvey [1,*], Helen F. Dacre [1], Helen N. Webster [2,3], Isabelle A. Taylor [4], Sujan Khanal [5], Roy G. Grainger [4] and Michael C. Cooke [2]

[1] Department of Meteorology, University of Reading, Reading RG6 6BB, UK
[2] Met Office, FitzRoy Road, Exeter EX1 3PB, UK
[3] College of Engineering, Mathematics and Physical Sciences, University of Exeter, Exeter EX4 4QF, UK
[4] COMET, Sub-Department of Atmospheric, Oceanic and Planetary Physics, University of Oxford, Oxford OX1 3PU, UK
[5] Sub-Department of Atmospheric, Oceanic and Planetary Physics, University of Oxford, Oxford OX1 3PU, UK
* Correspondence: n.j.harvey@reading.ac.uk

Abstract: Volcanic ash can interact with the earth system on many temporal and spatial scales and is a significant hazard to aircraft. In the event of a volcanic eruption, fast and robust decisions need to be made by aviation authorities about which routes are safe to operate. Such decisions take into account forecasts of ash location issued by Volcanic Ash Advisory Centers (VAACs) which are informed by simulations from Volcanic Ash Transport and Dispersion (VATD) models. The estimation of the time-evolving vertical distribution of ash emissions for use in VATD simulations in real time is difficult which can lead to large uncertainty in these forecasts. This study presents a method for constraining the ash emission estimates by combining an inversion modeling technique with an ensemble of meteorological forecasts, resulting in an ensemble of ash emission estimates. These estimates of ash emissions can be used to produce a robust ash forecast consistent with observations. This new ensemble approach is applied to the 2011 eruption of the Icelandic volcano Grímsvötn. The resulting emission profiles each have a similar temporal evolution but there are differences in the magnitude of ash emitted at different heights. For this eruption, the impact of precipitation uncertainty (and the associated wet deposition of ash) on the estimate of the total amount of ash emitted is larger than the impact of the uncertainty in the wind fields. Despite the differences that are dominated by wet deposition uncertainty, the ensemble inversion provides confidence that the reduction of the unconstrained emissions (*a priori*), particularly above 4 km, is robust across all members. In this case, the use of posterior emission profiles greatly reduces the magnitude and extent of the forecast ash cloud. The ensemble of posterior emission profiles gives a range of ash column loadings much closer in agreement with a set of independent satellite retrievals in comparison to the *a priori* emissions. Furthermore, airspace containing volcanic ash concentrations deemed to be associated with the highest risk (likelihood of exceeding a high concentration threshold) to aviation are reduced by over 85%. Such improvements could have large implications in emergency response situations. Future research will focus on quantifying the impact of uncertainty in precipitation forecasts on wet deposition in other eruptions and developing an inversion system that makes use of the state-of-the-art meteorological ensembles which has the potential to be used in an operational setting.

Keywords: dispersion modeling; volcanic ash; inversion modeling; meteorological ensemble; wet deposition; risk framework

1. Introduction

Volcanic ash, released into the atmosphere when a volcano explosively erupts, provides a significant hazard to aircraft as it can cause engines to malfunction and visibility can be reduced by external corrosion. It can also cause permanent engine damage which leads to high maintenance costs [1]. Grounding and re-routing of aircraft during an eruption also comes with a large economic cost. For example, the eruption at the summit of the Icelandic volcano Eyjafjallajökull in April 2010 disrupted European airspace for 13 days, grounded over 95,000 flights and is estimated to have cost the airline industry over £1 billion [2]. The aim of this paper is to present a method that optimally combines satellite retrievals and ensemble numerical weather prediction simulations to produce improved ash forecasts that can be used by the aviation industry during future volcanic eruptions.

In the event of an eruption, one of the nine worldwide Volcanic Ash Advisory Centers (VAACs) issue hazard maps forecasting horizontal ash coverage in three vertically integrated layers of the atmosphere at 6 h intervals to a maximum forecast lead time of 18 h. The ash forecasts show the maximum expected extent of the ash cloud within 3 flight level ranges but contain no quantitative information about ash concentration. These forecasts aid the aviation community in making decisions to minimize the risk of encountering ash during their flight operations. Before the 2010 Eyjafjallajökull eruption, the International Civil Aviation Organization (ICAO) guidelines were to avoid all ash [3]. However due to the unprecedented disruption in 2010, the United Kingdom's Civil Aviation Authority (CAA), in consultation with Rolls-Royce, the UK Met Office, international and European regulators, and aviation experts developed quantitative peak concentration limits [4,5]. The UK Met Office, who host the London VAAC, also began producing quantitative peak concentration forecasts for the North Atlantic and European areas during an eruption using the Numerical Atmospheric-Dispersion Modeling Environment (NAME), a Lagrangian particle dispersion model. The concentration thresholds that are currently used in the production of these forecasts are high (>4000 µg m^{-3}), medium (2000–4000 µg m^{-3}) and low (200–2000 µg m^{-3}) ash contamination. Before aircraft are permitted to fly in regions of medium and high ash contamination, operators are required to have a safety risk assessment approved by their national aviation authority [6,7]. ICAO guidelines are periodically under review and therefore there is the potential for procedural change in the future. A possible change is the need for all VAACs to issue quantitative ash concentration forecasts with predefined concentration thresholds. This is currently difficult due to the large uncertainty associated with volcanic emissions, particularly for remote locations with limited in situ observations.

The forecasting of ash location following an eruption is performed by the VAACs using volcanic ash transport and dispersion (VATD) models and is strongly dependent on information about the eruption that is used for initialization e.g., [8–10]. Information about the height of the plume, the mass eruption rate, the eruption duration and the particle size distribution is required. Eruption duration can be monitored by satellite or using estimates from observers in the locality and there are a variety of techniques to estimate the height of the plume e.g., [11]. Empirical relationships based on past eruptions e.g., [12,13] are typically used to estimate a mass eruption rate from the reported plume height. These empirical relationships do not account for the influence of the meteorological situation (e.g., wind bent plumes [14]) and a uniform vertical distribution of ash from the volcano vent to the reported plume height is typically assumed by the London VAAC.

VATD models also use meteorological information as input, including forecast 3-dimensional wind fields, precipitation and meteorological cloud location from numerical weather prediction models as input. This meteorological information governs the transport and dispersion of the ash cloud and the removal of ash from the atmosphere via deposition. The effect of smaller scale motions, not represented in the input meteorology, is parametrized within the VATD model. Current operational VATD models use only one realization of the meteorological situation, known as a deterministic forecast. Uncertainties in the synoptic situation are therefore not considered in the resulting forecasts and this can result in errors in the forecast ash cloud position [8]. This variability can be represented if an ensemble of meteorological forecasts are used as input for VATD models. Despite being advocated by the volcanic ash community as a way for accounting for wind and precipitation uncertainty [15,16], ensemble meteorology is not routinely used to produce ash forecasts. Barriers to the use of the state-of-the-art ensemble science include the need for forecasts to be produced within a short time window and the ability to present ensemble results in a way that can be easily used by decision makers [17].

A small number of studies have investigated the impact of ensemble meteorology on volcanic ash forecasts. Dare et al. [18] found that there is better agreement with satellite observations if an ensemble is used compared to a single realization of the meteorological situation for lead times greater than 12 h. More recent work by Zidikheri et al. [19] found that using an ensemble of dispersion model simulations with different meteorological fields obtained from an ensemble meteorological forecast model and different values of ash source parameters gave increased Brier skill scores at all lead times compared to a deterministic forecast. This is relevant as the current volcanic ash advisories and graphics produced by the VAACs are issued out to 18 h. Studies by Stefanescu et al. [20] and Madankan et al. [21] concluded that at longer lead times (48 h) there can be a large spread in predicted ash concentrations within forecasts made using ensemble meteorology. This large spread often occurs when ash particles encounter regions of large horizontal flow separation in the atmosphere. Nearby ash particle trajectories can rapidly diverge, leading to a reduction in the forecast accuracy of deterministic forecasts that do not represent uncertainty in wind fields at the synoptic scale [22]. Precipitation uncertainty can also impact ash forecast accuracy. In the case study presented in Langmann et al. [23] it was found that wet deposition, the removal of ash through microphysical processes, can remove up to 23% of ash, consistent with [18] which found that wet deposition removed 1–30% of ash, depending on the season.

Satellite observations show the extent of the ash cloud and can give estimates of ash column loading, ash cloud top height and effective ash radius e.g., [24,25]. Satellite retrievals can be combined with a dispersion model using an inversion technique to give time-evolving estimates of mass eruption rate, and the vertical distribution of the ash emissions from the eruptive plume. These quantities are not directly retrievable from satellites. There are numerous published methodologies that use inversion modeling to estimate ash [26–31] and sulfur dioxide (SO_2) [32–38] source parameters for volcanic eruptions. For example, the technique described in [29] is used in an operational framework by the London VAAC, where posterior estimates of emissions can be determined in near-real time as more satellite retrievals become available. However, all these methods make use of deterministic meteorological information, therefore errors in the wind fields or location of precipitation will lead to errors in the estimated source emissions.

This study brings together inversion modeling and the use of an ensemble of meteorological forecasts to give an ensemble of the most probable source emission estimates of volcanic ash that will undergo long range transport and therefore a robust ash forecast constrained by observations. This follows work by Zidikheri et al. [19] that addresses the problem of verifying and calibrating ensemble-based probabilistic volcanic ash forecasts using dispersion model simulations with different meteorological fields obtained from an ensemble meteorological forecast model and different values of ash source parameters. Zidikheri and Lucas [39] use an inversion modeling procedure involving an ensemble of meteorological forecasts to estimate fine ash mass emission rates and other source parameters for 14 eruption case studies. The ensemble forecasts used here as meteorological input to NAME are produced using the European Centre for Medium-Range Weather Forecasts (ECMWF) ensemble prediction system [40]. The NAME dispersion simulations will be used with satellite retrievals in the Inversion Technique for Emission Modeling (InTEM) system [29] that has been developed at the UK Met Office, to produce an ensemble of volcanic ash source emission terms. The impact of using the ensemble of emission source profiles on the resulting forecasts of ash location will be investigated for the 2011 Grímsvötn eruption, with a particular focus on regions where medium and high levels of ash contamination are predicted as these are areas that aircraft may be prohibited from entering.

The methods used to create the ash forecasts presented in this study are described in Section 2. Section 3 describes the details of the 2011 Grímsvötn eruption, the meteorological situation, the satellite observations of the ash and SO_2 cloud and the issued VAAC graphics. The impact of the use of ensemble meteorology on volcanic emission estimates and on flight planning decisions is presented in Sections 4 and 5. The summary and conclusions of the study are in Section 6.

2. Methods and Data

2.1. Ensemble of Meteorological Forecasts

In this study, NAME is driven by a set of bespoke ensemble meteorological datasets over the Grímsvötn eruption period produced using the ECMWF Integrated Forecast System (IFS) (cycle 41r1). To account for uncertainty in the initial meteorological fields and therefore the resulting numerical weather prediction (NWP) forecasts, the ECMWF IFS Ensemble Prediction System (EPS) was used to produce a 20-member meteorological ensemble. The EPS uses the singular-vector approach [41] to perturb initial conditions in the meteorology and a stochastic physics scheme [42] to account for model uncertainty. These global forecasts are initialized every 24 h between 0000 UTC 21 May and 0000 UTC 26 May 2011 and have a forecast lead time of 72 h. Data are extracted from the ECMWF archive at $0.25° \times 0.25°$ on a regular latitude/longitude grid, and the precipitation, surface stresses and sensible heat flux fields were post-processed so they can be used as input for NAME.

2.2. Satellite Observations

2.2.1. SEVIRI

The Spinning Enhanced Visible and InfraRed Imager (SEVIRI) is mounted on the geosynchronous Meteosat Second Generation (MSG) satellite. It has 12 spectral channels and provides high temporal (15 min) and spatial (3 km resolution at the equator) observations. The high temporal and spatial resolution makes these observations ideally suited to evaluate the transport of volcanic ash following an eruption. The volcanic ash measurements used in this paper are retrieved using the algorithm of [24] which uses three long-wave window channels centered at 8.7, 10.8, and 12.0 μm to discriminate between meteorological cloud and ash cloud. Where ash is detected this algorithm determines the ash column loading. These pixels are flagged as containing ash. If a pixel is free from both ash and meteorological cloud then it is flagged as a clear sky pixel. Pixels that neither have detectable ash nor are flagged as clear skies are unclassified.

Further processing is performed to regrid the retrieved column loadings on to a grid of $0.375°$ latitude by $0.5625°$ longitude (approximately 40 km $\times$ 40 km in mid-latitudes) and averaged over 1 h. This is to match the resolution of the NAME ash concentration output and to reduce data volumes. If 50% or more satellite pixels in a grid box contain ash or more than 90% of pixels are classified as ash or clear skies, then the grid box is selected for use in the InTEM inversion. If all classified pixels within a grid box are flagged as clear sky pixels then the grid box is deemed to be a clear sky observation. Otherwise, the grid box is deemed to be an ash grid observation with the column loading in this grid box given by the mean of all the classified pixels (including clear skies). More information about the processing of the SEVIRI retrievals can be found in [43]. These processed retrievals are used with InTEM as an observational constraint on the inverted emissions profile.

2.2.2. IASI

The Infrared Atmospheric Sounding Interferometer (IASI) is in a sun-synchronous polar orbit on Metop-A, Metop-B and Metop-C [44]. It has a swath width of 2200 km with 12 km circular pixels at nadir. It achieves near global coverage every 12 h. It has 8461 channels with wavelengths in 645 to $2760\ cm^{-1}$ which covers 3 SO_2 absorption bands [45] and the broad v-shaped absorption feature associated with volcanic ash at 750 and 1250 cm^{-1} [46]. There are several retrievals for SO_2 and ash developed for the IASI instrument e.g., [46–53]. The retrievals of SO_2 and ash for the Grímsvötn eruption presented in Figure 1 are determined using optimal estimation schemes developed by [50,52].

The ash retrieval used an ash refractive index measured by Reed et al. [54] from a sample of ash from the Grímsvötn eruption. IASI retrievals of ash and SO_2 are used in the description of the Grímsvötn 2011 case study investigated in this paper (Section 3).

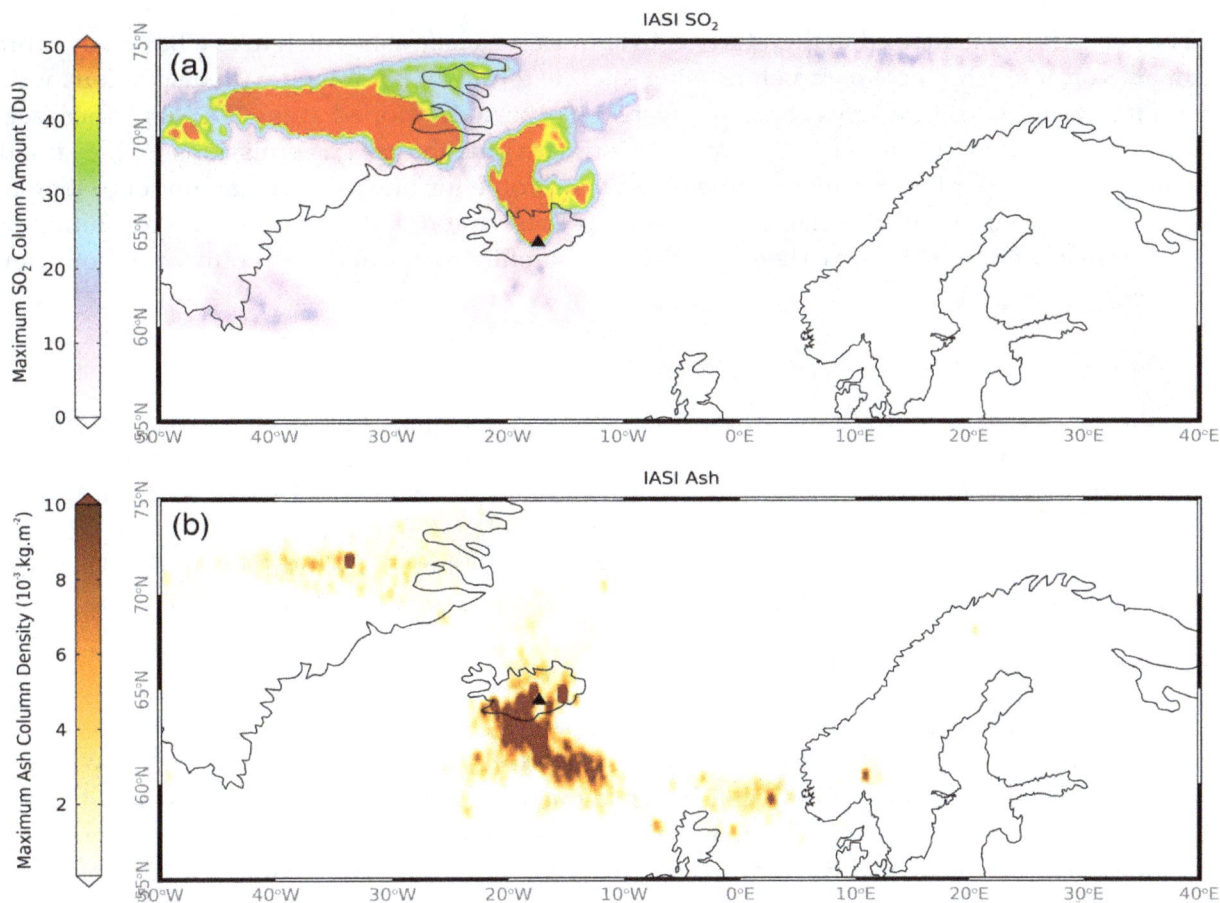

Figure 1. (**a**) Maximum observed SO_2 column amount and (**b**) Maximum observed ash column loading for the period between 21 and 26 May 2011 inclusive. The IASI retrieval outputs were gridded for each orbit and, similar to Moxnes et al. (2014), the maximum value observed during all the overpasses in the given period is shown.

2.2.3. MODIS

The Moderate Resolution Imaging Spectroradiometer (MODIS) is mounted on the NASA Terra and Aqua satellites. It is in a near-polar, sun-synchronous orbit. It has 36 spectral channels in the visible to infrared range (0.4 μm to 14.4 μm wavelength), a spatial resolution of 250–1000 m and global coverage every 1–2 days [55]. Here the Aqua MODIS calibrated and geolocated radiances data are used [56] and the Optimal Retrieval of Aerosol and Cloud (ORAC) retrieval algorithm is used to determine volcanic ash properties [57,58]. MODIS ash retrievals during the Grímsvötn eruption are used as independent satellite retrievals to evaluate NAME simulations of ash location and column loading.

2.3. VATD Model: NAME

To simulate the dispersion of volcanic ash, the VATD model NAME [59] was used. NAME includes parameterizations of turbulence, sedimentation, dry deposition and wet deposition, which are required to simulate the dispersion and removal of volcanic ash. Ash particles are typically assumed to have a density of 2300 kg m^{-3}, (although in reality density depends on chemical composition, porosity and grain size). Please note that aggregation of ash particles, near source plume rise and processes driven by the eruption dynamics (e.g., [14]) are not explicitly modeled in the operational configuration used by the London VAAC [16]. The particle size distribution used is based on data from [60].

2.4. Inversion System: InTEM

The inversion system used in this study is InTEM for volcanic ash which uses a Bayesian approach to estimate volcanic ash source parameters using satellite retrievals combined with dispersion modeling and an *a priori* estimate of the emission. This system has been developed at the UK Met Office (see [29] for full details) and was originally developed to estimate greenhouse gas emissions [61]. Using these input data, it provides a best estimate of the emissions profile for fine ash that can undergo long range dispersion. This emission profile has a chosen vertical resolution of 4 km and a time resolution of 3 h. The emissions profile can either be determined using satellite retrievals of ash only or of both ash and clear skies.

2.4.1. VATD Simulations

NAME simulations representing a nominal release rate (1 g s^{-1}) from each possible source term component (4 km height range and 3-hourly time period) are conducted. Model predictions of ash column loads can be easily determined for an arbitrary emission profile by a linear combination of these nominal simulations. Only ash particles with diameters less than 30 μm are included in these simulations. This assumption is similar to that made in Stohl et al. [34] that state SEVIRI retrievals have a preferential sensitivity to ash with particle diameters from 2 to 32 μm.

2.4.2. Estimate of the Source Term *a Priori*

An *a priori* estimate of the source term is used to ensure that the inverted source term is not over fitted to the satellite observations but is also guided by known information concerning the eruption (e.g. eruption time, maximum plume height). The mean and error covariance matrix of the *a priori* are estimated using a stochastic model that includes correlations between errors in the *a priori* source term emissions at different heights and times. A uniform emission profile from the volcano vent to the plume height is assumed and errors in the observed plume height (assumed to be ±2 km), in the assumed uniform emission profile and in the empirical relationship used to determine the mass eruption rate for a given plume height [12] are considered. A full description of the determination of the *a priori* source term used in the inversion process is given in [43]. In this study, the *a priori* source term is based on radar-recorded plume height which is combined with expert interpretation of other observations by the Icelandic Met Office State Volcano Observatory to provide an estimation of plume height.

2.4.3. The Inversion Algorithm

The inversion scheme yields a best estimate of the emission profile using the NAME simulations, SEVIRI retrievals and *a priori* estimate described above. The posterior distribution of emissions is Gaussian, and the best estimate of the emissions is taken as the peak of this distribution subject to a non-negative constraint. This best estimate of the emissions can then be used in forecasting the volcanic ash cloud. To find this peak, a cost function, which represents the simultaneous fit of the model ash loadings to the satellite retrievals and of the emissions to the *a priori* estimate, is minimized. The minimization is performed using the Lawson and Hanson non-negative least squares algorithm [62]. This algorithm is iterative (but fast) and converges completely in a finite number of iterations. Webster et al. [63] states that inversion run time for using 88791 satellite observations until 00:00 UTC on 31 May 2011 from the 2011 Grímsvötn eruption was 1 min 5 s. A full description of the inversion method is given in [29,43]. Please note that it is possible that there are significant uncertainties introduced by selecting the peak of posterior distribution but that is not the focus of this study.

3. Grímsvötn 2011: Case Study Description

The case study chosen to demonstrate the ensemble inversion method is the 2011 Grímsvötn eruption. Grímsvötn is located at 64.42° N, 17.33° W and has a vent height of 1719 m above sea level (asl). It is one of Iceland's most active volcanoes and its most recent explosive eruption started at 1913 UTC on 21 May 2011 and had an initial eruptive plume height of 20–25 km asl [11]. The eruption lasted approximately 3 days and ended at 0230 UTC on 25 May 2011. Throughout the eruption period the eruptive plume height varied substantially between 2 and 20 km asl [11]. There is some evidence to suggest that there was a (or several) partial column collapse(s) leading to the compression of the lower part of the ash column, and a mechanism for driving a gravity current of ash outwards from the eruption column [64]. It is possible that this column collapse, which cannot currently be represented in VATD models, led to the SO_2 and ash clouds experiencing different atmospheric wind speeds and directions leading to errors in the forecasting of ash in northern Europe. This separation is clearly seen in Figure 1 which shows the maximum retrieved SO_2 (Figure 1a) and ash column loadings (Figure 1b), from IASI, travelling in different directions away from Grímsvötn.

Figure 2 shows the synoptic situation at 0000 UTC on 22 May 2011. There is a mature low-pressure system to the south-east of Iceland (992 hPa), with another developing cyclone following in the north Atlantic (1005 hPa). At low levels, the flow around these features would transport low level ash and SO_2, emitted from the eruption, south-eastwards towards the UK. Analysis of the Skew-T thermodynamic chart from Keflavik radiosonde station at 0000 UTC on 22 May (not shown) indicates a large amount of vertical wind shear. The upper-level wind direction suggests that ash and SO_2 emitted above approximately 9 km would be transported northwards—the opposite direction to low level material. The meteorological situation remains similar for the duration of the eruption. The deepest system during the eruption had a central pressure of 977 hPa and moved north of Scotland overnight on 23/24 May. It was associated with strong north-westerly winds near the surface and heavy precipitation [65]. Beckett et al. [16] states that NWP forecast errors associated with these low-pressure systems led to the operational NAME simulations forecasting the ash cloud further south than observed and [66] suggests that precipitation over most of northern and central Europe might have caused a significant removal of ash due to wet deposition. The use of ensemble meteorology in this study will enable the investigation of uncertainties in the ash cloud position and in ash column loads associated with errors in the wind and precipitation fields.

Figure 3 shows two volcanic ash advisory graphics for (a) 1800 UTC on 23 May 2011 and (b) 0000 UTC on 25 May taken from the London VAAC: Volcanic ash advisories and graphics archive [67]. These graphics show an ash cloud extending northwards and then eastwards towards the Arctic and another branch extending south, travelling cyclonically towards western Europe. When bringing together the information from the IASI satellite retrievals (Figure 1) and the Met Office surface analysis chart (Figure 2), it seems likely that this north-eastward travelling branch of the forecast cloud may have been dominated by SO_2 emitted at a greater height than the ash. This is consistent with the findings of Moxnes et al. [38] and Cooke et al. [68] and surface observations of ash in many regions of Europe [65,66,69–71]. Pelley et al. [29] shows that the InTEM inversion, using a single realization of the meteorological situation, can distinguish between ash emitted at different altitudes and hence produces an emission profile that emits ash at lower altitudes than the *a priori* vertical distribution. This emission profile results in a forecast ash cloud that is consistent with the location of the ash, not SO_2. In this study, the uncertainty in the posterior emission profile will be estimated by performing an ensemble of inversions using an ensemble of meteorological conditions.

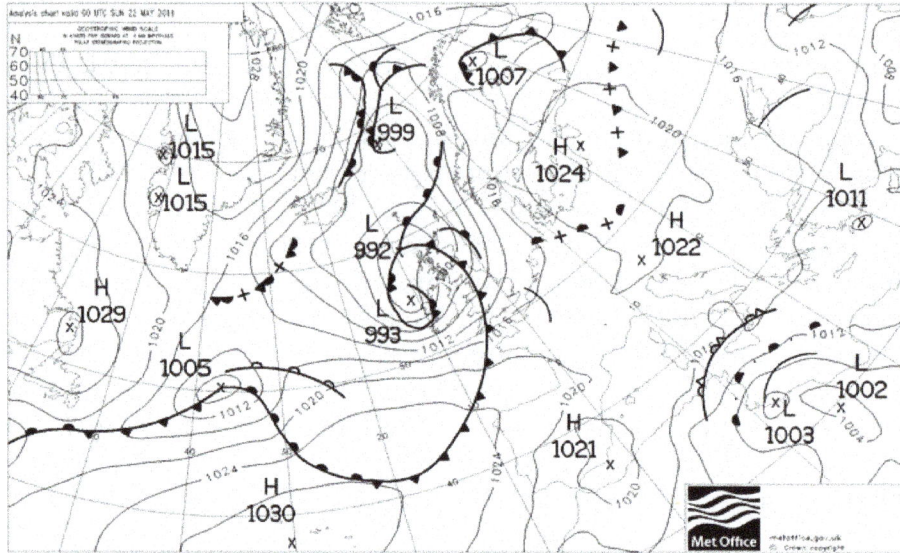

Figure 2. UK Met Office surface analysis chart at 0000 UTC on 22 May 2011. Mean sea level pressure isobars overlaid with active surface fronts (solid black lines with filled symbols), decaying surface fronts (crosses with filled symbols) upper-level fronts (solid black lines with unfilled symbols) and upper-level troughs (solid black lines).

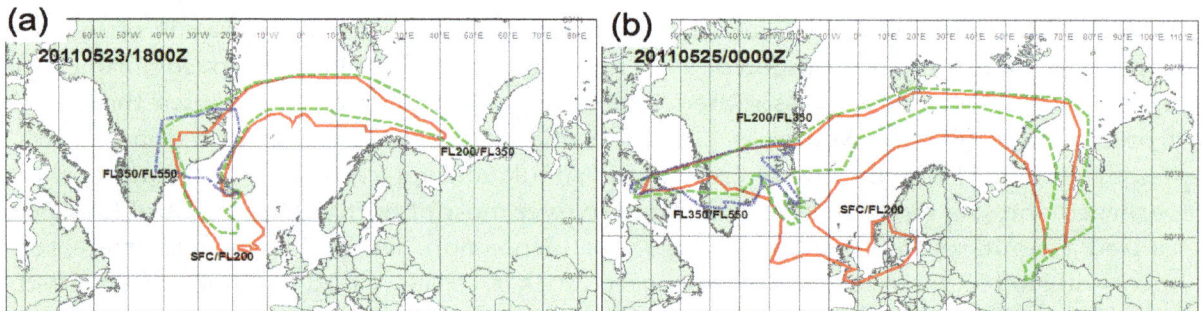

Figure 3. Volcanic ash advisory graphics issued by the London VAAC at (**a**) 1800 UTC on 23 May 2011 and (**b**) 0000 UTC on 25 May 2011 taken from the London VAAC: Volcanic ash advisories and graphics archive [67]. Contours show the outermost extent of the ash cloud in 3 layers of the atmosphere: surface—flight level (FL) 200 (red), FL200-350 (green) and FL350-550 (blue).

4. Results

4.1. Posterior Inversion Estimates

Figure 4 shows the posterior height-time ash emission rates (g/h) obtained using InTEM for the Grímsvötn eruption (21–25 May 2011) using each of the EPS meteorological ensemble members with SEVIRI retrievals of ash and clear skies. Each panel shows the ash emission rates determined by InTEM using a single member of the EPS ensemble and the emission rates have a temporal resolution of 3 h and a vertical resolution of 4 km. All members have high emission rates between 16 and 20 km above vent level (avl) shortly after the start of the eruption. These high emission rates are similar to the *a priori* emission (Figure 5a) and are not directly informed by observations. From 1200 UTC on 22 May, ash emissions are largely confined to between 0–4 km avl for the remaining eruption period. Although the vertical emission profiles are similar, there are differences in the magnitude of ash emitted at different heights. For example, member 13 has a continuous emission of ash between

0–4 km whereas member 10 has times when there is no emission of ash at all at this height level. These differences lead to a range of total emissions that vary by a factor of approximately 1.5 over the entire eruption (shown in Figure 6 as red circles) with an ensemble mean value of 2.00×10^{12} g. There is also a range, 17,754–20,268, in the number of observations which impact the inversion between ensemble members. This represents the variability between the ensemble meteorology that is used in the VATD model simulations used in the inversion process.

The ensemble of posterior emission profiles is comparable to those found in [38,63]. Figure 5a shows the *a priori* emission profile regridded on to the height-time grid used in the inversion. The emission rates are higher at all heights and times when compared to the posterior estimates. The *a priori* total emission is 32×10^{12} g, which is approximately 16 times larger than any of the posterior estimates (shown in Figure 6). Figure 5b shows the ensemble mean posterior emission profile which indicates the large difference between total emissions in the *a priori* and posterior is mainly due to the inversion reducing emissions above 4 km avl in the posterior estimates.

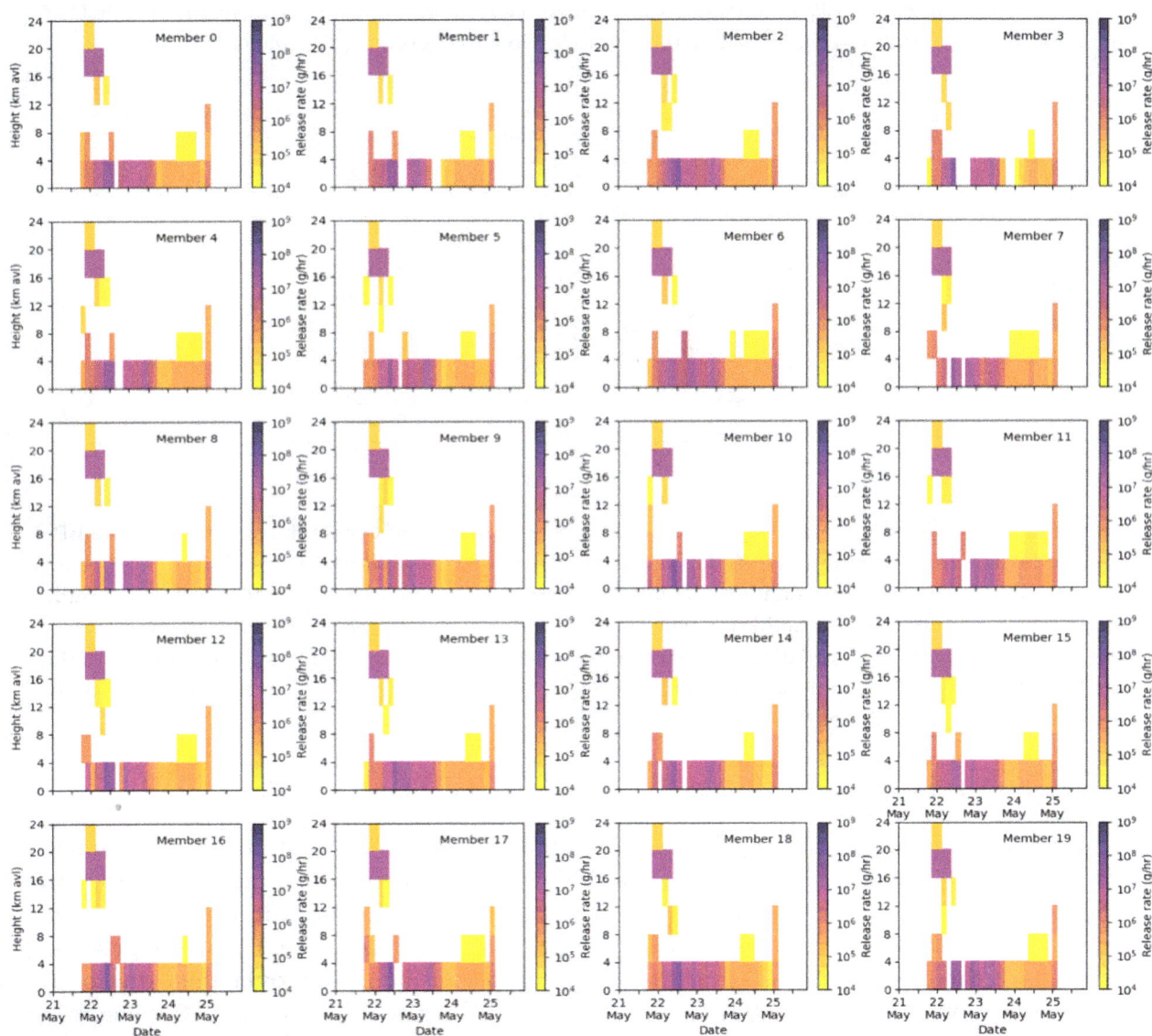

Figure 4. Posterior emission profiles (g/h) estimated by InTEM for the 2011 Grímsvötn eruption for each member of the ECMWF EPS ensemble using SEVIRI retrievals of ash and clear skies. Note the logarithmic color scale.

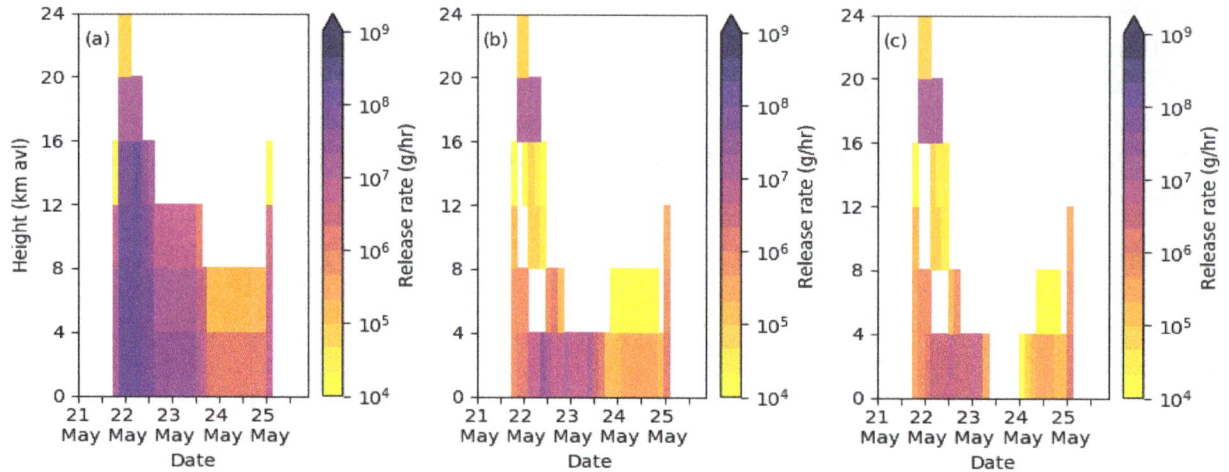

Figure 5. (**a**) *a priori* (**b**) ensemble mean posterior with wet deposition represented (**c**) ensemble mean posterior without wet deposition represented time-height emission profile (g/h) determined by InTEM for the Grímsvötn eruption (21–25 May 2011). Note the log scale used for the release rate.

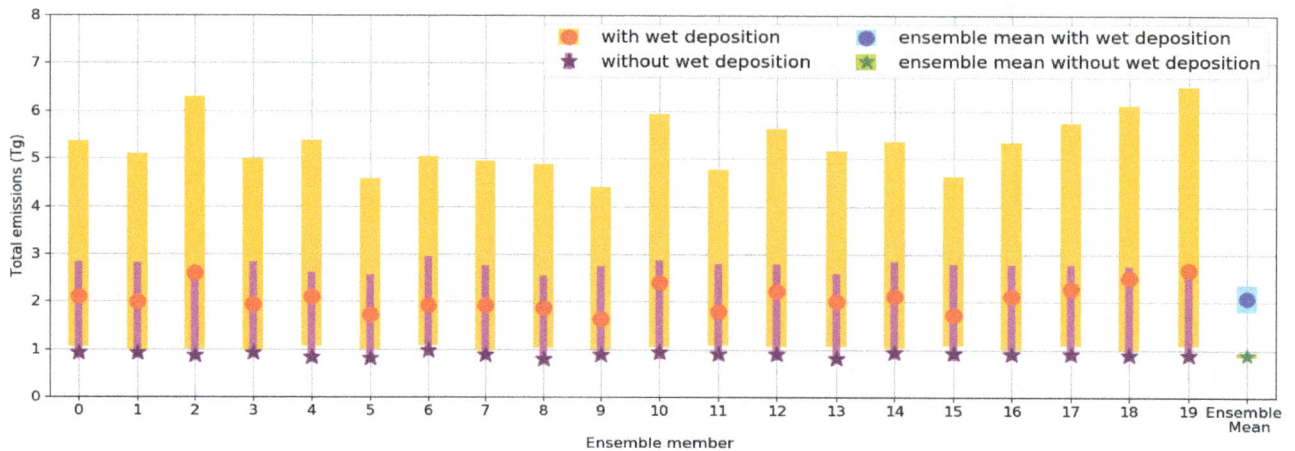

Figure 6. Total ash emitted during the Grímsvötn eruption for each ensemble member in the EPS ensemble determined using InTEM. Red circles and orange bars indicate the peak in the posterior distribution and range (±one standard deviation) of the total ash emitted for simulations that include wet deposition. Purple stars and lilac bars indicate the peak of the posterior distribution and range of the total ash emitted for simulations that do not include wet deposition. The final column shows the mean and standard deviation for the ensemble emissions with (blue circle and cyan bar) and without wet deposition (green star and light green bar).

4.2. Quantifying Wet Deposition Uncertainty

As stated in Section 3, during the eruption there were several extratropical cyclones that transited the North Atlantic (as shown in Figure 2) which could impact ash location forecasts through small forecast errors in the NWP wind and precipitation location. The location of precipitation and meteorological cloud is important as ash is removed from the atmosphere by wet deposition [72]. Figure 7 shows the ensemble mean posterior wet deposition integrated over the duration of the Grímsvötn eruption (21–25 May 2011) from NAME simulations driven by the EPS meteorological ensemble. Wet deposition occurs over a large area with the largest values to the south and south-east of Iceland. The removal of ash by precipitation impacts atmospheric ash concentrations as less ash remains to be transported. To investigate the relative importance of potential wind and precipitation uncertainty, the InTEM inversion system was run again with the EPS meteorological ensemble but using

NAME simulations without wet deposition processes being represented. This removes differences in the inversion that occur due to differences in the location and magnitude of forecast precipitation.

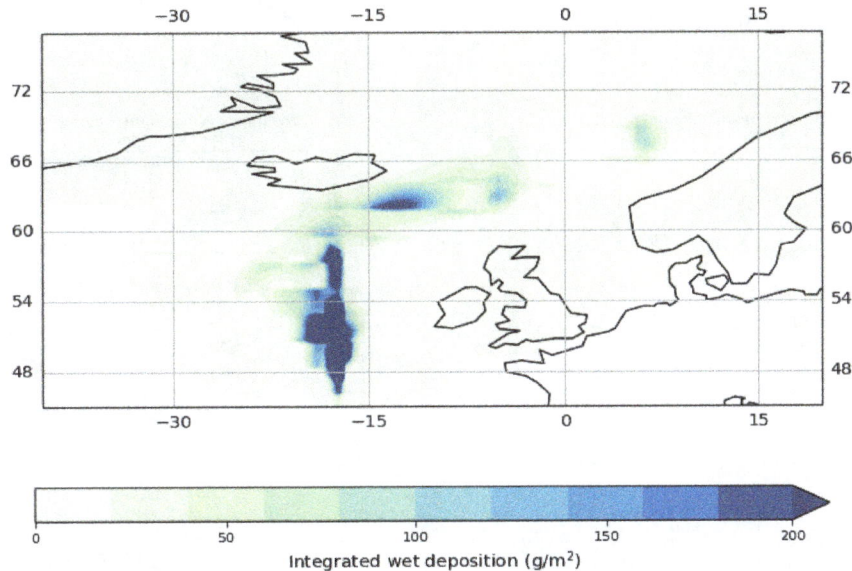

Figure 7. Ensemble mean wet deposition integrated over the duration of the Grímsvötn eruption (21–25 May 2011).

The ensemble of posterior emission profiles from InTEM without wet deposition represented are qualitatively similar to those shown in Figure 4 but with smaller emission rates. This is expected as, to match to the SEVIRI observations, less ash needs to be released if ash is not removed through wet deposition in the NAME simulations. The ensemble mean emission profile without wet deposition represented is shown in Figure 5. Please note that no ash was emitted for 15 h on 23 May for all members of the meteorological ensemble. Figure 6 shows the total emission rates over the whole eruption period for the posterior emissions with wet deposition processes represented (red circles show the mean, with the uncertainty ($\pm$one standard deviation) indicated by an orange bar) and without wet deposition processes represented (purple stars show the mean, with the uncertainty ($\pm$one standard deviation) indicated by lilac bars) for each ensemble member. The range of mean values for the ensemble without wet deposition is 0.81–0.97×10^{12} g which is over a factor of 2 less than the range of total emissions for the ensemble with wet deposition processes represented (1.72–2.66×10^{12} g). This suggests that in this case, wet deposition has a significant impact on the posterior emission rates. In both sets of inversions, the uncertainty of solutions for the total emission is large and skewed (yellow and pink bars in Figure 6) due to the non-negative constraint.

The ensemble mean total emission and standard deviation of the mean is shown in the last column of Figure 6. The ensemble mean total emission for the posterior emissions with wet deposition represented is 2.08×10^{12} g with a standard deviation of 0.28×10^{12} g compared to 0.89×10^{12} g with a standard deviation of 0.04×10^{12} g for the posterior emissions without wet deposition processes. The large separation of the ensemble means shows that the impact of wet deposition processes in the uncertainty on the total ash emitted is greater than uncertainty from the variability in the winds for this eruption.

4.3. Ensemble InTEM Ash Forecasts

The previous discussion in Sections 4.1 and 4.2 focused on the differences in posterior emission profiles; however VATD model forecasts of the ash cloud are used to inform VAAC graphics and advisories. This section analyses NAME simulations driven with the EPS meteorological ensemble both using the peak of the posterior distribution of emissions (shown in Figure 4) and *a priori* emissions and evaluates these simulations against independent satellite retrievals of volcanic ash from MODIS. Please note that the results presented use satellite retrievals for the whole eruption period (21–25 May 2011) and not just those retrievals available before 1800 UTC 23 May 2011 (Figure 8) and 0000 UTC 25 May 2011 (Figure 9).

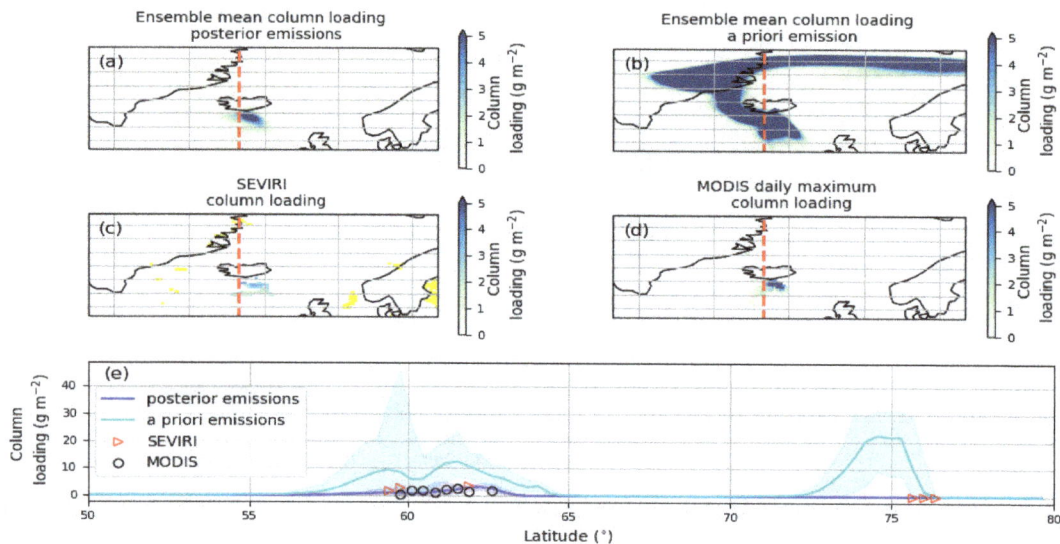

Figure 8. (a) Ensemble mean ash column loading at 1800 UTC 23 May 2011 for simulations run with emission profiles determined using InTEM with matching ensemble meteorology at 1800 UTC 23 May 2011, (b) Ensemble mean ash column loading at 1800 UTC 23 May 2011 for simulations run with *a priori* emission profile and ensemble meteorology, (c) SEVIRI ash column loading (yellow shading indicates grid boxes that are classified as clear sky) (d) Daily MODIS maximum ash column loading for 23 May 2011, (e) Ash column loading profile along the cross section indicated by the red dashed line in panels (a–d) for the posterior ensemble (blue), *a priori* ensemble (cyan), SEVIRI ash column loading (red triangles) and MODIS ash column loading (black circles). The shading indicates the range of column loading in the ensemble.

Figure 8a,b show the spatial extent of the ash cloud at 1800 UTC 23 May 2011 for the ensemble of VATD model simulations with posterior emissions with wet deposition considered (Figure 8a) and *a priori* emissions (Figure 8b) , while Figure 8c,d show SEVIRI (Figure 8c) and MODIS (Figure 8d) satellite retrievals. To ensure a fair comparison, these are presented as ensemble mean column loading but note that the VATD model simulations here use the default NAME particle size distribution which represents a range of particle sizes (0.1–100 μm). Please note that the difference between this particle size distribution and what is found using by the satellite retrieval may contribute to discrepancies between the simulated and retrieved ash column loadings.

There is a large difference between the magnitude of the simulated ash clouds in Figure 8a,b. The ash cloud shown in Figure 8a is the mean of the simulations with posterior emissions. It extends south-east from the volcano with the maximum column loading values (greater than 4 g m^{-2}) located to the south of Grímsvötn. This is quantitatively similar to both the SEVIRI retrievals used in the inversion process (Figure 8c) and the independent MODIS retrievals (Figure 8d). The mean ash cloud extent from the *a priori* ensemble has a much larger extent when considering column loadings greater than 4 g m^{-2}, both to the south-west and to the north-west with a filament extending from Greenland towards northern Scandinavia. As expected, this is very similar to the VAAC issued graphic shown in Figure 3. The region of highest column loading, 44.5 g m^{-2}, is over Greenland which is also consistent with the VAAC graphic (Figure 3). The region of highest mean column loading in the simulations with posterior emissions is located to the south-east of Grímsvötn with a magnitude of 7.8 g m^{-2}.

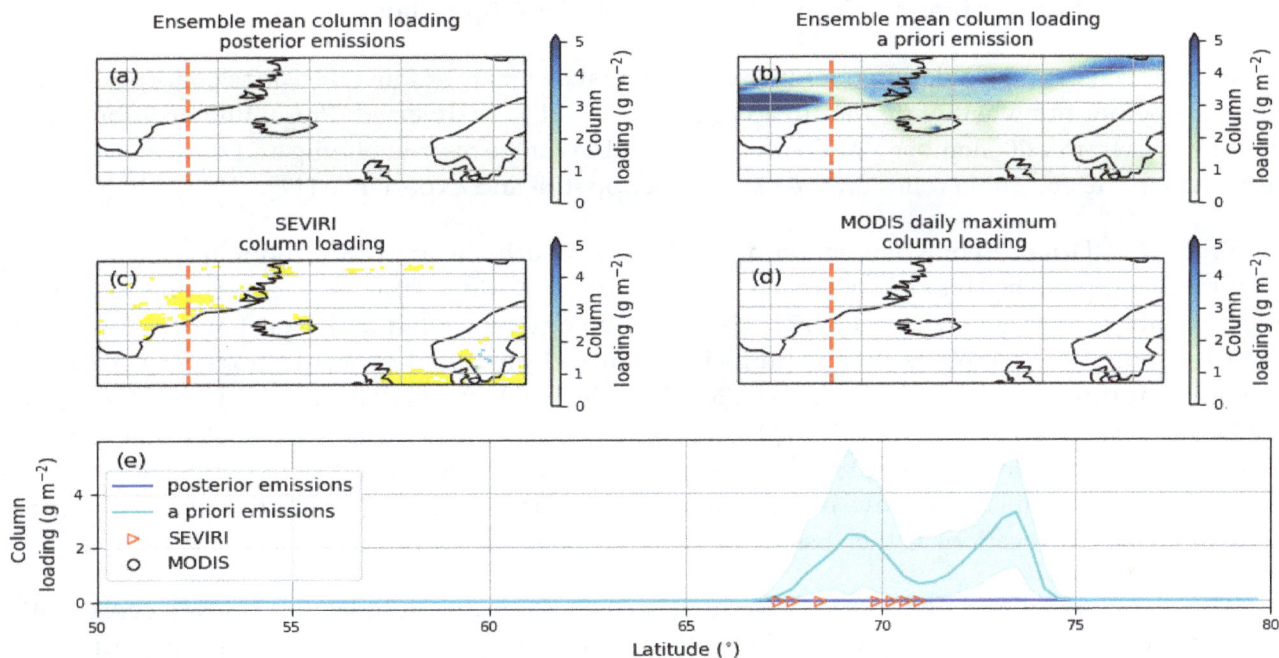

Figure 9. As Figure 8 for 0000 UTC on 25 May 2011. Note MODIS (panel d) did not retrieve any ash on this day.

Figure 9a,b show a similar difference between the simulated ash clouds at 0000 UTC on 25 May 2011. This is 30 h later than the ash cloud shown in Figure 8 and before the end of the eruption at 0240 UTC on 25 May. The ash cloud from the *a priori* simulations is the most similar to the VAAC issued graphic (Figure 3b). On this day, MODIS did not detect any ash and there are a small number of SEVIRI pixels classified as ash over Scandinavia and clear sky pixels over Greenland and between the UK and mainland Europe. There may have been ash present at other locations at this time that could not be detected due to the presence of meteorological cloud.

Producing an ensemble of VATD simulations allows assessment of the ensemble spread. Figure 8e shows the range of posterior column loading values for the 20 posterior emission simulations (blue line and shading) and 20 *a priori* ensemble simulations (cyan line and shading) along the cross section shown in panels (a–d) as a red dashed line. The shading represents the uncertainty in ash column loading due to the meteorological ensemble. Satellite retrievals of ash column loading from SEVIRI (red triangles) and MODIS (black circles) are shown for comparison. Between 55–65 degrees north, the posterior ensemble spread encompasses both the MODIS and SEVIRI retrievals. The *a priori* ensemble has a much higher mean magnitude than the posterior emissions (by up to 200%) and the ensemble spread does not encompass the satellite observations. At 75 degrees north the *a priori* ensemble has a mean column loading of 21 g m^{-2}. However, the posterior emission ensemble predicts a very small amount of ash in this location which better matches the location of SEVIRI clear sky grid boxes at this time. Meteorological cloud obscures much of the domain so obtaining a contiguous retrieval of ashy and clear sky pixels is not possible at this time.

Figure 9e shows the same variables as Figure 8 at 0000 UTC 25 May 2011 along the cross section highlighted in panels (a–d) as a red dashed line. At this time there is no satellite retrieved ash along the cross section but there is a large number of SEVIRI grid boxes that are classified as clear sky (yellow shading). The magnitude of the column loading in the VATD simulations is approximately 10% of the column loadings shown in Figure 8. Along the cross section the mean column loading in the posterior emission ensemble is very small which agrees better with the location of SEVIRI clear sky observations. However, the *a priori* ensemble predicts ash between 65–75° N, with a large spread in forecast ash column loadings. This spread shows the variability that can be introduced by using an ensemble of meteorological conditions with the same ash emission profile.

5. Application of Probabilistic Volcanic Ash Forecasts for Flight Planning

During an eruption, aviation operators need to make fast and robust decisions concerning which flights to operate and whether flights already in the air need to be diverted. Presenting decision makers with graphics from a 20 member VATD forecast ensemble can be overwhelming and how they interpret the spread in the ensemble relies on the user's risk appetite and experience [17].

Prata et al. [10] present a new method for visualizing ash concentration along flight paths using a risk-matrix approach. This risk approach is used by the UK Met Office when presenting forecasts of severe weather to the general public [73]. The methodology presented by Prata et al., determines the fraction of ensemble members that result in VATD model predicted concentrations over 3 different thresholds at three flight levels (surface-FL200, FL200-FL350 and FL350-FL550). The concentration thresholds used are 200–2000 $\mu g\, m^{-3}$, 2000–4000 $\mu g\, m^{-3}$ and > 4000 $\mu g\, m^{-3}$. The risk of encountering ash is the likelihood (or probability of exceedance) multiplied by impact (where the concentration ranges above are deemed to be low, medium and high impact respectively). In Prata et al. [10] the risk of flying in specific locations is then assigned to be low, medium or high.

The overall risk presented is the maximum risk over the three flight levels. This is a conservative estimate but consistent with current UK regulations that state that where possible, aircraft should laterally avoid forecasted areas of ash rather than over- or under-fly [7]. Please note that ICAO guidelines state that, for the purposes of flight planning, operators should treat the horizontal and vertical limits of the Danger Area to be overflown as it would mountainous terrain [3]. Prata et al. [10] suggest possible actions for the user for each of the risk levels, although these are a subset of the actions a flight planner might take. These range from checking updated forecasts, to loading more fuel and performing engine checks and considering other routes. The risk-matrix graphics reduce the state-of-the-art ensemble information into an easy to use decision making tool that can be used to make fast and scientifically robust decisions.

Figure 10 shows the risk determined using the Prata et al. [10] approach using both the posterior emissions ensemble with wet deposition considered and the *a priori* ensemble at 1800 UTC 23 May 2011 and 0000 UTC 25 May 2011. These ensembles represent the uncertainty in ash location and magnitude due to uncertainty in the meteorological situation. At both times, the region of forecasted risk is greatly reduced when the posterior emission ensemble is used compared to the *a priori* ensemble. At 1800 UTC 23 May, the forecasted risk area is reduced by 72%, with the highest risk area (blue) reduced by 88%. Similarly, at 0000 UTC 25 May, the forecasted risk area is reduced by 74%, with the highest risk area reduced by 97%. The area of forecasted risk determined by the *a priori* ensemble has a large extent to the north of Iceland with a further branch extending over the north of the UK into Scandinavia (Figure 10d). The posterior risk, also has a branch that extends from Iceland into Scandinavia (Figure 10b), although the risk diagnosed in this extension towards Europe is mostly at the medium level (turquoise). Please note that when calculating the likelihood of exceeding concentration thresholds, only uncertainty arising from the meteorology is estimated in this study.

At both times shown in Figure 10, if this risk-matrix approach is taken, for this eruption the disruption to airline operations has the potential to be greatly reduced if the posterior emissions were used. Please note that to be consistent with [10], the ash concentration fields output by NAME were multiplied by a factor of 10, known as the "peak-to-mean" factor. This factor accounts for peak concentrations that are not resolved by the NAME modeling [74]. This is similar to the operational approach of the London VAAC at the time of the Grímsvötn eruption, although the peak-to-mean-ratio is no longer applied operationally [16].

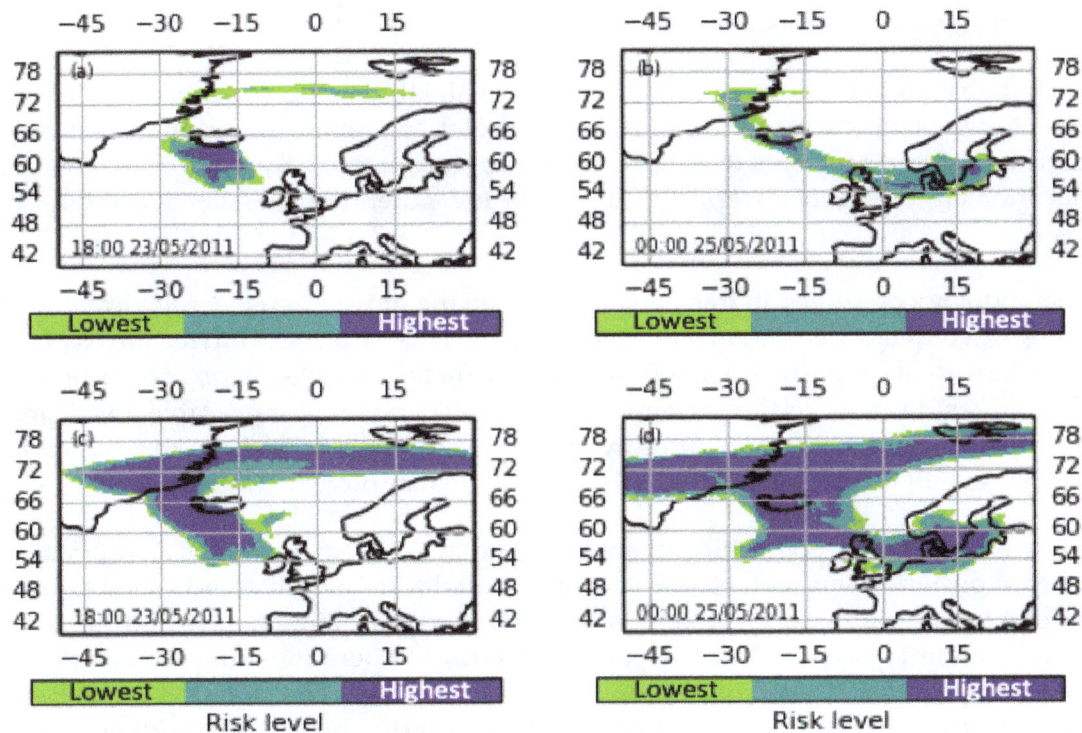

Figure 10. Overall ash concentration risk map at (**a,c**) 1800 UTC on 23 May 2011, (**b,d**) 0000 UTC on 25 May 2011 for (**a,b**) simulations with posterior emissions with matching ensemble meteorology and (**c,d**) simulations with *a priori* emissions with ensemble meteorology. Green shading indicates the lowest level of risk, turquoise shading indicates mid-level risk and purple indicates the highest level of risk.

6. Summary and Conclusions

In the event of an emergency, aviation authorities need to make fast and robust decisions. These decisions take into account the VAAC forecasts which are informed by deterministic VATD model simulations. Two important inputs to these simulations are the ash emission source description and the driving NWP meteorological forecast. The determination of ash emissions in real time is, however, difficult. This study combines optimally estimated ash emission profiles using state-of-the-art inversion techniques with an ensemble of meteorological forecasts for the 2011 eruption of the Icelandic volcano Grímsvötn. Also determined is the dependence of emission estimates on wind and precipitation field uncertainty.

For this case study, the InTEM posterior ash emission rates are substantially reduced compared to the *a priori* emission profile that would be used if an inversion were not performed. The ensemble of posterior emission profiles, produced using a range of plausible meteorological situations, are similar but there are differences in the magnitude of the ash emitted at different heights. This leads to a large range of values (1.72–2.66×10^{12} g) for the total amount of ash (in the size range 0.1–100 μm) emitted over the eruption period. In the case study presented here, the inclusion of wet deposition processes, and the variability in the cloud and precipitation fields, has a greater impact on the uncertainty in the total amount of ash emitted than the variability of the winds. Despite the differences that are dominated by wet deposition uncertainty, the ensemble inversion provides confidence that the reduction of the *a priori* emissions, particularly above 4 km, is robust across all members. That is because, in this case, the emission profile and changing wind direction with height dominate this aspect of the simulations. The VATD model forecasts using the posterior emission ensemble have ash clouds with much lower column loadings compared to the *a priori* ensemble simulations. The posterior emission ensemble ash clouds are also a better match to the independent MODIS satellite retrievals of ash and have a much smaller range of column loadings.

The risk-matrix methodology outlined in Prata et al. [10] has been applied to the ensemble of forecast ash clouds obtained using the *a priori* and posterior emissions. In this case study, the use of the posterior emissions reduces the region of highest forecast risk by up to 94%. This could have large implications in emergency response situations and potentially reduce disruption to the civil flight plans.

The methodology presented in this paper focuses on the impact of meteorological uncertainty on the forecasting of volcanic ash location and concentration. There are many other sources of uncertainty that can contribute to uncertainty in ash forecasts which are not included here, such as uncertainties in the source (e.g., ash density) and uncertainties in the VATD model, for example, the representation of free tropospheric turbulence and unresolved mesoscale motions and the representation of wet deposition, that would ideally be included in an operational approach (as in [10]).

To be able to use an ensemble inversion approach in an emergency response situation, the ensemble of posterior emission profiles and associated forecast ash clouds need to be produced in a timely manner. The ensemble method is well suited to parallelization with different ensemble members run simultaneously on different cores. The current operational implementation of InTEM uses an iterative procedure whereby the posterior emissions can be updated during an ongoing eruption as more observations become available and the eruption progresses. Further work is needed to determine how the ensemble members should be combined to produce the best possible ash forecasts in this iterative framework.

7. Data Statement

The SEVIRI satellite data, NAME simulation and InTEM output are available in the University of Reading Research Data Archive at http://dx.doi.org/10.17864/1947.260. Further information about the data supporting these findings and requests for access to the data can be directed to n.j.harvey@reading.ac.uk. The Oxford IASI volcanic products can be obtained by contacting Isabelle Taylor (isabelle.taylor@physics.ox.ac.uk). For InTEM and NAME license enquiries, please contact the Met Office (atmospheric.dispersion@metoffice.gov.uk).

Author Contributions: Conceptualization, N.J.H. and H.F.D.; methodology, N.J.H. and H.F.D.; software, N.J.H. and H.N.W.; validation, N.J.H., investigation, N.J.H.; resources, N.J.H., S.K., M.C.C. and I.A.T.; writing—original draft preparation, N.J.H.; writing—review and editing, N.J.H., H.F.D., H.N.W., I.A.T. and R.G.G.; visualization, N.J.H. and I.A.T.; funding acquisition, H.F.D. and R.G.G. All authors have read and agreed to the published version of the manuscript.

Acknowledgments: We would like to acknowledge EUMETSAT and CEDA for access to the IASI level 1c spectra [75], and ECMWF and CEDA for the meteorological profiles used within the IASI retrievals [76]). We thank Keith Bevan (Lancaster University), and David Thomson (UK Met Office) for useful discussions on the methodology and Matt Hort (UK Met Office) for feedback on the manuscript.

References

1.   Casadevall, T.J. The 1989–1990 eruption of Redoubt Volcano, Alaska: Impacts on aircraft operations. *J. Volcanol. Geotherm. Res.* **1994**, *62*, 301–316. [CrossRef]

2.   Mazzocchi, M.; Hansstein, F.; Ragona, M. The volcanic ash cloud and its financial impact on the European airline industry. In *CESifo Forum*; No 2:92-100; Ifo Institut für Wirtschaftsforschung an der Universität München: München, Germany, 2010.

3. *Doc 9691 AN/954: Manual on Volcanic Ash, Radioactive Material and Toxic Chemical Clouds*, 2nd ed.; International Civil Aviation Organization: Montréal, QC, Canada, 2007. Available online: https://skybrary.aero/bookshelf/books/2997.pdf (accessed on 5 February 2020).

4. Witham, C.; Webster, H.; Hort, M.; Jones, A.; Thomson, D. Modelling concentrations of volcanic ash encountered by aircraft in past eruptions. *Atmos. Environ.* **2012**, *48*, 219–229. [CrossRef]

5. Clarkson, R.J.; Majewicz, E.J.; Mack, P. A re-evaluation of the 2010 quantitative understanding of the effects volcanic ash has on gas turbine engines. *Proc. Inst. Mech. Eng. Part G J. Aerosp. Eng.* **2016**, *230*, 2274–2291. [CrossRef]

6. European Commission. *Volcano Grimsvötn: How Is the European Response Different to the Eyjafjallajökull Eruption Last Year? Frequently Asked Questions*; European Commission: Brussels, Belgium, 2011.

7. UK Civil Aviation Authority. *CAP1236: Guidance Regarding Flight Operations in the Vicinity of Volcanic Ash*; UK Civil Aviation Authority: London, UK, 2017.

8. Dacre, H.; Grant, A.; Hogan, R.; Belcher, S.; Thomson, D.; Devenish, B.; Marenco, F.; Hort, M.; Haywood, J.M.; Ansmann, A.; et al. Evaluating the structure and magnitude of the ash plume during the initial phase of the 2010 Eyjafjallajökull eruption using lidar observations and NAME simulations. *J. Geophys. Res.* **2011**, *116*, [CrossRef]

9. Harvey, N.J.; Huntley, N.; Dacre, H.F.; Goldstein, M.; Thomson, D.; Webster, H. Multi-level emulation of a·volcanic ash transport and dispersion model to quantify sensitivity to uncertain parameters. *Nat. Hazards Earth Syst. Sci.* **2018**, *18*, 41–63. [CrossRef]

10. Prata, A.T.; Dacre, H.F.; Irvine, E.A.; Mathieu, E.; Shine, K.P.; Clarkson, R.J. Calculating and communicating ensemble-based volcanic ash dosage and concentration risk for aviation. *Meteorol. Appl.* **2019**, *26*, 253–266. [CrossRef]

11. Petersen, G.N.; Bjornsson, H.; Arason, P.; von Löwis, S. Two weather radar time series of the altitude of the volcanic plume during the May 2011 eruption of Grimsvotn, Iceland. *Earth Syst. Sci. Data* **2012**, *4*, 121–127, [CrossRef]

12. Mastin, L.; Guffanti, M.; Servranckx, R.; Webley, P.; Barsotti, S.; Dean, K.; Durant, A.; Ewert, J.; Neri, A.; Rose, W.; et al. A multidisciplinary effort to assign realistic source parameters to models of volcanic ash-cloud transport and dispersion during eruptions. *J. Volcanol. Geotherm. Res.* **2009**, *186*, 10–21. [CrossRef]

13. Sparks, R.S.J.; Bursik, M.; Carey, S.; Gilbert, J.; Glaze, L.; Sigurdsson, H.; Woods, A. *Volcanic Plumes*; Wiley: Chichester, UK, 1997.

14. Woodhouse, M.J.; Hogg, A.J.; Phillips, J.C.; Sparks, R.S.J. Interaction between volcanic plumes and wind during the 2010 Eyjafjallajökull eruption, Iceland. *J. Geophys. Res.* **2013**, *118*, 92–109, [CrossRef]

15. Bonadonna, C.; Folch, A.; Loughlin, S.; Puempel, H. Future developments in modelling and monitoring of volcanic ash clouds: Outcomes from the first IAVCEI-WMO workshop on Ash Dispersal Forecast and Civil Aviation. *Bull. Volcanol.* **2012**, *74*, 1–10. [CrossRef]

16. Beckett, F.M.; Witham, C.S.; Leadbetter, S.J.; Crocker, R.; Webster, H.N.; Hort, M.C.; Jones, A.R.; Devenish, B.J.; Thomson, D.J. Atmospheric Dispersion Modelling at the London VAAC: A Review of Developments since the 2010 Eyjafjallajökull Volcano Ash Cloud. *Atmosphere* **2020**, *11*, 352. [CrossRef]

17. Mulder, K.J.; Lickiss, M.; Harvey, N.; Black, A.; Charlton-Perez, A.; Dacre, H.; McCloy, R. Visualizing Volcanic Ash Forecasts: Scientist and Stakeholder Decisions Using Different Graphical Representations and Conflicting Forecasts. *Weather. Clim. Soc.* **2017**, *9*, 333–348. [CrossRef]

18. Dare, R.A.; Potts, R.J.; Wain, A.G. Modelling wet deposition in simulations of volcanic ash dispersion from hypothetical eruptions of Merapi, Indonesia. *Atmos. Environ.* **2016**, *143*, 190–201. [CrossRef]

19. Zidikheri, M.J.; Lucas, C.; Potts, R.J. Quantitative verification and calibration of volcanic ash ensemble forecasts using satellite data. *J. Geophys. Res. Atmos.* **2018**, *123*, 4135–4156. [CrossRef]

20. Stefanescu, E.; Patra, A.; Bursik, M.; Madankan, R.; Pouget, S.; Jones, M.; Singla, P.; Singh, T.; Pitman, E.; Pavolonis, M.; et al. Temporal, probabilistic mapping of ash clouds using wind field stochastic variability and uncertain eruption source parameters: Example of the 14 April 2010 Eyjafjallajökull eruption. *J. Adv. Model. Earth Syst.* **2014**, *6*, 1173–1184. [CrossRef]

21. Madankan, R.; Pouget, S.; Singla, P.; Bursik, M.; Dehn, J.; Jones, M.; Patra, A.; Pavolonis, M.; Pitman, E.B.; Singh, T.; et al. Computation of probabilistic hazard maps and source parameter estimation for volcanic ash transport and dispersion. *J. Comput. Phys.* **2014**, *271*, 39–59. [CrossRef]

22. Dacre, H.F.; Harvey, N.J. Characterizing the Atmospheric Conditions Leading to Large Error Growth in Volcanic Ash Cloud Forecasts. *J. Appl. Meteorol. Climatol.* **2018**, *57*, 1011–1019. [CrossRef]

23. Langmann, B.; Zakšek, K.; Hort, M. Atmospheric distribution and removal of volcanic ash after the eruption of Kasatochi volcano: A regional model study. *J. Geophys. Res. Atmos.* **2010**, *115*. [CrossRef]

24. Francis, P.N.; Cooke, M.C.; Saunders, R.W. Retrieval of physical properties of volcanic ash using Meteosat: A case study from the 2010 Eyjafjallajökull eruption. *J. Geophys. Res.* **2012**, *117*, [CrossRef]

25. Pavolonis, M.J.; Heidinger, A.K.; Sieglaff, J. Automated retrievals of volcanic ash and dust cloud properties from upwelling infrared measurements. *J. Geophys. Res. Atmos.* **2013**, *118*, 1436–1458, [CrossRef]

26. Kristiansen, N.I.; Stohl, A.; Prata, A.J.; Bukowiecki, N.; Dacre, H.; Eckhardt, S.; Henne, S.; Hort, M.C.; Johnson, B.T.; Marenco, F.; et al. Performance assessment of a volcanic ash transport model mini-ensemble used for inverse modeling of the 2010 Eyjafjallajökull eruption. *J. Geophys. Res.* **2012**, *117*. [CrossRef]

27. Schmehl, K.J.; Haupt, S.E.; Pavolonis, M.J. A genetic algorithm variational approach to data assimilation and application to volcanic emissions. *Pure Appl. Geophys.* **2012**, *169*, 519–537. [CrossRef]

28. Denlinger, R.P.; Pavolonis, M.; Sieglaff, J. A robust method to forecast volcanic ash clouds. *J. Geophys. Res.* **2012**, *117*, [CrossRef]

29. Pelley, R.E.; Cooke, M.C.; Manning, A.J.; Thomson, D.J.; Witham, C.S.; Hort, M.C. *Initial Implementation of an Inversion Techniques for Estimating Volcanic Ash Source Parameters in Near Real Time Using Satellite Retrievals*; Forecasting Research Technical Report No. 644; Met Office: Exeter, UK, 2015.

30. Zidikheri, M.J.; Lucas, C.; Potts, R.J. Toward quantitative forecasts of volcanic ash dispersal: Using satellite retrievals for optimal estimation of source terms. *J. Geophys. Res. Atmos.* **2017**, *122*, 8187–8206. [CrossRef]

31. Zidikheri, M.J.; Lucas, C.; Potts, R.J. Estimation of optimal dispersion model source parameters using satellite detections of volcanic ash. *J. Geophys. Res. Atmos.* **2017**, *122*, 8207–8232. [CrossRef]

32. Eckhardt, S.; Prata, A.; Seibert, P.; Stebel, K.; Stohl, A. Estimation of the vertical profile of sulfur dioxide injection into the atmosphere by a volcanic eruption using satellite column measurements and inverse transport modeling. *Atmos. Chem. Phys.* **2008**, *8*, 3881–3897. [CrossRef]

33. Kristiansen, N.; Stohl, A.; Prata, A.; Richter, A.; Eckhardt, S.; Seibert, P.; Hoffmann, A.; Ritter, C.; Bitar, L.; Duck, T.; et al. Remote sensing and inverse transport modeling of the Kasatochi eruption sulfur dioxide cloud. *J. Geophys. Res. Atmos.* **2010**, *115*. [CrossRef]

34. Stohl, A.; Prata, A.; Eckhardt, S.; Clarisse, L.; Durant, A.; Henne, S.; Kristiansen, N.; Minikin, A.; Schumann, U.; Seibert, P.; et al. Determination of time-and height-resolved volcanic ash emissions and their use for quantitative ash dispersion modeling: The 2010 Eyjafjallajökull eruption. *Atmos. Chem. Phys.* **2011**, *11*, 4333–4351. [CrossRef]

35. Seibert, P.; Kristiansen, N.I.; Richter, A.; Eckhardt, S.; Prata, A.J.; Stohl, A. Uncertainties in the inverse modelling of sulphur dioxide eruption profiles. *Geomat. Nat. Hazards Risk* **2011**, *2*, 201–216. [CrossRef]

36. Boichu, M.; Menut, L.; Khvorostyanov, D.; Clarisse, L.; Clerbaux, C.; Turquety, S.; Coheur, P.F. Inverting for volcanic SO_2 flux at high temporal resolution using spaceborne plume imagery and chemistry-transport modelling: The 2010 Eyjafjallajökull eruption case-study. *Atmos. Chem. Phys.* **2013**, *13*, 8569–8584. [CrossRef]

37. Zidikheri, M.J.; Potts, R.J. A simple inversion method for determining optimal dispersion model parameters from satellite detections of volcanic sulfur dioxide. *J. Geophys. Res. Atmos.* **2015**, *120*, 9702–9717. [CrossRef]

38. Moxnes, E.; Kristiansen, N.; Stohl, A.; Clarisse, L.; Durant, A.; Weber, K.; Vogel, A. Separation of ash and sulfur dioxide during the 2011 Grímsvötn eruption. *J. Geophys. Res. Atmos.* **2014**, *119*, 7477–7501. [CrossRef]

39. Zidikheri, M.J.; Lucas, C. Using Satellite Data to Determine Empirical Relationships between Volcanic Ash Source Parameters. *Atmosphere* **2020**, *11*, 342, [CrossRef]

40. Molteni, F.; Buizza, R.; Palmer, T.N.; Petroliagis, T. The ECMWF Ensemble Prediction System: Methodology and validation. *Q. J. R. Meteorol. Soc.* **1996**, *122*, 73–119, [CrossRef]

41. Buizza, R.; Palmer, T.N. The singular-vector structure of the atmospheric global circulation. *J. Atmos. Sci.* **1995**, *52*, 1434–1456. [CrossRef]

42. Buizza, R.; Milleer, M.; Palmer, T.N. Stochastic representation of model uncertainties in the ECMWF ensemble prediction system. *Q. J. R. Meteorol. Soc.* **1999**, *125*, 2887–2908. [CrossRef]

43. Thomson, D.J.; Webster, H.N.; Cooke, M.C. *Developments in the Met Office InTEM Volcanic Ash Source Estimation System Part 1: Concepts*; Met Office: Exeter, UK, 2017.

44. Clerbaux, C.; Boynard, A.; Clarisse, L.; George, M.; Hadji-Lazaro, J.; Herbin, H.; Hurtmans, D.; Pommier, M.; Razavi, A.; Turquety, S.; et al. Monitoring of atmospheric composition using the thermal infrared IASI/MetOp sounder. *Atmos. Chem. Phys.* **2009**, *9*, 6041–6054, [CrossRef]

45. Theys, N.; Campion, R.; Clarisse, L.; Brenot, H.; van Gent, J.; Dils, B.; Corradini, S.; Merucci, L.; Coheur, P.F.; Van Roozendael, M.; et al. Volcanic SO_2 fluxes derived from satellite data: A survey using OMI, GOME-2, IASI and MODIS. *Atmos. Chem. Phys.* **2013**, *13*, 5945–5968, [CrossRef]

46. Clarisse, L.; Prata, F.; Lacour, J.L.; Hurtmans, D.; Clerbaux, C.; Coheur, P.F. A correlation method for volcanic ash detection using hyperspectral infrared measurements. *Geophys. Res. Lett.* **2010**, *37*, [CrossRef]

47. Clarisse, L.; Coheur, P.F.; Prata, A.; Hurtmans, D.; Razavi, A.; Phulpin, T.; Hadji-Lazaro, J.; Clerbaux, C. Tracking and quantifying volcanic SO_2 with IASI, the September 2007 eruption at Jebel at Tair. *Atmos. Chem. Phys.* **2008**, *8*, 7723–7734, [CrossRef]

48. Clarisse, L.; Hurtmans, D.; Prata, A.J.; Karagulian, F.; Clerbaux, C.; Mazière, M.D.; Coheur, P.F. Retrieving radius, concentration, optical depth, and mass of different types of aerosols from high-resolution infrared nadir spectra. *Appl. Opt.* **2010**, *49*, 3713–3722, [CrossRef] [PubMed]

49. Clarisse, L.; Hurtmans, D.; Clerbaux, C.; Hadji-Lazaro, J.; Ngadi, Y.; Coheur, P.F. Retrieval of sulphur dioxide from the infrared atmospheric sounding interferometer (IASI). *Atmos. Meas. Tech.* **2012**, *5*, 581–594, [CrossRef]

50. Carboni, E.; Grainger, R.; Walker, J.; Dudhia, A.; Siddans, R. A new scheme for sulphur dioxide retrieval from IASI measurements: application to the Eyjafjallajökull eruption of April and May 2010. *Atmos. Chem. Phys.* **2012**, *12*, 11417–11434, [CrossRef]

51. Walker, J.C.; Carboni, E.; Dudhia, A.; Grainger, R.G. Improved detection of sulphur dioxide in volcanic plumes using satellite-based hyperspectral infrared measurements: Application to the Eyjafjallajökull 2010 eruption. *J. Geophys. Res. Atmos.* **2012**, *117*. [CrossRef]

52. Ventress, L.J.; McGarragh, G.; Carboni, E.; Smith, A.J.; Grainger, R.G. Retrieval of ash properties from IASI measurements. *Atmos. Meas. Tech.* **2016**, *9*, 5407–5422, [CrossRef]

53. Taylor, I.A.; Carboni, E.; Ventress, L.J.; Mather, T.A.; Grainger, R.G. An adaptation of the CO_2 slicing technique for the Infrared Atmospheric Sounding Interferometer to obtain the height of tropospheric volcanic ash clouds. *Atmos. Meas. Tech.* **2019**, *12*, 3853–3883, [CrossRef]

54. Reed, B.E.; Peters, D.M.; McPheat, R.; Grainger, R.G. The Complex Refractive Index of Volcanic Ash Aerosol Retrieved From Spectral Mass Extinction. *J. Geophys. Res. Atmos.* **2018**, *123*, 1339–1350, [CrossRef]

55. Barnes, W.L.; Pagano, T.S.; Salomonson, V.V. Prelaunch characteristics of the Moderate Resolution Imaging Spectroradiometer (MODIS) on EOS-AM1. *IEEE Trans. Geosci. Remote Sens.* **1998**, *36*, 1088–1100. [CrossRef]

56. MODIS Characterization Support Team (MCST). *MODIS 1 km Calibrated Radiances Product*; Goddard Space Flight Center: Greenbelt, MD, USA, 2017. Available online: http://dx.doi.org/10.5067/MODIS/MYD021KM.061 (accessed on 1 April 2020).

57. Poulsen, C.; Siddans, R.; Thomas, G.; Sayer, A.; Grainger, R.; Campmany, E.; Dean, S.; Arnold, C.; Watts, P. Cloud retrievals from satellite data using optimal estimation: Evaluation and application to ATSR. *Atmos. Meas. Tech.* **2012**, *5*, 1889. [CrossRef]

58. McGarragh, G.R.; Poulsen, C.A.; Thomas, G.E.; Povey, A.C.; Sus, O.; Stapelberg, S.; Schlundt, C.; Proud, S.; Christensen, M.W.; Stengel, M.; et al. The Community Cloud retrieval for CLimate (CC4CL)—Part 2: The optimal estimation approach. *Atmos. Meas. Tech.* **2018**, *11*, 3397–3431. [CrossRef]

59. Jones, A.; Thomson, D.; Hort, M.; Devenish, B. The UK Met Office's next-generation atmospheric dispersion model, NAME III. In *Air Pollution Modeling and Its Application XVII*; Springer: Boston, MA, USA, 2007; pp. 580–589.

60. Hobbs, P.V.; Radke, L.F.; Lyons, J.H.; Ferek, R.J.; Coffman, D.J.; Casadevall, T.J. Airborne measurements of particle and gas emissions from the 1990 volcanic eruptions of Mount Redoubt. *J. Geophys. Res.* **1991**, *96*, 18735–18752. [CrossRef]

61. Manning, A.J.; O'Doherty, S.; Jones, A.R.; Simmonds, P.G.; Derwent, R.G. Estimating UK methane and nitrous oxide emissions from 1990 to 2007 using an inversion modeling approach. *J. Geophys. Res. Atmos.* **2011**, *116*, [CrossRef]

62. Lawson, C.L.; Hanson, R.J. *Solving Least Squares Problems*; Prentice-Hall: Englewood Cliffs, NJ, USA, 1974.

63. Webster, H.N.; Thomson, D.J.; Cooke, M.C. *Developments in the Met Office InTEM Volcanic Ash Source Estimation System Part 2: Results*; Met Office: Exeter, UK, 2017.

64. Prata, F.; Woodhouse, M.; Huppert, H.E.; Prata, A.; Thordarson, T.; Carn, S. Atmospheric processes affecting the separation of volcanic ash and SO_2 in volcanic eruptions: Inferences from the May 2011 Grímsvötn eruption. *Atmos. Chem. Phys.* **2017**, *17*, 10709–10732. [CrossRef]

65. Stevenson, J.A.; Loughlin, S.C.; Font, A.; Fuller, G.W.; MacLeod, A.; Oliver, I.W.; Jackson, B.; Horwell, C.J.; Thordarson, T.; Dawson, I. UK monitoring and deposition of tephra from the May 2011 eruption of Grímsvötn, Iceland. *J. Appl. Volcanol.* **2013**, *2*, 3. [CrossRef]

66. Tesche, M.; Glantz, P.; Johansson, C.; Norman, M.; Hiebsch, A.; Ansmann, A.; Althausen, D.; Engelmann, R.; Seifert, P. Volcanic ash over Scandinavia originating from the Grímsvötn eruptions in May 2011. *J. Geophys. Res.* **2012**, *117*, doi:10.1029/2011JD017090. [CrossRef]

67. UK Met Office. London VAAC: Volcanic ash Advisories and Graphics Archive. 2020. Available online: https://www.metoffice.gov.uk/services/transport/aviation/regulated/vaac/advisories/archive (accessed on 1 April 2020).

68. Cooke, M.C.; Francis, P.N.; Millington, S.; Saunders, R.; Witham, C. Detection of the Grímsvötn 2011 volcanic eruption plumes using infrared satellite measurements. *Atmos. Sci. Lett.* **2014**, *15*, 321–327. [CrossRef]

69. Kerminen, V.M.; Niemi, J.; Timonen, H.; Aurela, M.; Frey, A.; Carbone, S.; Saarikoski, S.; Teinilä, K.; Hakkarainen, J.; Tamminen, J.; et al. Characterization of a volcanic ash episode in southern Finland caused by the Grimsvotn eruption in Iceland in May 2011. *Atmos. Chem. Phys.* **2011**, *11*, 12227. [CrossRef]

70. Cazacu, M.; Timofte, A.; Talianu, C.; Nicolae, D.; Danila, M.; Unga, F.; Dimitriu, D.; Gurlui, S. Grímsvötn volcano: Atmospheric volcanic ash cloud investigations, modelling-forecast and experimental environmental approach upon the Romanian area. *J. Optoelectron. Adv. Mater.* **2012**, *14*, 517.

71. Kvietkus, K.; Šakalys, J.; Didžbalis, J.; Garbarienė, I.; Špirkauskaitė, N.; Remeikis, V. Atmospheric aerosol episodes over Lithuania after the May 2011 volcano eruption at Grimsvötn, Iceland. *Atmos. Res.* **2013**, *122*, 93–101. [CrossRef]

72. Pruppacher, H.R.; Klett, J.D. Microphysics of clouds and precipitation. *Nature* **1980**, *284*, 88. [CrossRef]

73. Neal, R.A.; Boyle, P.; Grahame, N.; Mylne, K.; Sharpe, M. Ensemble based first guess support towards a risk-based severe weather warning service. *Meteorol. Appl.* **2014**, *21*, 563–577. [CrossRef]

74. Webster, H.; Thomson, D.; Johnson, B.; Heard, I.; Turnbull, K.; Marenco, F.; Kristiansen, N.; Dorsey, J.; Minikin, A.; Weinzierl, B.; et al. Operational prediction of ash concentrations in the distal volcanic cloud from the 2010 Eyjafjallajökull eruption. *J. Geophys. Res.* **2012**, *117*, doi:10.1029/2011JD016790 [CrossRef]

75. EUMETSAT. *IASI: Atmospheric Sounding Level 1C Data Products*; NERC Earth Observation Data Centre: Oxford, UK, 2009. Available online: https://catalogue.ceda.ac.uk/uuid/ea46600afc4559827f31dbfbb8894c2e (accessed on 27 August 2020).

76. European Centre for Medium-Range Weather Forecasts. *ECMWF Operational Regular Gridded Data at 1.125 Degrees Resolution*; NCAS British Atmospheric Data Centre: Oxford, UK, 2012. Available online: https://catalogue.ceda.ac.uk/uuid/a67f1b4d9db7b1528b800ed48198bdac (accessed on 27 August 2020).

2

Multi-Model Ensemble Sub-Seasonal Forecasting of Precipitation over the Maritime Continent in Boreal Summer

Yan Wang [1,2,3], Hong-Li Ren [2,3,*], Fang Zhou [4], Joshua-Xiouhua Fu [5], Quan-Liang Chen [1], Jie Wu [3], Wei-Hua Jie [3] and Pei-Qun Zhang [3]

[1] Plateau Atmosphere and Environment Key Laboratory of Sichuan Province, College of Atmospheric Science, Chengdu University of Information Technology, Chengdu 610225, China; wangy_0618@126.com (Y.W.); chenql@cuit.edu.cn (Q.-L.C.)

[2] State Key Laboratory of Severe Weather, Chinese Academy of Meteorological Sciences, Beijing 100081, China

[3] Laboratory for Climate Studies & CMA-NJU Joint Laboratory for Climate Prediction Studies, National Climate Center, China Meteorological Administration, Beijing 100081, China; wujie@cma.gov.cn (J.W.); jiewh@cma.gov.cn (W.-H.J.); zhangpq@cma.gov.cn (P.-Q.Z.)

[4] Climate Change Research Center, Institute of Atmospheric Physics, and Nansen–Zhu International Research Centre, Chinese Academy of Sciences, Beijing 100029, China; zhouf@cma.gov.cn

[5] Department of Atmospheric and Oceanic Sciences & Institute of Atmospheric Sciences, Fudan University, Shanghai 200433, China; fuxh@fudan.edu.cn

* Correspondence: renhl@cma.gov.cn

Abstract: The Maritime Continent (MC) is a critical region with unique geographical conditions and significant monsoon activities that plays a vital role in global climate variation. In this study, the weekly prediction of precipitation over the MC during boreal summer (from May to September) was analyzed using the 12-year reforecasts data from five Sub-seasonal to Seasonal (S2S) models, including the China Meteorological Administration (CMA), the European Centre for Medium-Range Weather Forecasts (ECMWF), Environment and Climate Change Canada (ECCC), the National Centers for Environmental Prediction (NCEP), and the Met Office (UKMO). The result shows that, compared with the individual models, our newly derived median multi-model ensemble (MME) can significantly improve the prediction skill of sub-seasonal precipitation in the MC. Both the Temporal Correlation Coefficient (TCC) skill and the Pattern Correlation Coefficient (PCC) skill reached 0.6 in lead week 1, dropped the following week, did not exceed 0.2 in lead week 3, and then lost their significance. The results show higher prediction skill near the Equator than in the north at 10° N. It is difficult to make effective predictions with the models beyond three weeks. The prediction ability of the median MME improves significantly as the total number of model members increases. The prediction performance of the median MME depends not only on the diversity of models but also on the number of model members. Moreover, the prediction skill is particularly sensitive to the intensity and phase of Boreal Summer Intraseasonal Oscillation 1 (BSISO1) with the highest skills appearing at initial phases 1 and 5.

Keywords: Maritime Continent; multi-model ensemble; sub-seasonal prediction; precipitation

1. Introduction

The Maritime Continent (MC), located at the junction of the Indian Ocean and the Pacific Ocean, consists of several islands and shallow seas [1]. It is an essential area in the Northern and Southern Hemispheres that interacts through the cross-equatorial flow. Asian–Australian monsoon activity is closely related to atmospheric changes in this area [2]. Due to the distinct geographical conditions and remarkable monsoon activities, MC plays a critical role in surrounding and even global climate variations [3]. Convection in the MC region features multi-scale spatial-temporal activities [4]. Since the 1960s, many studies have been conducted on the climatic features and impacts of the MC region [5–9].

The sub-seasonal to seasonal (S2S) prediction (the so called "desert of predictability") is an urgent problem that necessarily needs to be solved. On one hand, S2S fills in the gap between medium-to-long range weather forecast and seasonal prediction. On the other hand, the improvement of S2S forecast for extreme events will contribute to the prevention of climatic disaster and economic loss [10]. S2S prediction lies between weather forecast and climate prediction, in which initial conditions as well as boundary conditions are extremely important [11–14]. The study [15] has shown that there are several potential sources of predictability that contribute to this timescale, such as the intraseasonal oscillation (ISO), El Nino Southern Oscillation (ENSO), soil moisture, snow cover, sea ice, stratosphere–troposphere interaction, etc. [16–20]. Madden–Julian Oscillation (MJO) is the primary intraseasonal mode of organized convective activity in tropical regions, and it significantly influences on the global monsoon and also has a considerable impact on the middle and high latitudes [21–23]. In addition, the Boreal Summer Intraseasonal Oscillation (BSISO) has become the dominant mode during boreal summer [24,25].

The S2S Prediction Project and associated database [26] have already been jointly established by the World Weather Research Programme (WWRP)/World Climate Research Programme (WCRP) to promote S2S research and operational prediction [27]. Studies based on S2S model outputs have been carried out recently. The MJO prediction skill of the majority of S2S models has exceeded 20 days. Although models have significantly improved the predictions of MJO and major climatic phenomena [28,29], they tend to underestimate the intensity and propagation speed of MJO, and it is difficult to propagate MJO across the MC [30]. Some studies also show model assessment for monsoons, extreme events, and tropical cyclones (TCs). It is shown that models have some capability in reproducing the atmospheric dynamics that influence monsoon variability, predicting the onset, evolution, and decay of some large-scale extreme events and predicting the occurrence of basin-wide TCs. The eye-catching performance of the European Centre for Medium-Range Weather Forecasts (ECMWF) model is no surprise [31–33]. Some previous studies have focused on the comparison of multi-model performances and the evaluation of predictability sources [34–38].

The multi-model ensemble (MME) is found to be an effective approach to improve weather [39,40] and climate prediction [41,42]. MME can greatly reduce the uncertainty of the forecast and increases its reliability compared to single-model ensemble prediction [43,44]. To better promote the development of MME forecast for short-term climate (seasonal to interannual) prediction, several research projects, such as ENSEMBLES [45], the US National Multi-Model Ensemble (NMME) [46], and China MME (CMME) [47] have been launched. The implementations of the MME technique have effectively improved the prediction of weather (e.g., tropical cyclones) and climate variability on both seasonal [48–51] and sub-seasonal timescales [52,53]. With the S2S database, Specq et al. [54] first attempted to assess the balanced MME forecast skill of precipitation over the southwest tropical Pacific at the sub-seasonal timescale.

Compared to well-implemented weather and seasonal forecasts, less attention has been paid to sub-seasonal prediction in general, much less the precipitation in the MC. Studying precipitation in the MC contributes to raising awareness of climate evolution and provides scientific reference for precipitation prediction. This paper aims at constructing MME forecast and revealing the predictability of sub-seasonal precipitation during boreal summer in the MC. This paper is organized as follows: models, data, and methods are introduced in Section 2; in Section 3, we focus on the prediction skill of individual models and the MME; the association of MME prediction with MJO and BSISO is analyzed in Section 4. The summary and discussion are given in Section 5.

2. Models, Data, and Methods

2.1. Data

Data products generated from five S2S models are used in this study, namely the China Meteorological Administration (CMA), ECMWF, Environment and Climate Change Canada (ECCC), the National Centers for Environmental Prediction (NCEP), and the Met Office (UKMO). Boreal summer is defined as May–June–July–August–September (MJJAS). Noting that the S2S database was not produced following a consistent protocol, basic information on the 5 models is shown in Table 1.

Table 1. Information on the five Sub-seasonal to seasonal models used in this study.

Model	Resolution	Model Top	Time Range	Reforecast Type	Period	Frequency	Size
CMA	1° × 1° L40	0.5 hPa	Day 1–60	Fix	1994–2014	Daily	4
UKMO	0.5° × 0.8° L85	85 km	Day 1–60	On the fly	1993–2015	4/month	3
ECMWF	0.25° × 0.25°/0.5° × 0.5° L91	0.01 hPa	Day 1–46	On the fly	1996–2015	2/week	11
ECCC	0.45° × 0.45° L40	2h Pa	Day 1–32	On the fly	1995–2014	Weekly	4
NCEP	1° × 1° L64	0.02 Pa	Day 1–44	Fix	1999–2010	Daily	4

The analysis period of the five model outputs is unified to 1999–2010 MJJAS. Each month consists of four start dates; the start dates of May are shown in Figure 1. The observational daily precipitation data were provided by the Global Precipitation Climatology Project (GPCP) [55] with a horizontal resolution of 1° × 1° latitude/longitude. As part of the WCRP, the GPCP aims to observe and estimate global precipitation. It is based on more than 6000 routine surface observation stations and integrates the satellite observational results. For the purpose of comparing the results of models and observation in the MC, model and observational data were consistently interpolated onto the 2.5° × 2.5° grid by using the bilinear interpolation. In order to examine the relationship between sub-seasonal precipitation and ISO in the MC, the Real-time Multivariate MJO (RMM) index [56,57] and the BSISO index [58] were also used. The MC domain was taken as 90° E–150° E, 10° S–20° N.

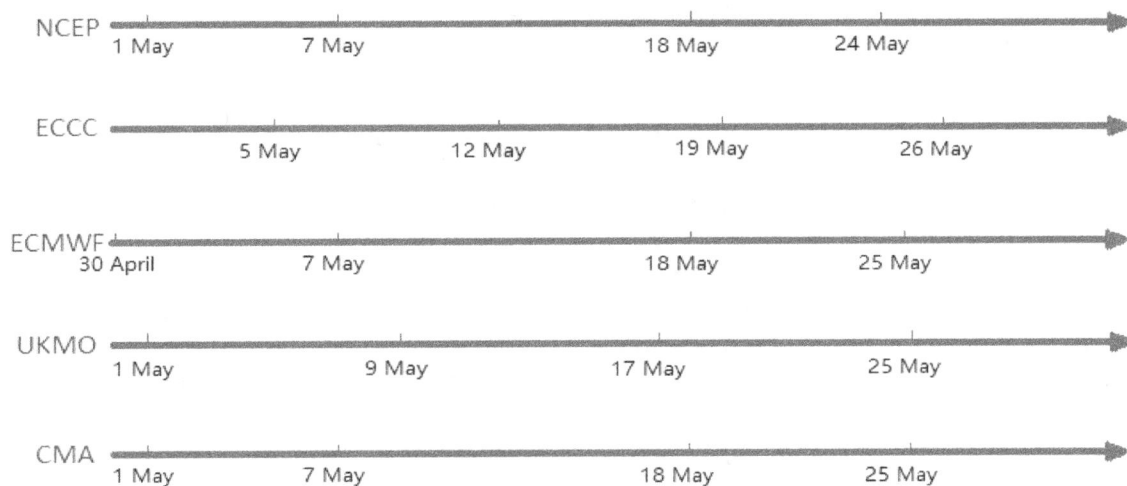

Figure 1. Start dates of the five models used in May.

2.2. Construction of MME

As can be seen from Table 1 and Figure 1, it is obvious that the five S2S models do not follow an agreed-upon protocol; there is also no standard construction of the MME due to the start dates of each model being different. Thus, the construction of an S2S-MME is quite necessary but a challenging issue. This study tries to make such an S2S-MME possible by constructing a median MME instead of traditional arithmetic MME (e.g., Ren et al. [47]), also referring to the work of Specq et al. [54]. We think that median MME is a better method in terms of different S2S models. The construction of the MME refers to the start date of the CMA. Table 2 is an example of constructing MME with three different start dates (30 April, 1 May, 5 May) for the forecast of the target week (6–12 May). In this example, the CMA's lead-a-week forecast for 6 to 12 May can choose 1 May as the start date. For the ECCC, the nearest start date is 5 May. Therefore, a given forecast target week may correspond to different start dates for different models.

Table 2. Example of constructing a multi-model ensemble (MME) with the first start date.

Model	Start Date	Lead Days	Target Week 1
CMA	1 May	Lead 5 to 11	6–12 May
UKMO	1 May	Lead 5 to 11	6–12 May
ECMWF	30 April	Lead 6 to 12	6–12 May
ECCC	5 May	Lead 1 to 7	6–12 May
NCEP	1 May	Lead 5 to 11	6–12 May

The number of ensemble members is not exactly the same for each model (CMA, ECCC, and NCEP each have 4 members, ECMWF has 11 members, and UKMO has 3 members); therefore, a uniform number of members was selected to construct MME. In this study, we considered constructing MMEs of 15, 10, and 5 members (including 3, 2, and 1 from each model), respectively. The median performance ensemble of each model was selected to construct the MME. For example, CMA has four ensemble members (a, b, c, d), and thus we can get four ensembles (abc, abd, acd, bcd) constructed by three members. We chose the median of four ensembles as the ensemble of CMA, and the MME was composed of the respective median ensemble of five models.

2.3. Forecasting Skill Metrics

In this paper, WMO-recommended standards are used to evaluate prediction performance [59]. The Temporal Correlation Coefficient (TCC) is used to measure the prediction skill. TCC can represent the temporal agreement of model forecast statistical significance and is able to acquire a distribution of prediction skill by calculating the TCC of each grid point. The calculation formation is given as:

$$TCC_i = \frac{\sum_{j=1}^{N}(x_{i,j} - \overline{x}_i)(y_j - \overline{y})}{\sqrt{\sum_{j=1}^{N}(x_{i,j} - \overline{x}_i)^2}\sqrt{\sum_{j=1}^{N}(y_j - \overline{y})^2}}, \tag{1}$$

where i denotes the start date and j denotes the target week. $x_{i,j}$ represents the predicted week j mean anomalies starting on date i, and $\overline{x}_i$ represents the average of predicted mean anomalies starting on date i of each week. y_i expresses the observation anomalies in week j, and $\overline{y}$ expresses the average of y_j. N is the number of time samples. The range of TCC is between -1 and 1, and the closer the TCC is to 1, the higher the forecasting skill.

Unlike TCC, the Pattern Correlation Coefficient (PCC) reflects the spatial agreement of the model forecast and acquires a variation of prediction skill by calculating PCC of different times. The calculation formula is as follows:

$$PCC_j = \frac{\sum_{i=1}^{M} \Delta x_{i,j} \Delta y_{i,j}}{\sqrt{\sum_{j=1}^{M} \Delta x_{i,j}^2}\sqrt{\sum_{j=1}^{M} \Delta y_{i,j}^2}}, \tag{2}$$

where j denotes the target week and i denotes the grid point. $\Delta x_{i,j}$ and $\Delta y_{i,j}$ represent the predicted and observational week j mean anomalies on grid i, respectively. M is the total number of the grid point over the MC. The PCC also ranges from -1 to 1. A larger PCC indicates a more consistent spatial pattern between model forecast and observation.

3. Prediction Skills of Individual Models and MME

The TCC skills of the five individual models and the MME are shown in Figure 2. The ensemble member of each model is unified for three members. The result shows that the TCC skill of each model decreases with the increase in lead time. During lead week 1, TCC of most of the MC regions is greater than 0.4. After lead week 1, prediction skill rapidly drops. The prediction skill of the MME is significantly larger than the single model, and most regions can still reach above 0.3 in lead week 2. Compared with the single model, the MME still shows significant skills in lead week 3, and all models lose their prediction ability in lead week 4. The performance of the ECMWF and NCEP is relatively better than the other three models. Furthermore, the results show that prediction skill by the Equator tends to be higher than that north of the Equator.

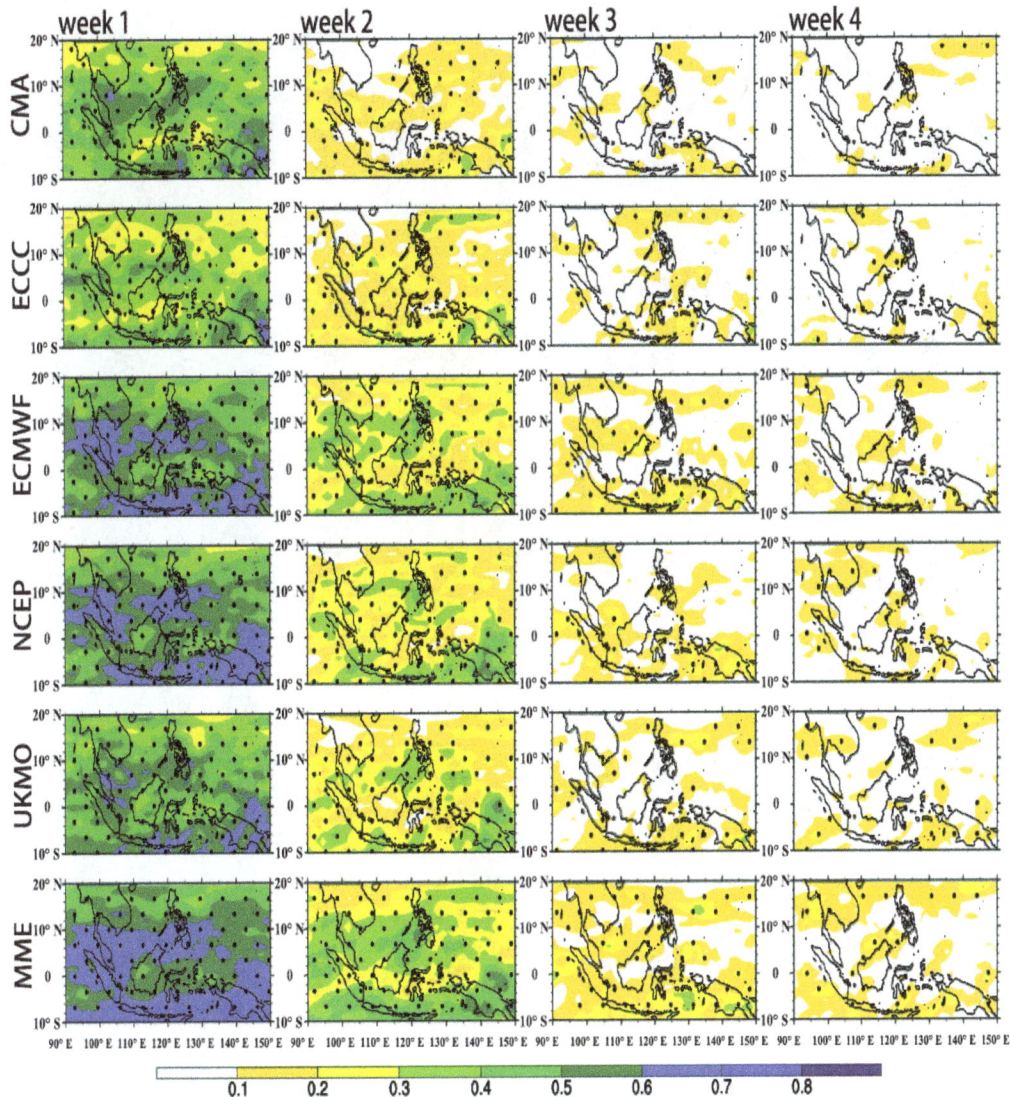

Figure 2. Spatial distribution of temporal correlation coefficient (TCC) scores of precipitation anomalies in May–June–July–August–September (MJJAS) for China Meteorological Administration (CMA), Environment and Climate Change Canada (ECCC), European Centre for Medium-Range Weather Forecasts (ECMWF), National Centers for Environmental Prediction (NCEP), the UK Met Office (UKMO), and the 15-member MME (from top to bottom). The 1st, 2nd, 3rd, and 4th columns are predictions for lead weeks 1–4, respectively. Areas exceeding the 95% confidence level are dotted.

The prediction skill of the MC is significantly higher than those of the middle and high latitudes [60]. As shown in the MME, high-skill regions are concentrated near the Equator over ocean;

Sumatra, Kalimantan and the Philippines have higher skill over land. There is no significant difference through analyzing the spatial average of the TCC skill over the ocean and the land. Interestingly, the skill over ocean is more significant after lead week 3, which indicates that the skill over ocean is higher compared to that over land.

Figure 3 shows the interannual variations of the PCC skills for the five individual models and the MME. Similar to TCC, the PCC skill of each model decreases with the increase in lead time, and the prediction skill of the MME is relatively higher than individual models. The PCC skills of all models drop significantly in lead week 3 but are still positive for most cases, even for lead week 4.

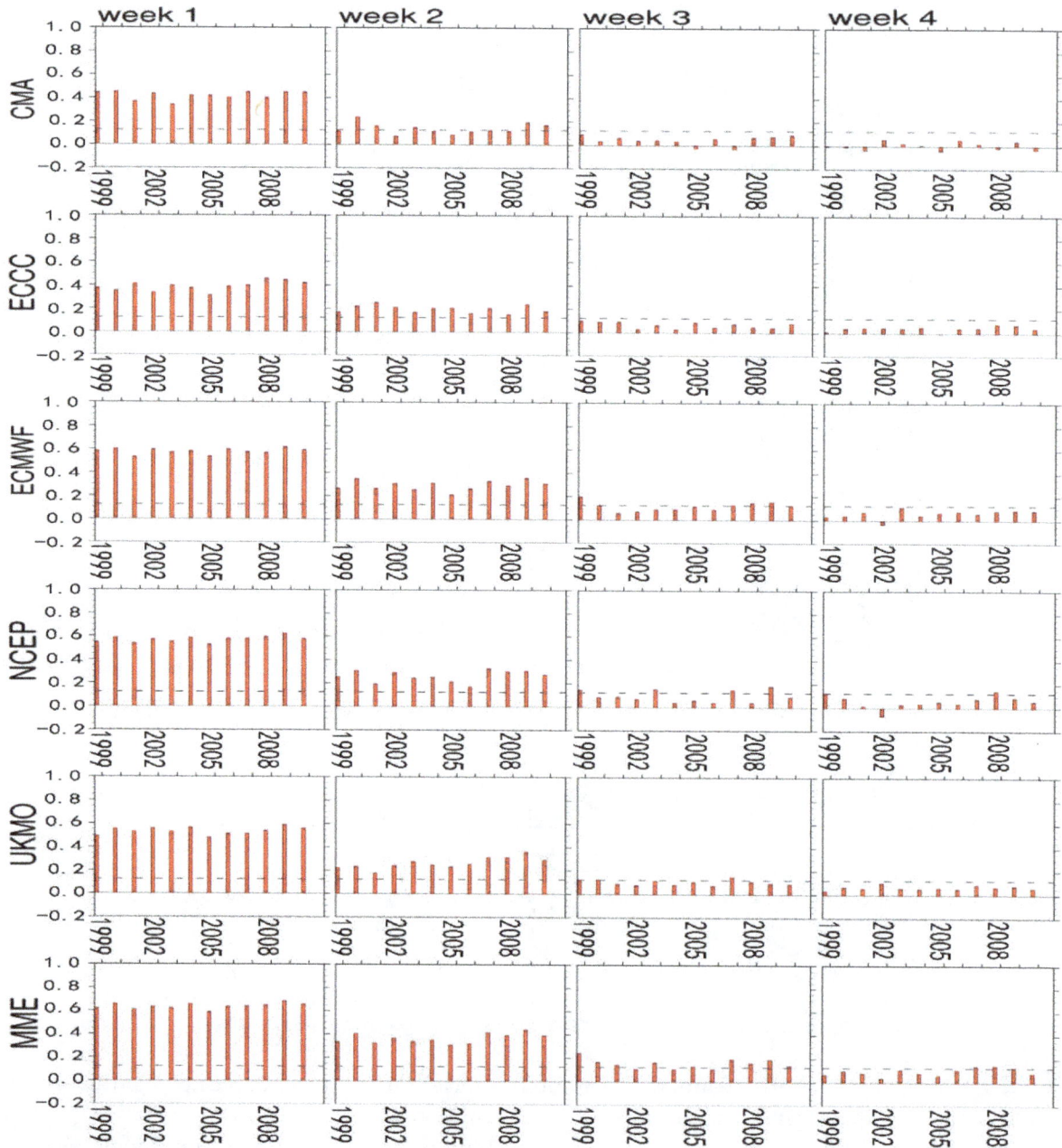

Figure 3. Yearly averaged Pattern Correlation Coefficient (PCC) scores of precipitation anomalies in MJJAS for CMA, ECCC, ECMWF, NCEP, UKMO, and the 15-member MME (from top to bottom). The 1st, 2nd, 3rd, and 4th columns are predictions for lead weeks 1–4, respectively. The dashed black line represents the 95% significance threshold.

Focusing on the overall prediction skill in the MC, the spatial average (90° E–150° E, 10° S–20° N) of the TCC and time average (1999–2010) of the PCC are shown in Figure 4. The skills are relatively high in lead week 1 and rapidly drop in lead week 2. The ECMWF has the best performance, with the NCEP and UKMO very close behind, followed by the CMA and ECCC models. MME can effectively improve the forecast skill and is still significant in lead week 3, although the skills of all models are not significant. The forecast skill after 3 weeks decreases slowly and is not statistically significant as the lead week increases. This indicates that the MME lost its prediction skill after three weeks. In addition, the time average of the PCC is relatively higher than the spatial average of the TCC, indicating that the model is better at capturing the spatial distribution characteristics of precipitation relative to its temporal evolution.

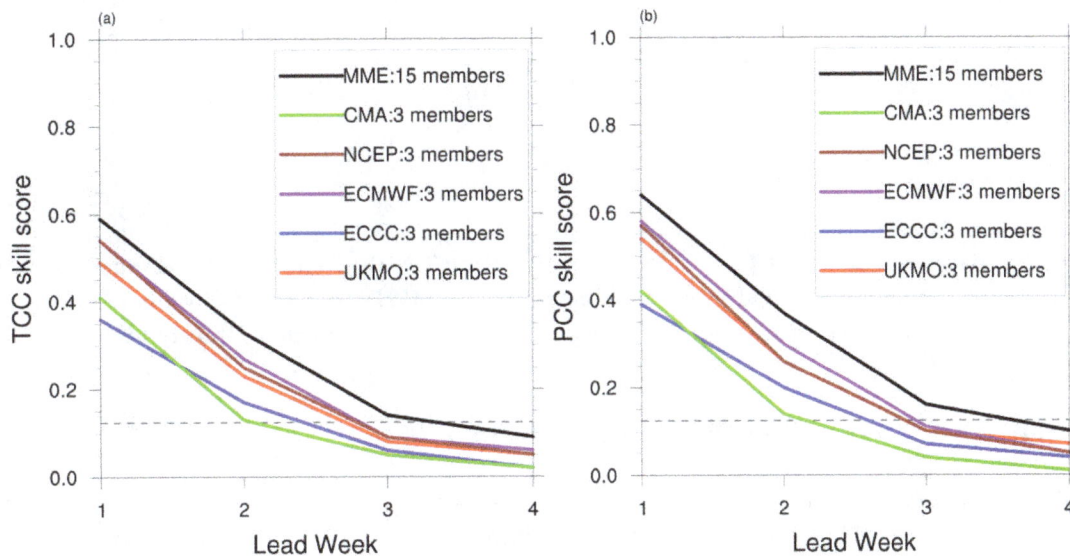

Figure 4. Spatial average of the TCC (**a**) and time average of the PCC (**b**) over the MC. CMA, NCEP, ECMWF, ECCC, UKMO, and the 15-member MME are, respectively, plotted in green, brown, purple, blue and red. The dashed black line represents the 95% significance threshold.

To explore the impact of ensemble members on the prediction skill of the MME, MMEs of 5, 10, and 15 members were constructed. We call this the median MME because the MMEs are constructed with five models, as shown in Figure 5. The prediction skill of the MME relies on ensemble members, with more members denoting higher skill when the number of models is constant, which is reflected by both the TCC and PCC skills. To compare the prediction ability of the MME with that of the individual model under the same number of members, the ECMWF model of 10 members ensemble was also plotted. Although the ECMWF model has excellent performance (consistent with the results in Figures 2–4), the MME can still improve the prediction ability.

In order to explore the impact of the diversity of models on the prediction ability of the MME, we constructed two types of 15-member MMEs for comparison. The ECMWF model with 11 members and the NCEP model with four members were selected to construct a 15-member MME, known as the optimal MME. As shown in Figure 5, impressively, the median MME skill is larger than the optimal MME. The diversity of models significantly contributes to the performance of the MME than the single model with better performance. This shows that the number and the diversity of models with a certain number of ensemble members needed to be taken into consideration when constructing an MME. Some suitable ensemble methods also have a great impact on improving the prediction performance of the MME, which is also worth further exploring.

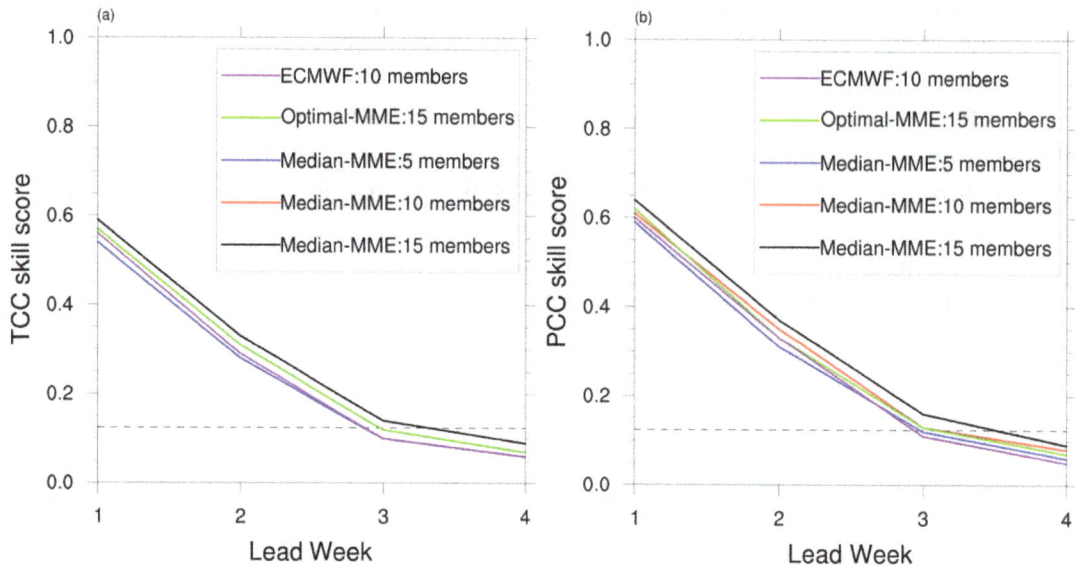

Figure 5. Spatial average of the TCC (**a**) and time average of the PCC (**b**); 5-member, 10-member, 15-member median MMEs, 15-member optimal MME, and 10-member ECMWF are, respectively, plotted in blue, red, black, green and purple. The dashed black line represents the 95% significance threshold.

4. Association of MME Prediction with MJO and BSISO

ISO plays a crucial role in the prediction of sub-seasonal timescales. Madden and Julian [61,62] defined the tropical atmospheric low-frequency oscillation MJO as the most significant ISO in tropical regions, especially in winter. The MJO convection signal is strongest over the tropical Indian Ocean and Western Pacific. It propagates eastward with a speed of 5 m s^{-1}, circulating the globe within a period of 30–60 days. The MJO index [56] can visually reflect the position and propagation of MJO according to the two-dimensional spatial phase diagram composed of RMM.

In addition to the eastward propagation, ISO in the boreal summer also has a significant northward propagating component, which has been popularly called BSISO [63]. The BSISO index proposed by Lee et al. [25] include BSISO1 and BSISO2 indices, which describe both the eastward and northward movements of BSISO within a period of 10–60 days. The BSISO1 index describes the typical northward and the eastward propagation within a period of 30 to 60 days.

The PCC skill of precipitation in lead week 1 was scattered against the amplitudes of BSISO1 and MJO to explore the correlation between the prediction skill of the 15-member median MME and the ISO intensity by linear regression. As shown in Figure 6, the prediction skill of the MME is weakly positively correlated with MJO intensity, and the correlation coefficient is around 0.12. Compared with MJO, BSISO1 has a prominent correlation relationship with a correlation coefficient of 0.37. It shows that the amplitude of BSISO1 has a significant influence on weekly precipitation prediction; in other words, BSISO1 provides important sub-seasonal predictability for precipitation in the MC.

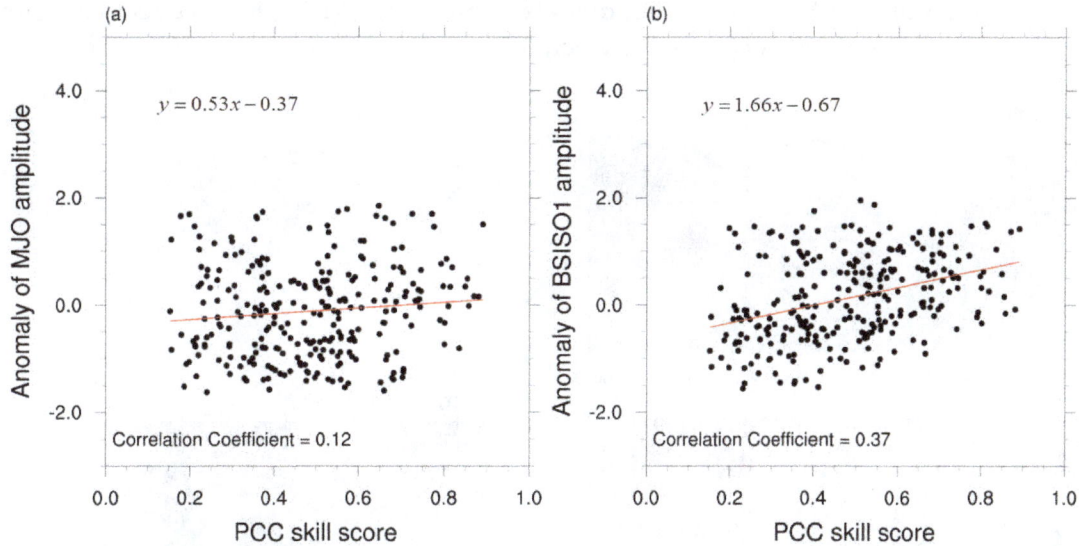

Figure 6. Scatter plots of the PCC of 15-member median MME prediction in lead week 1 versus the anomalies of (**a**) Madden–Julian Oscillation (MJO) amplitude and (**b**) Boreal Summer Intraseasonal Oscillation 1 (BSISO1) amplitude.

In order to further explore the relationship between BSISO1 phase and MME prediction, the prediction skill for different BSISO1 phases is displayed in Figure 7. The PCC skill for each phase is the highest in lead week 1, and skill decreases with the increase in lead week. The skill of phase 1 and phase 5 is relatively high and declines slowly after lead week 2, indicating that there is relatively high predictability when initiated in phase 1 and phase 5. The skill initiated in phase 3 is the lowest.

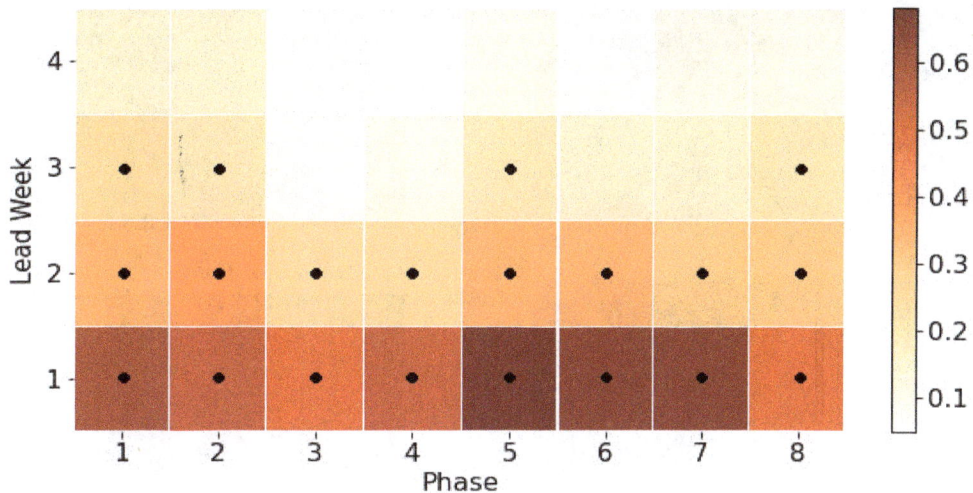

Figure 7. Averaged PCC of 15-member median MME prediction as a function of lead week and initial BSISO1 phase; results exceeding the 95% confidence level are dotted.

Since phase 1 and phase 5 of BSISO1 have relatively high predictability, the two phases with the prominent prediction skill are as shown in Figures 8 and 9 skill of phase 1 is higher). In phase 1 (Figure 8), during the 1–4 weeks of observation, there is a rain belt between the west of Sumatra and Kalimantan Island, and the dry area north of 10° N shrinks after week 1. There is a trend of eastward movement over time; this trend can be basically simulated in the first two weeks, and there is some simulation capability for the dry regions of the eastern Philippines, but it is difficult to simulate the west of the Philippines (Figure 8f,g). Furthermore, models in median MME failed to effectively simulate the propagation and position of convection after week 3.

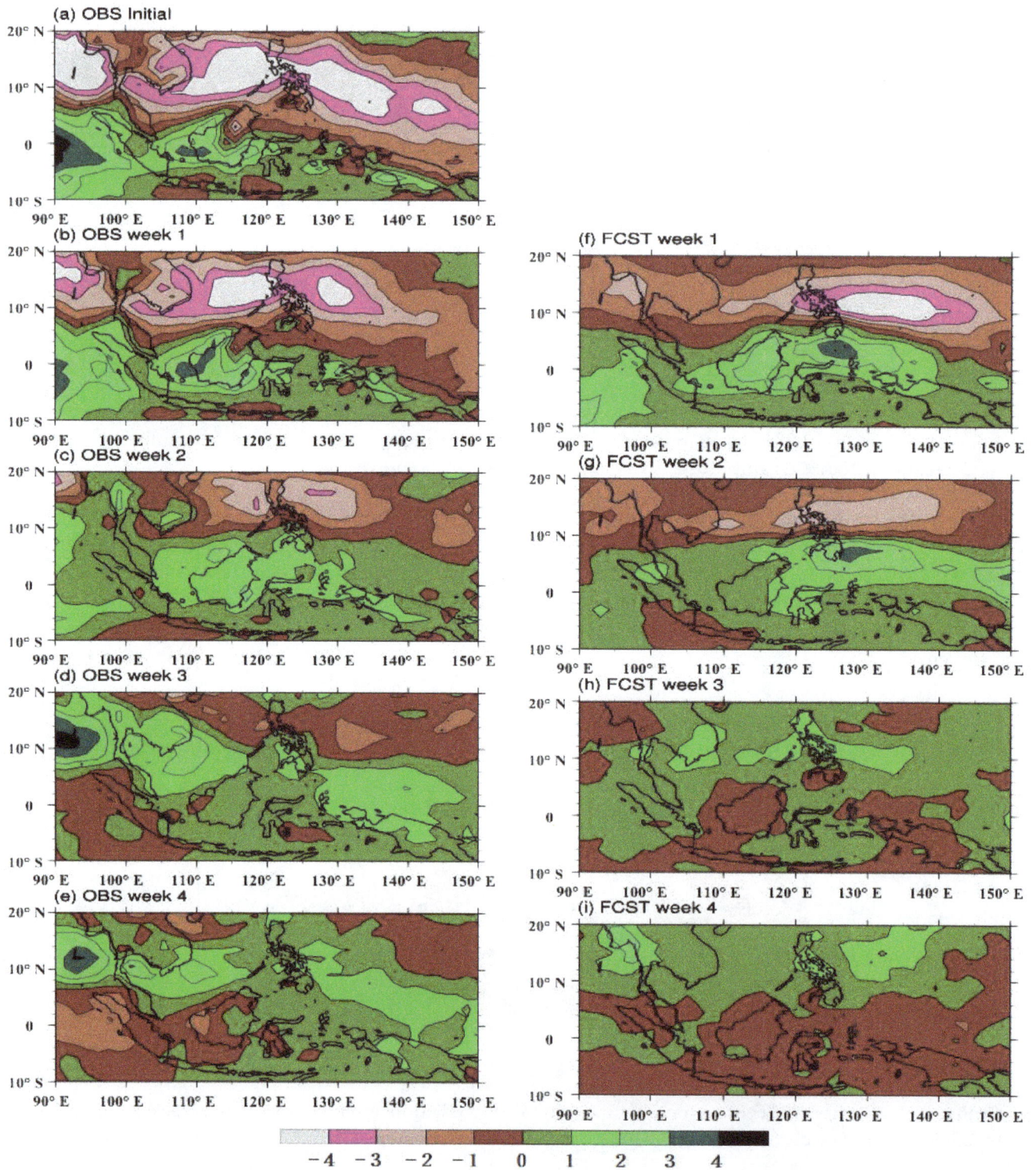

Figure 8. Composite observation (left panels) and (median MME) forecast precipitation anomalies (mm/day) from initial conditions of BSISO1 phase 1. (**a**) Observed initial conditions, (**b–e**) observed average for weeks 1–4, and (**f–i**) forecast average for weeks 1–4.

In phase 5 (Figure 9), the convection occurs near the Bay of Bengal to the South China Sea and the northwest Pacific region in the north of New Guinea. There are significant dry regions in Kalimantan

Island and west of Sumatra. Convection tends to weaken and propagates toward the northeast with the increase in lead week. However, the simulation ability is weak and lasts for just two weeks. The models in the median MME can capture the convection cells in mainland Southeast Asia, near the South China Sea and east of the Philippines, but, in the Indochina Peninsula, the models tend to underestimate precipitation in lead week 1 and overestimate precipitation in lead week 2. Moreover, the capture of the eastward signal lags for models in the median MME (Figure 9c,g).

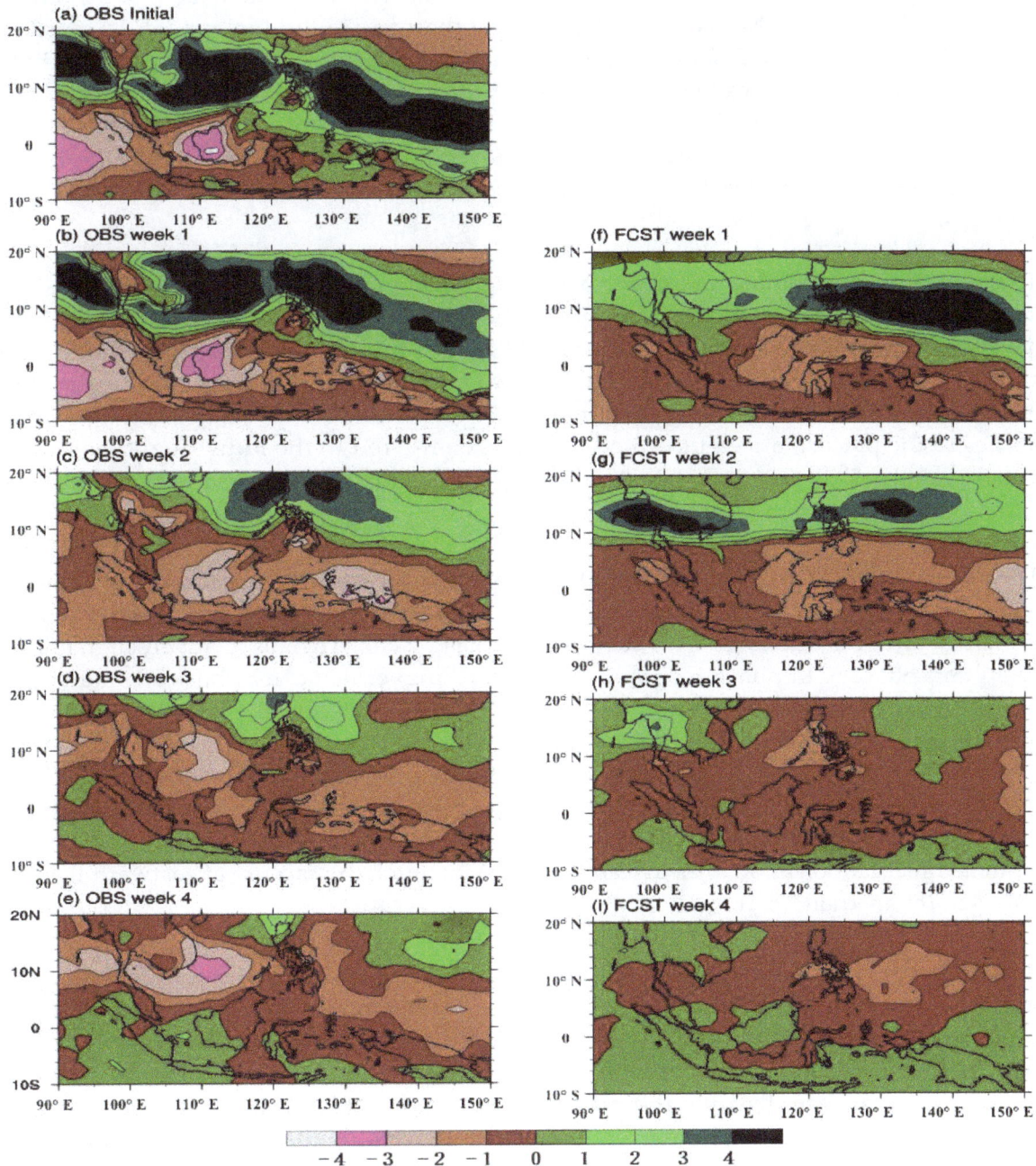

Figure 9. Composite observation (left panels) and (median MME) forecast precipitation anomalies (mm/day) from initial conditions of BSISO1 phase 5. (**a**) Observed initial conditions, (**b–e**) observed average for weeks 1–4, and (**f–i**) forecast average for weeks 1–4.

5. Summary and Discussions

In this paper, 12-year (1999–2010) reforecasts from five S2S models, including CMA, ECMWF, ECCC, NCEP, and UKMO, are used to study the sub-seasonal predictability of the precipitation of the

MC during the boreal summer. Based on the 20 start dates (May–September), a median MME was constructed, and the multi-week prediction of precipitation in the MC was evaluated based on TCC and PCC skill scores. The relationship between prediction skill in this region and the MJO and BSISO is examined.

It is found that there are significant differences between the abilities of the five individual models and their Multi-model ensemble. The prediction skills of all models decrease with the increasing the lead week. The performances of ECMWF and NCEP are relatively better than those of the other models. The prediction skill of the MME is significantly larger than any individual models. According to the spatial distribution of the TCC skill, the prediction skill is higher in low latitudes than that in high latitudes, and the skill over ocean is more significant than that over land after lead week 3. PCC skills exhibit significantly interannual variations. Both skill measures are relatively high in lead week 1, then rapidly drop in lead week 2. In lead week 3, the skills of all models are not significant except for the MME. Therefore, the models have no useful prediction skill for summer precipitation in the MC after 3 weeks.

The number of ensemble members has a significant impact on the prediction skill of the median MME, with more members denoting higher skill when the number of models is constant. When ensemble members are constant, median MME can improve the prediction ability compared with the best-performing model. It is obvious in both that the improvement in prediction skill comes from the error elimination of the individual model. Furthermore, the diversity of models significantly contributes to the performance of the median MME compared to the individual model with better performance, so the diversity of models should be given priority to construct the MME, and there is no doubt that the importance of both is worth exploring further.

The intensity of ISO has a significant influence on weekly precipitation prediction; this characteristic is especially true with the BSISO. Precipitation prediction skills for different BSISO1 phases are also quite different, and skill decreases with the increase in lead week. There is relatively high predictability in phase 1 and phase 5. In phase 1, a trend of eastward movement can be basically simulated in the first two weeks, but models in the median MME failed to effectively simulate the propagation and position of convection after week 3. In phase 5, models in the median MME can basically simulate the convection cells in mainland Southeast Asia, near the South China Sea and east of the Philippines, within 2 weeks, based on a composite analysis for 1-4 weeks of precipitation anomalies. In the Indochina Peninsula, the models tend to underestimate precipitation anomalies in lead week 1 and overestimate precipitation anomalies in lead week 2. Moreover, the capture of eastward the signal lags for models in the median MME.

Although we have demonstrated that the MME is an effective way to improve the prediction skill of sub-seasonal precipitation in the MC, a considerable amount of further research is warranted. The construction of the MME requires the participation of more S2S models. The proportion of excellent models and the diversity of models require further research when constructing the MME. The key physical processes [64] and initial value problems [65] of the S2S models still need in-depth diagnosis and analysis. Considering the interaction between sub-seasonal scale and other scale signals is also an important direction to improve short-term climate prediction [66]. Furthermore, applying dynamic-statistical prediction methods [67–71] on the base of the MME is necessary for the improvement of S2S prediction.

Author Contributions: Conceptualization, H.-L.R., J.-X.F. and Q.-L.C.; methodology, H.-L.R. and J.-X.F.; data curation, W.-H.J.; writing—original draft preparation, Y.W.; writing—review and editing, H.-L.R., F.Z., J.-X.F. and J.W.; visualization, Y.W.; project administration, H.-L.R. and P.-Q.Z.; funding acquisition, H.-L.R. and J.W. All authors have read and agreed to the published version of the manuscript.

References

1. Ramage, C.S. Role of tropical "Maritime Continent" in the atmospheric circulation. *Mon. Weather Rev.* **1968**, *96*, 365–370. [CrossRef]
2. McBride, J.L. Indonesia, Papua New Guinea, and tropical Australia: The southern hemisphere monsoon, in meteorology of the Southern Hemisphere. *Am. Meteorol. Soc.* **1998**, *27*, 89–100. [CrossRef]
3. Neale, R.; Slingo, J. The Maritime continent and its role in the global climate: A GCM study. *J. Clim.* **2003**, *16*, 834–848. [CrossRef]
4. Meehl, G.A. The annual cycle and interannual variability in the tropical Pacific and Indian ocean regions. *Mon. Weather Rev.* **1987**, *115*, 27–50. [CrossRef]
5. McBride, J.L.; Haylock, M.R.; Nicholls, N. Relationships between the Maritime Continent heat source and the EI Nino-Southern Oscillation phenomenon. *J. Clim.* **2003**, *16*, 2905–2914. [CrossRef]
6. Chang, C.P.; Harr, P.A.; Chen, H.J. Synoptic disturbances over the equatorial South China sea and Western Maritime Continent during boreal winter. *Mon. Weather Rev.* **2005**, *133*, 489–503. [CrossRef]
7. Qian, W.H.; Lee, D.K. Seasonal march of Asian summer monsoon. *Int. J. Climatol.* **2000**, *20*, 1371–1386. [CrossRef]
8. Qian, J.H. Why precipitation is mostly concentrated over islands in the Maritime Continent. *J. Atmos. Sci.* **2008**, *65*, 1428–1441. [CrossRef]
9. Wang, B.; Ding, Q.H. Global monsoon: Dominant mode of annual variation in the tropics. *Dyn. Atmos. Ocean.* **2008**, *44*, 165–183. [CrossRef]
10. Materia, S.; Muñoz, Á.G.; Álvarez-Castro, M.C.; Mason, S.J.; Vitart, F.; Gualdi, S. Multimodel Subseasonal Forecasts of Spring Cold Spells: Potential Value for the Hazelnut Agribusiness. *Weather. Forecast.* **2020**, *35*, 237–254. [CrossRef]
11. National Research Council; Division on Earth and Life Studies; Board on Atmospheric Sciences and Climate; Committee on Assessment of Intraseasonal to Interannual Climate Prediction and Predictability. *Assessment of Intraseasonal to Interannual Climate Prediction and Predictability*; The National Academies Press: Washington, DC, USA, 2010. [CrossRef]
12. Fu, X.; Wang, B.; Waliser, D.; Tao, L. Impact of atmosphere-ocean coipling on the predictability of monsoon intraseasonal oscillation. *J. Atmos. Sci.* **2007**, *64*, 157–174. [CrossRef]
13. Fu, X.; Yang, B.; Bao, Q.; Wang, B. Sea surface temperature feedback extends the predictability of tropical intraseasonal oscillation. *Mon. Wea. Rev.* **2008**, *136*, 577–596. [CrossRef]
14. Fu, X.; Wang, B.; Bao, Q.; Liu, P.; Lee, J.Y. Impacts of initial conditions on monsoon intraseasonal forecasting. *Geo. Res. Lett.* **2009**, *36*, L08801. [CrossRef]
15. Vitart, F.; Robertson, A.W.; S2S Steering Group. *Sub-Seasonal to Seasonal Prediction: Linking Weather and Climate. Seamless Prediction of the Earth System: from Minutes to Months (WMO-No. 1156)*; WMO: Geneva, Switzerland, 2015; pp. 385–401.
16. Waliser, D.E.; Sperber, K.; Hendon, H.; Kim, D.; Maloney, E.; Wheeler, M.; Weickmann, K.; Zhang, C.; Donner, L.; Gottschalck, J.; et al. MJO simulation diagnostics. *J. Clim.* **2009**, *22*, 3006–3030. [CrossRef]
17. Koster, R.D.; Mahanama, S.; Yamada, T.; Balsamo, G.; Berg, A.A.; Boisserie, M.; Dirmeyer, P.A.; Doblas-Reyes, F.J.; Drewitt, G.; Gordon, C.T.; et al. The Contribution of Land Surface Initialization to Subseasonal Forecast Skill: First Results from a multi-model experiment. *Geo. Res. Lett.* **2010**, *37*, L02402. [CrossRef]
18. Deser, C.; Tomas, R.A.; Peng, S. The transient atmospheric circulation response to North Atlantic SST and sea ice anomalies. *J. Clim.* **2007**, *20*, 4751–4767. [CrossRef]
19. Holland, M.M.; Bailey, D.A.; Vavrus, S. Inherent sea ice predictability in the rapidly changing Arctic environment of the Community Climate System Model, version 3. *Clim. Dyn.* **2011**, *36*, 1239–1253. [CrossRef]
20. Baldwin, M.P.; Dunkerton, T.J. Stratospheric harbingers of anomalous weather regimes. *Science* **2001**, *244*, 581–584. [CrossRef]
21. Li, T.; Wang, B. A review on the Western North Pacific monsoon: Synoptic-to-interannual variabilities. *Terr. Atmos. Ocean. Sci.* **2005**, *16*, 285–314. [CrossRef]
22. Li, K.; Yu, W.; Li, T.; Murty, V.S.N.; Khokiattiwong, S.; Adi, T.R.; Budi, S. Structures and mechanisms of the first-branch northward-propagating intraseasonal oscillation over the tropical Indian Ocean. *Clim. Dyn.* **2013**, *40*, 1707–1720. [CrossRef]

23. Waliser, D.E. Predictability and Forecasting. In *Intraseasonal variability of the Atmosphere-Ocean Climate System*, 2nd ed.; Springer: Berlin/Heidelberg, Germany, 2011; pp. 433–468.

24. Suhas, E.; Neena, J.M.; Goswami, B.N. An Indian monsoon intraseasonal oscillations (MISO) index for real time monitoring and forecast verification. *Clim. Dyn.* **2013**, *40*, 2605–2616. [CrossRef]

25. Lee, J.Y.; Wang, B.; Wheeler, M.C.; Fu, X.; Waliser, D.E.; Kang, I.-S. Real-time multivariate indices for the boreal summer intraseasonal oscillation over the Asian summer monsoon region. *Clim. Dyn.* **2013**, *40*, 493–509. [CrossRef]

26. Homepage of S2S Project. Available online: http://www.s2sprediction.net (accessed on 18 October 2019).

27. Vitart, F.; Ardilouze, C.; Bonet, A.; Brookshaw, A.; Chen, M.; Codorean, C.; Déqué, M.; Ferranti, L.; Fucile, E.; Fuentes, M.; et al. The Subseasonal to Seasonal (S2S) Prediction Project Database. *Bull. Am. Meteorol. Soc.* **2017**, *98*, 163–173. [CrossRef]

28. Vitart, F. Madden-Julian oscillation prediction and teleconnections in the S2S database. *Q. J. R. Meteorol. Soc.* **2017**, *143*, 2210–2220. [CrossRef]

29. Wu, J.; Ren, H.L.; Zuo, J.; Zhao, C.; Chen, L.; Li, Q. MJO prediction skill, predictability, and teleconnection impacts in the beijing climate centeratmospheric general circulation model. *Dyn. Atmos. Ocean.* **2016**, *75*, 78–90. [CrossRef]

30. Lim, Y.; Son, S.W.; Kim, D. MJO Prediction Skill of the Subseasonal-to-Seasonal Prediction Models. *J. Clim.* **2018**, *31*, 4075–4094. [CrossRef]

31. Olaniyan, E.; Adefisan, E.A.; Oni, F.; Afiesimama, E.; Balogun, A.A.; Lawal, K.A.A. Evaluation of the ECMWF Sub-seasonal to Seasonal Precipitation Forecasts during the Peak of West Africa Monsoon in Nigeria. *Front. Environ. Sci.* **2018**, *6*, 4. [CrossRef]

32. Vitart, F.; Robertson, A.W. The sub-seasonal to seasonal prediction project (S2S) and the Prediction of extreme events. *Clim. Atmospheric Sci.* **2018**, *3*, 1–7. [CrossRef]

33. Lee, C.Y.; Camargo, S.J.; Vitart, F.; Sobel, A.H.; Tippett, M.K. Subseasonal Tropical Cyclone Genesis Prediction and MJO in the S2S Dataset. *Weather Forecast.* **2018**, *33*, 967–988. [CrossRef]

34. Liu, R.F.; Wang, W. Multi-week prediction of South-East Asia rainfall variability during boreal summer in CFSv2. *Clim. Dyn.* **2015**, *45*, 493–509. [CrossRef]

35. Jie, W.H.; Vitart, F.; Wu, T.; Liu, X. Simulations of the Asian summer monsoon in the sub-seasonal to seasonal prediction project (S2S) database. *Q. J. R. Meteorol. Soc.* **2017**, *143*, 2282–2298. [CrossRef]

36. Li, C.C.; Ren, H.L.; Zhou, F.; Li, S.; Fu, J.-X.; Li, G. Multi-pentad Prediction of Precipitation Variability over Southeast Asia during Boreal Summer Using BCC_CSM1.2. Dyn. *Atmos. Ocean.* **2018**, *82*, 20–36. [CrossRef]

37. Zhao, C.; Ren, H.L.; Song, L.; Wu, J. Madden-Julian oscillation simulated in BCC climate models. *Dyn. Atmos. Ocean.* **2015**, *72*, 88–101. [CrossRef]

38. Wu, J.; Ren, H.L.; Lu, B.; Zhang, P.Q.; Zhao, C.B.; Liu, X.W. Effects of moisture initialization on MJO and its teleconnection prediction in BCC subseasonal coupled model. *J. Geophys. Res.* **2020**, *125*, e2019JD031537. [CrossRef]

39. Zheng, M.; Chang, E.K.; Colle, B.A.; Luo, Y.; Zhu, Y. Applying fuzzy clustering to a multimodel ensemble for US East Coast winter storms: Scenario identification and forecast verification. *Weather Forecast.* **2017**, *32*, 881–903. [CrossRef]

40. Zheng, M.; Chang, E.K.; Colle, B.A. Evaluating US East Coast Winter Storms in a Multimodel Ensemble Using EOF and Clustering Approaches. *Mon. Weather Rev.* **2019**, *147*, 967–1987. [CrossRef]

41. Krishnamurti, T.N.; Kishtawal, C.M.; LaRow, T.E.; Bachiochi, D.R.; Zhang, Z.; Williford, C.E.; Gadgil, S.; Surendran, S. Improved weather and seasonal climate forecasts from multimodel superensemble. *Science* **1999**, *285*, 1548–1550. [CrossRef]

42. Palmer, T.N.; Branković, Č.; Richardson, D.S. A probability and decision-model analysis of PROVOST seasonal multi-model ensemble integrations. *Q. J. R. Meteorol. Soc.* **2000**, *126*, 2013–2033. [CrossRef]

43. Peng, P.T.; Kumar, A.; Dool, H.V.D.; Barnston, A.G. An analysis of multimodel ensemble predictions for seasonal climate anomalies. *J. Geophys. Res.* **2002**, *107*, 4710. [CrossRef]

44. Doblas-Reyes, F.J.; Hagedorn, R.; Palmer, T.N. The rationale behind the success of multi-model ensembles in seasonal forecasting-II. Calibration and combination. *Tellus A* **2005**, *57*, 234–252. [CrossRef]

45. Hewitt, C.D. The ENSEMBLES project: Providing ensemble-based predictions of climate changes and their impacts. *EGGS Newslett.* **2005**, *13*, 22–25.

46. Kirtman, B.P.; Min, D.; Infanti, J.M.; Kinter, J.L.; Paolino, D.A.; Zhang, Q.; van den Dool, H.; Saha, S.; Mendez, M.P.; Becher, E.; et al. The North American Multimodel Ensemble: Phase-1 Seasonal-to-Interannual Prediction; Phase-2 toward Developing Intraseasonal Prediction. *Bull. Am. Meteorol. Soc.* **2014**, *95*, 585–601. [CrossRef]

47. Ren, H.L.; Wu, Y.J.; Bao, Q.; Ma, J.; Liu, C.; Wan, J.; Li, Q.; Wu, X.; Liu, Y.; Tian, B.; et al. The China Multi-Model Ensemble Prediction System and Its Application to Flood-Season Prediction in 2018. *J. Meteor. Res.* **2019**, *33*, 540–552. [CrossRef]

48. Krishnamurti, T.N.; Kishtawal, C.M.; Zhang, Z.; LaRow, T.; Bachiochi, D.; Williford, E.; Gadgil, S.; Surendran, S. Multimodel Ensemble Forecasts for Weather and Seasonal Climate. *J. Clim.* **2000**, *13*, 4196–4216. [CrossRef]

49. Kotal, S.D.; Bhowmik, S.K.R. A multimodel ensemble (MME) technique for cyclone track prediction over the North Indian Sea. *Geofizika* **2011**, *28*, 275–291.

50. Kirtman, B.P.; Min, D. Multimodel ensemble ENSO prediction with CCSM and CFS. Mon. *Weather Rev.* **2009**, *137*, 2908–2930. [CrossRef]

51. Bougeault, P.; Burridge, D.; Chen, D.H.; Ebert, B.; Mylne, K.; Nicolau, J.; Park, Y.-Y.; Raoult, B.; Schuster, D.; Dias, P.S.; et al. The THORPEX Interactive Grand Global Ensemble. *Bull. Am. Meteorol. Soc.* **2010**, *91*, 1059–1072. [CrossRef]

52. Vigaud, N.; Robertson, A.W.; Tippett, M.K. Multimodel Ensembling of Subseasonal Precipitation Forecasts over North America. *Mon. Weather Rev.* **2017**, *145*, 3913–3928. [CrossRef]

53. Kathy, P.; Ben, P.K.; Emily, B.; Dan, C.C.; Emerson, L.; Robert, B.; Ray, B.; Timothy, D.; Dughong, M.; Yuejian, Z.; et al. The Subseasonal Experiment (SubX): A Multimodel Subseasonal Prediction Experiment. Bull. *Am. Meteorol. Soc.* **2019**, *100*, 2043–2060. [CrossRef]

54. Specq, D.; Batte, L.; Déqué, M.; Ardilouze, C. Multimodel forecasting of precipitation at subseasonal timescales over the southwest tropical Pacific. *Earth Space Sci.* **2020**, in press. [CrossRef]

55. Homepage of NEWS. Available online: http://precip.gsfc.nasa.gov (accessed on 20 October 2019).

56. Wheeler, M.C.; Hendon, H.H. An all-season real-time multivariate MJO Index: Development of an Index for monitoring and prediction. *Mon. Weather Rev.* **2004**, *132*, 1917–1932. [CrossRef]

57. MJO Monitoring. Available online: http://www.bom.gov.au/climate/mjo/#tabs=MJO~{}%20phase (accessed on 21 October 2019).

58. BSISO Monitoring. Available online: http://www.apcc21.org/ser/moni.do?lang=en (accessed on 22 October 2019).

59. WMO. *Standardised Verification System (SVS) for Long-Range Forecasts (LRF): New Attachment II-8 to the Manual on the GDPFS (WMO-No. 485)*; WMO: Geneva, Switzerland, 2006; Volume I.

60. Li, S.; Robertson, A.W. Evaluation of submonthly precipitation forecast skill from global ensemble prediction systems. *Mon. Wea. Rev.* **2015**, *143*, 2871–2889. [CrossRef]

61. Madden, R.A.; Julian, P.R. Detection of a 40-50 Day Oscillation in the Zonal Wind in the Tropical Pacific. *J. Atmos. Sci.* **1971**, *28*, 702–708. [CrossRef]

62. Madden, R.A.; Julian, P.R. Description of global-scale circulation cells in tropics with a 40-50 day period. *J. Atmos. Sci.* **1972**, *29*, 1109–1123. [CrossRef]

63. Wang, B.; Xie, X. A model for the boreal summer intraseasonal oscillation. *J. Atmos. Sci.* **1997**, *54*, 72–86. [CrossRef]

64. Han, J.; Pan, H.L. Revision of convection and vertical diffusion schemes in the NCEP global forecast system. *Weather Forecast.* **2011**, *26*, 520–533. [CrossRef]

65. Fu, X.; Wang, B.; Lee, J.Y. Sensitivity of Dynamical Intraseasonal Prediction Skills to Different Initial Conditions. *Mon. Weather Rev.* **2011**, *139*, 2572–2592. [CrossRef]

66. Hendon, H.H.; Wheeler, M.C.; Zhang, C.D. Seasonal Dependence of the MJO ENSO Relationship. *J. Clim.* **2007**, *20*, 531–543. [CrossRef]

67. Feddersen, H.; Andersen, U. A method for statistical downscaling of seasonal ensemble predictions. *Tellus A* **2005**, *57*, 398–408. [CrossRef]

68. Kang, H.W.; An, K.H.; Park, C.K.; Solis, A.L.S.; Stitthichivapak, K.; Kang, H. Multimodel output statistical downscaling prediction of precipitation in the Philippines and Thailand. *Geo. Res. Lett.* **2007**, *34*, L15710. [CrossRef]

69. Kang, H.W.; Park, C.K.; Hameed, S.N.; Ashok, K. Statistical downscaling of precipitation in Korea using multimodel output variables as predictors. *Mon. Weather Rev.* **2009**, *137*, 1928–1938. [CrossRef]

70. Liu, Y.; Ren, H.L. A hybrid statistical downscaling model for prediction of winter precipitation in China. *Int. J. Climatol.* **2015**, *35*, 1309–1321. [CrossRef]

71. Liu, Y.; Ren, H.L. Improving ENSO prediction in CFSv2 with an analogue-based correction method. *Int. J. Climatol.* **2017**, *37*, 5035–5046. [CrossRef]

3

Increasing Neurons or Deepening Layers in Forecasting Maximum Temperature Time Series?

Trang Thi Kieu Tran [1], Taesam Lee [1,*]⊕ and Jong-Suk Kim [2,*]⊕

[1] Department of Civil Engineering, ERI, Gyeongsang National University, 501 Jinju-daero, Jinju 660-701, Korea; trangtran281292@gmail.com
[2] State Key Laboratory of Water Resources and Hydropower Engineering Science, Wuhan University, Wuhan 430072, China
*   Correspondence: tae3lee@gnu.ac.kr (T.L.); jongsuk@whu.edu.cn (J.-S.K.)

Abstract: Weather forecasting, especially that of extreme climatic events, has gained considerable attention among researchers due to their impacts on natural ecosystems and human life. The applicability of artificial neural networks (ANNs) in non-linear process forecasting has significantly contributed to hydro-climatology. The efficiency of neural network functions depends on the network structure and parameters. This study proposed a new approach to forecasting a one-day-ahead maximum temperature time series for South Korea to discuss the relationship between network specifications and performance by employing various scenarios for the number of parameters and hidden layers in the ANN model. Specifically, a different number of trainable parameters (i.e., the total number of weights and bias) and distinctive numbers of hidden layers were compared for system-performance effects. If the parameter sizes were too large, the root mean square error (RMSE) would be generally increased, and the model's ability was impaired. Besides, too many hidden layers would reduce the system prediction if the number of parameters was high. The number of parameters and hidden layers affected the performance of ANN models for time series forecasting competitively. The result showed that the five-hidden layer model with 49 parameters produced the smallest RMSE at most South Korean stations.

Keywords: artificial neural network; neurons; layers; temperature; South Korea; deep learning

1. Introduction

An artificial neural network (ANN) is a system for information processing inspired by biological neural networks. The key element of this network is the huge amount of highly interconnected processing nodes (neurons) that work together by a dynamic response to process the information. A neural network is useful for modeling the non-linear relation between the input and output of a system [1]. Compared to other machine learning methods such as autoregressive moving averages (ARMA), autoregressive integrated moving averages (ARIMA), and random forest (RF), the ANN model showed better performance in regression prediction problems [2–4]. According to Agrawal [5], the ANN model predicted rainfall events more accurately than the ARIMA model. In another work, ANNs have been applied to forecast monthly mean daily global solar radiation [6].

Furthermore, the ANN model has also been employed to forecast climatological and meteorological variables. Although it is known that the weather forecasting problem is challenging because of its chaotic and dynamic process, weather forecasting based on ANNs has been employing considerably in recent years due to the success of the ANN's ability. From some previous research, artificial neural networks have been shown as a promising method to forecast weather and time series data due to their capability of pattern recognition and generalization [7,8]. Smith et al. [9] developed an improved ANN to forecast the air temperature from 1 to 12 h ahead by increasing the number of samples in the training, adding additional seasonal variables, extending the duration of prior observations, and varying the number of hidden neurons in the network. Six hours of prior data were chosen as the inputs for the temperature prediction since a network with eight prior observations performed worse than the six hour network. Moreover, it is demonstrated that the models using one hidden layer with 40 neurons performed better than other models over repeated instantiations. In another study, the ANN models for the maximum as well as minimum temperature, and relative humidity forecasting were proposed by Sanjay Mathur [10] using time series analysis. The multilayer feedforward ANN model with a back-propagation algorithm was used to predict the weather conditions in the future, and it was found that the forecasting model could make a highly accurate prediction. The authors in [11] employed the ANN models to forecast air temperature, relative humidity, and soil temperature in India, showing that the ANN model was a robust tool to predict meteorological variables as it showed promising results with 91–96% accuracy for predictions of all cases. In this study, we also aimed to predict the air temperature one day ahead of past observations using the ANN model.

The effectiveness of a network-based approach depends on the architecture of the network and its parameters. All of these considerations are complex, and the configuration of a neural network structure depends on the problem [12]. If unsuitable network architecture and parameters are selected, the results may be undesirable. On the other hand, a proper design of network architecture and parameters can produce desirable results [13,14]. However, little investigation has been conducted on the effect of parameters and architecture on the model's performance. The selection of ANN architecture, consisting of input variables, the number of neurons in the hidden layer, and the number of hidden layers is a difficult task, so the structure of the network is usually determined by a trial and error approach and based on the experience of the modeler [5]. In another previous study, we compared one-hidden layer and multi-hidden layer ANN models in maximum temperature prediction at five stations in South Korea [15]. In addition, the genetic algorithm was applied to find the best architecture of models. It showed that the ANN with one hidden layer performed the most accurate forecasts. However, the effect of the number of hidden layers and neurons on the ANN's performance in the maximum temperature time series prediction is not sufficient. It may expect that the model performs worse when the number of parameters decreases. However, what happens if we further increase the number of tunable parameters? There are two competing effects. On the one hand, more parameters, which mean more neurons, become available, possibly allowing for better predictions. On the other hand, the higher the parameter number, the more overfitting the model is. Will the networks be robust if more trainable parameters than necessary are present? Is a one-hidden layer model always better than a multi-hidden layer model for maximum temperature forecasting in South Korea? Therefore, it is also apparently several problems related to the model proper architecture.

This paper proposed a new strategy that applied the ANNs using different learning parameters and hidden layers to empirically compare the prediction performance of daily maximum temperature time series. This study aimed to discuss the effect of parameters on the performance of ANN for temperature time series forecasting.

The rest of the paper is structured as follows. Section 2 describes the data and methodology used for the experiments. Section 3 describes the results, and the final section provides conclusions and directions for future work.

2. Data and Methods

In the current study, 55 weather stations that record maximum temperature in South Korea at the daily timescale were employed. Most stations have a data period of 40 years from 1976 to 2015, except for Andong station (1983–2015) and Chuncheon station (1988–2015).

Figure 1 presents the locations of the stations at which the data were recorded. The forecasting model for the maximum temperature was built based on the neural network. There were six neurons in the input layer, which corresponds to the number of previous days provided to the network for the prediction of the next maximum temperature value and one neuron in the output layer, respectively. The number of hidden layers and the number of hidden neurons are discussed. This study tested the performance of ANN models for one day ahead of the maximum temperature prediction using prior observations as inputs corresponding to three different cases of hidden layers; they were one, three, and five hidden layers, respectively. Besides, the following five levels of numbers of trainable parameters (i.e., the total number of its weights and biases) were selected for testing: 49, 113, 169, 353, and 1001. Combining the number of hidden layers and the number of parameters, Table 1 shows the model architectures. Besides, the configurations of 1-, 3- and 5-hidden layer ANN models with 49 learnable parameters are illustrated in Figure 2. It is noticed that the total number of trainable parameters was computed by summing the connections between layers and biases in every layer.

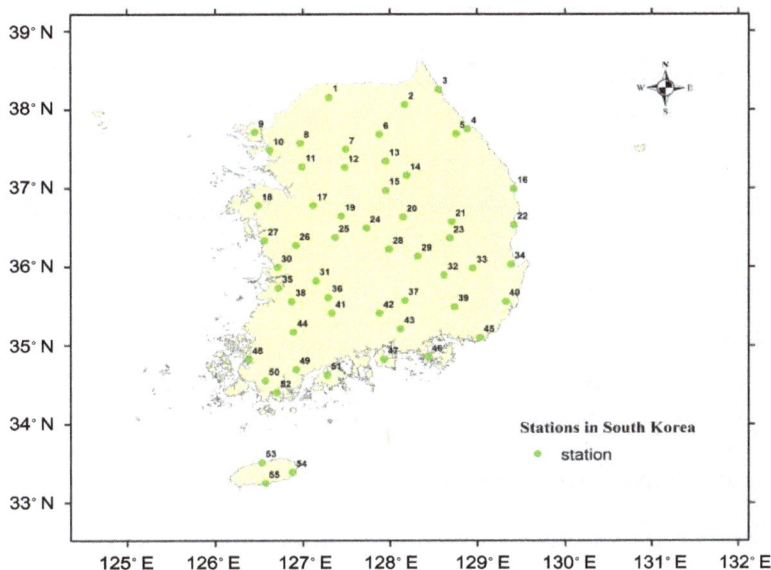

Figure 1. Map of the locations of the stations used in this study.

Table 1. Structure of the ANN models used for the study.

Number of Parameters	Number of Hidden Layers	Structure
49	1	6-6-1
	3	6-3-3-3-1
	5	6-5-1-1-1-1-1
113	1	6-14-1
	3	6-6-5-5-1
	5	6-4-4-4-4-4-1
169	1	6-21-1
	3	6-7-7-7-1
	5	6-6-6-4-4-6-1
353	1	6-44-1
	3	6-11-11-11-1
	5	6-8-8-8-8-8-1
1001	1	6-125-1
	3	6-20-20-20-1
	5	6-18-17-16-9-10-1

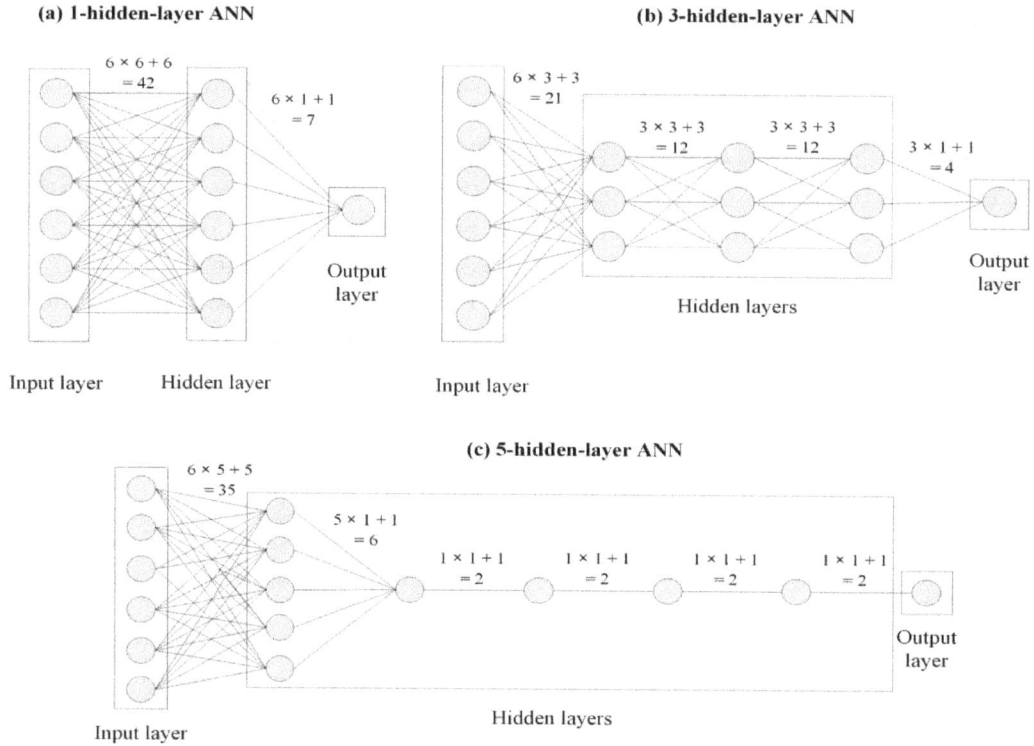

Figure 2. Schematic of the artificial neural network (ANN) with the total parameter number of 49 corresponding to (**a**) 1 hidden layer; (**b**) 3 hidden layers; and (**c**) 5 hidden layers.

To evaluate the effectiveness of each network, the root mean square error (RMSE) was used as a performance index. RMSE is calculated as

$$\text{RMSE} = \sqrt{\frac{1}{n}\sum_{i=1}^{n}(y_i - y'_i)^2} \tag{1}$$

where y_i is the observed data, y'_i is the predicted data, and n is the number of observations. The RMSE indicates a discrepancy between the observed and predicted data. The lower the RMSE, the more accurate the prediction is.

At each station, the data were subdivided into three parts: a training set consisting of 70% of the total data, a validation set using 20% of the total data, and a testing set containing 10% of the total data. The first two splits were used to train and examine the training performance, and the test set was used to evaluate the actual performance of the prediction. All models were trained and validated with the same training set and validation set, respectively. A popular technique to avoid the effect of overfitting the network on the training data is the early stopping method presented by Sarle [16]. Model overfitting is understood as the fitness of the model to the signal and noise that are usually present in the training sample. The possibility of overfitting depends on the size of the network, the number of training samples, and the data quality.

The maximum epoch for training was set at 1000. An epoch was described as one pass through all the data in the training set, and the weights of a network were updated after each epoch. The training was stopped when the error in the validation data reached its lowest value, or the training reached the maximum epoch, whichever came first. One iteration step during the ANN training usually works with a subset (call batch or mini-batch) of the available training data. The number of samples per batch (batch size) is a hyperparameter, defined as 100 in our case. Finally, the networks were evaluated based on the testing data. The ANN network was implemented using Keras [17], with TensorFlow [18] as the backend. According to Chen et al. [19], the design of neural networks has several aspects of concern. A model should have sufficient width and depth to capture the underlying pattern of the data. In contrast, a model should also be as simple as possible to avoid overfitting and high computational costs. However, the general trend of the number of parameters versus RMSE indeed provides some insights into selecting the proper structure of the model.

Data normalization is a preprocessing technique that transforms time series into a specified range. The quality of the data is guaranteed when normalized data are fed to a network. The MinMaxscaler technique was chosen for normalizing the data and making them in a range of [0, 1], which was defined as follows:

$$x' = \frac{x - x_{min}}{x_{max} - x_{min}} \tag{2}$$

where x' is the normalized data; x is the original data; and x_{max}, x_{min} are the maximum and minimum values of the data, respectively. The max and min used for standardization are calculated from the calibration period only. At the end of each algorithm, the outputs were denormalized into the original data to receive the final results.

An ANN consists of an input layer, one or more hidden layers of computation nodes, and an output layer. Each layer uses several neurons, and each neuron in a layer is connected to the neurons in the next layer with different weights, which change the value when it goes through that connection. The input layer receives the data one case at a time, and the signal will be transmitted through the hidden layers before arriving at the output layer, which is interpreted as the prediction or classification. The network weights are adjusted to minimize the output error based on the difference between the expected and target outputs. The error at the output layer propagates backward to the hidden layer until it reaches the input layer. The ANN models are used as an efficient tool to reveal a nonlinear relationship between the inputs and outputs [8]. Generally, the ANN model with three layers can be mathematically formulated as Lee et al. [20]:

$$y_k = f_2\left[\sum_{j=1}^{m} W_{kj} f_1\left(\sum_{i=1}^{n} W_{ji} x_i + b_j\right) + b_k\right] \tag{3}$$

where x_i is the input value to neuron i; y_k is the output at neuron k; f_1 and f_2 are the activation function for the hidden layer and output layer, respectively; n and m indicate the number of neurons in the input and hidden layers. W_{ji} is the weight between the input node i and hidden node j while W_{kj} is the weight between the hidden node j and output node k. b_j and b_k are the bias of the jth node in the hidden layer and the kth node in the output layer, respectively.

The weights in Equation (3) were adjusted to reduce the output error by calculating the difference between the predicted values and expected values using the back-propagation algorithm. This algorithm is executed in two specified stages, called forward and backward propagation. In the forward phase, the inputs were fed into the network and propagated to the hidden nodes at each layer until the generation of the output. In the backward phase, the difference between the true values and the estimated values or loss function was calculated by the network. The gradient of the loss function with respect to each weight can be computed and propagated backward to the hidden layer until it reaches the input layer [20].

In the current study, the 'tanh' or hyperbolic tangent activation function and an unthresholded linear function were used in the hidden layer and output layer, respectively. The range of the tanh function is from −1 to 1, and it is defined as follows:

$$\tanh(x) = \frac{e^x - e^{-x}}{e^x + e^{-x}} \tag{4}$$

It is noteworthy that there is no direct method well established for selecting the number of hidden nodes for an ANN model for a given problem. Thus, the common trial-and-error approach remains the most widely used method. Since ANN parameters are estimated by iterative procedures, which provide slightly different results each time they are run, we estimated each ANN 5 times and reported the mean and standard deviation errors in the figure.

The purpose of this study was not to find the best station-specific model, but to investigate the effects of hidden layers and trainable parameters on the performance of ANNs for maximum temperature modeling. We think that the sample size of 55 stations is large enough to infer some of the (average) properties of these factors to the ANN models.

3. Results

To empirically test the effect of the number of learnable parameters and hidden layers, we assessed and compared the model results obtained at 55 stations for five different parameters: 49, 113, 169, 353, and 1001, respectively. Moreover, we also tested the ANN models with different hidden layers (1, 3, and 5) having the same number of parameters at each station. Therefore, the mean and standard deviation were computed for the RMSE to analyze the impact of these factors on the ANN's performance. It is noted that all other modeling conditions, e.g., input data, activation function, number of epochs, and the batch size, were kept identical. After training and testing the datasets, the effects of the parameters and hidden layers of models were discussed.

3.1. Effect of the Number of Parameters

We first evaluated the performance of the ANN model by using the testing datasets. For each studied parameter, the prediction performance values were also presented as a rate of change in RMSE based on the original RMSE value obtained at 55 stations. This reference RMSE changes depending on the site used to study the impact of each parameter. Thus, as the results are depicted in Figures 3–5, the proposed ANN with 49 parameters consistently outperforms the other parameters at almost all stations in South Korea, since it produces the lowest change in error for different model configurations. Taking the single-hidden layer ANN as an example, the rate of change of the RMSE slightly increases with the extension of parameters for most sites (see Figure 3).

However, we can also observe that the increased parameter size of the ANN model made the error's change decrease lightly at Buan stations. Moreover, several sites have the lowest change values of RMSE at the parameter of 1001, such as Daegwallyeong, Gunsan, Hongcheon, and Tongyeong. These sites have an increasing trend of error when the number of tunable parameters in the network is raised from 49 to 169 (Hongcheon) or 353 (Daegwallyeong, Gunsan, and Tongyeong) before declining to the lowest point at 1001. Similarly, Figure 4 illustrates the general relationship between the total numbers of parameters versus the rate of change of RMSE on testing data in all 55 stations for three hidden layers. It can be noted from this figure that in the majority of stations, the rise of parameter numbers makes the performance of the model worse due to the increase in the change of RMSE. In contrast, few stations have the best results at the parameter of 133 (Pohang), 169 (Haenam), 353 (Buan and Yeongdeok), and 1001 (Daegwallyeong). Although the fluctuation of the RMSE's change rate for the three-hidden layer ANN, corresponding to various parameter sizes, varies from site to site, the ANN model with a structure of 49 trainable parameters still shows the best solution for predicting the maximum temperature one day ahead for most stations in South Korea.

In the case of five hidden layers, it can be observed from Figure 5 that the 49-parameter ANN model continues showing the smallest error in 52 out of the total 55 stations. In most cases, the increase in the number of parameters deteriorates the performance of the ANN model. However, it should be noticed that the model achieves the best result at the parameter of 353 in Buan and Gunsan stations while the smallest RMSE in Mokpo is obtained at the parameter of 119.

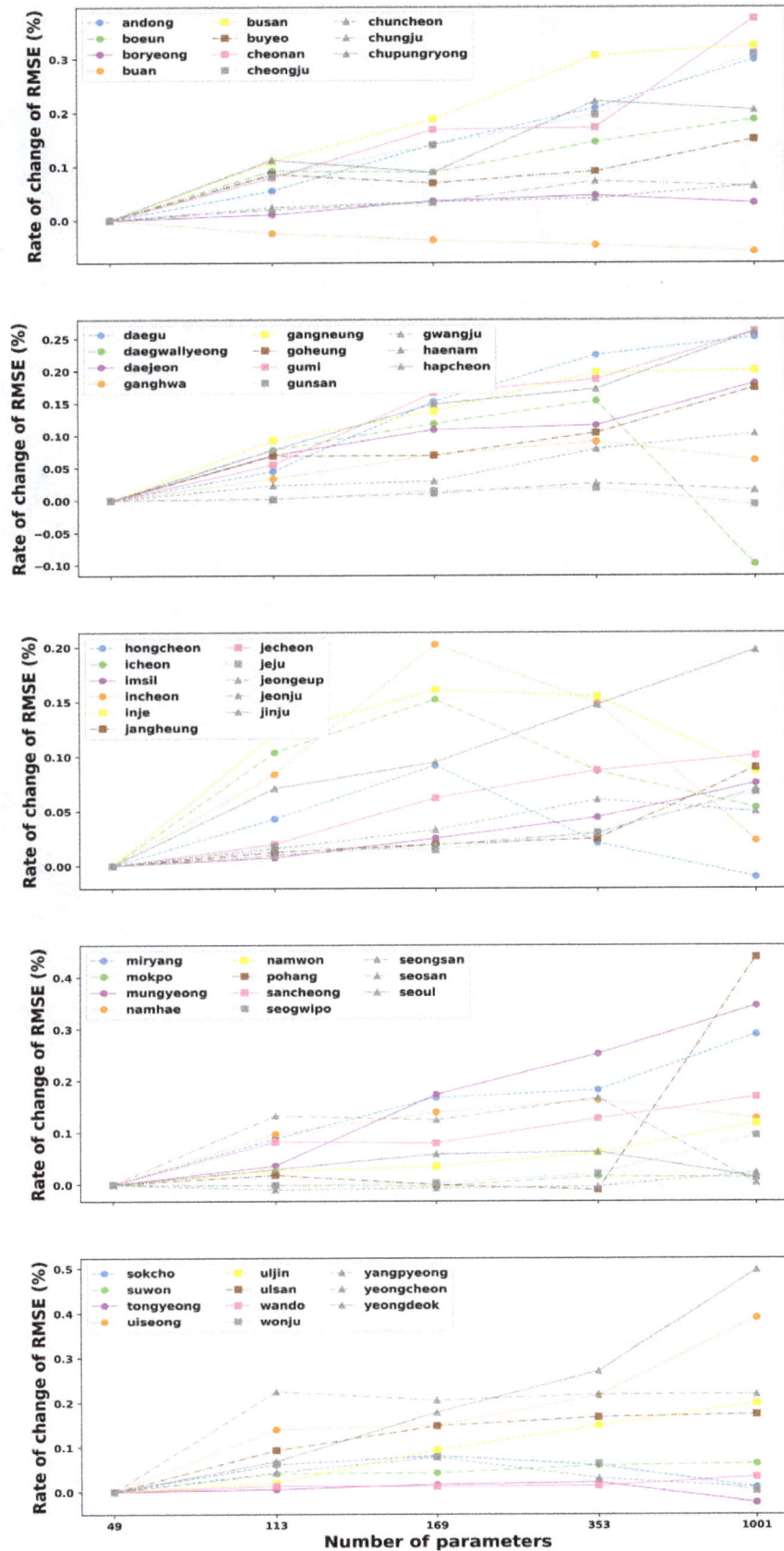

Figure 3. The rate of change of mean root mean square error (RMSE) of the 1-hidden layer ANN corresponding to 49, 113, 169, 353, and 1001 trainable parameters for the test data for all 55 stations in South Korea.

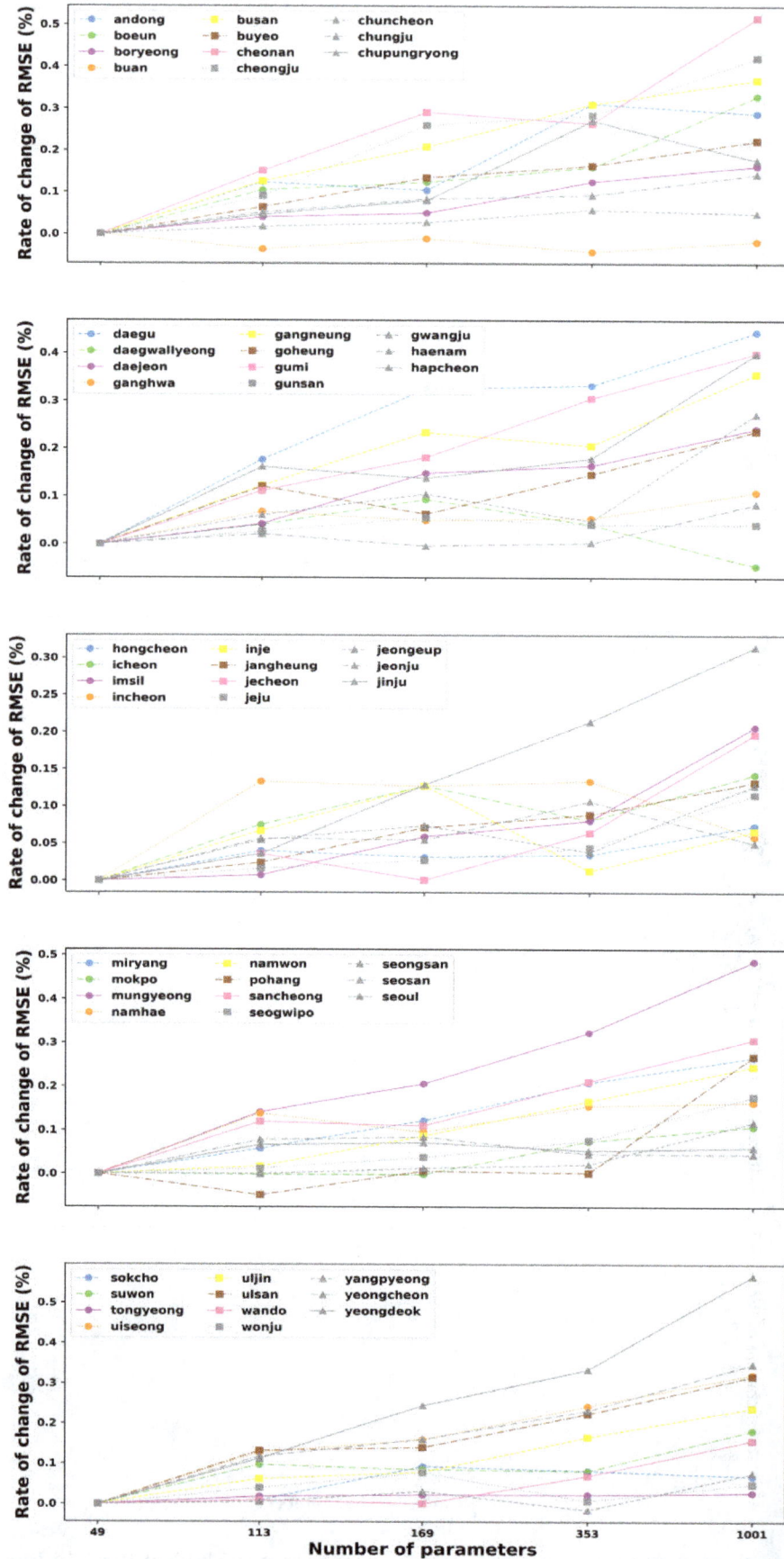

Figure 4. The rate of change of the mean RMSE of the 3-hidden layer ANN corresponding to 49, 113, 169, 353, and 1001 trainable parameters for the test data for all 55 stations in South Korea.

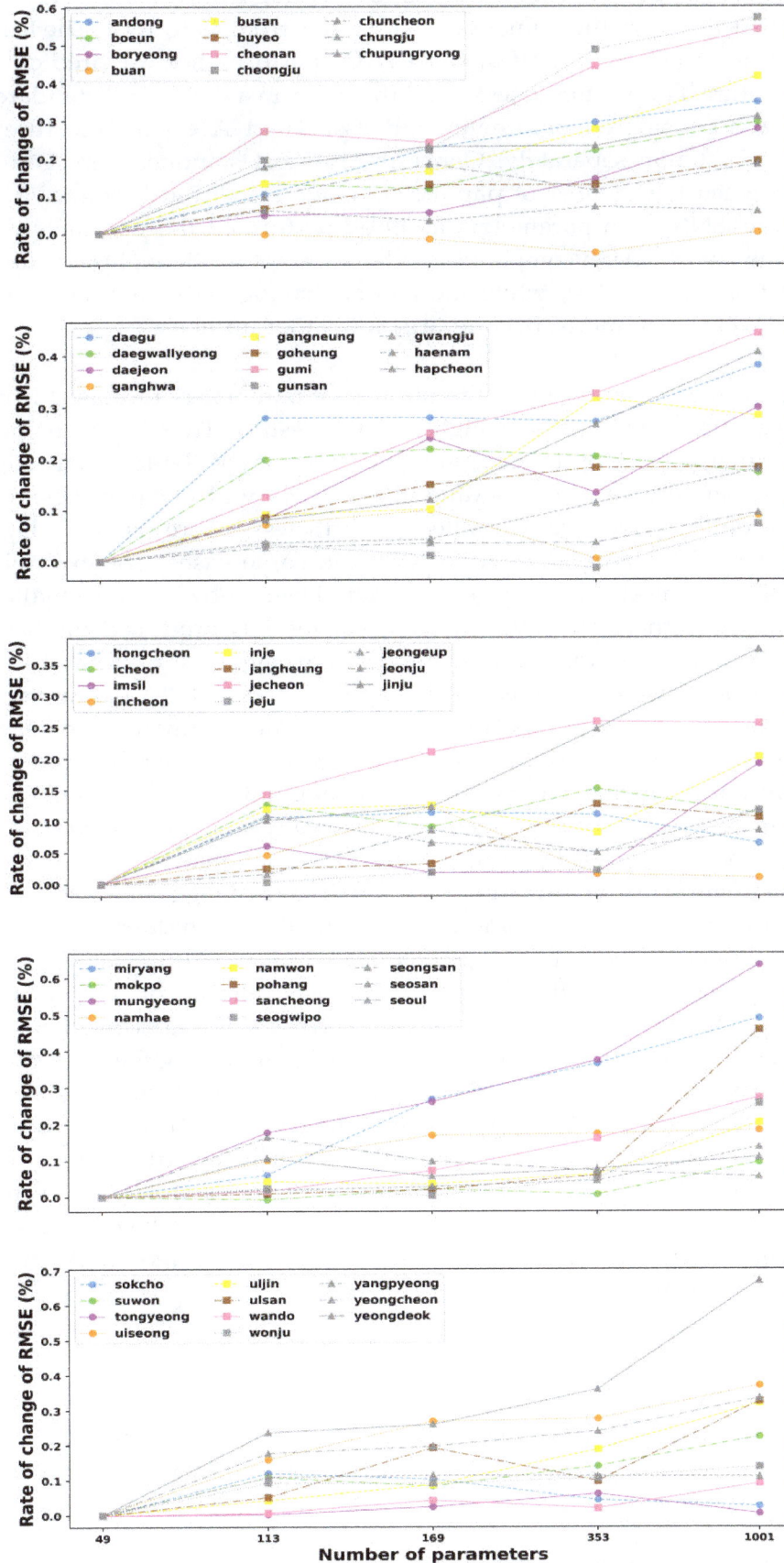

Figure 5. The rate of change of the mean RMSE of the 5-hidden layer ANN corresponding to 49, 113, 169, 353, and 1001 trainable parameters for the test data for all 55 stations in South Korea.

Figures S1–S4 (see Supplementary Materials) depict the RMSE of the ANN models with three different hidden layers that vary the number of parameters from 49 to 1001. The learnable parameter is an important parameter that may affect the ANN's performance for predicting the maximum temperature in the future. For a better assessment, the five-run average and standard deviation of the performances of each considered parameter and associated ANN architecture are depicted in those figures for each station separately. Generally, in most stations, there is a slight increasing trend of the RMSE when the number of parameters is increased. Furthermore, it can be seen from those figures that the ANN with parameters at 49 outperformed the parameters of other models, since it produced the lowest RMSE compared to others, except for Buan, Daegwallyeong (Figure S1), and Tongyeong stations (Figure S4), while the models having 1001 parameters yielded the worst results at around 50% of total stations, in comparison to other numbers of parameters.

The difference in performance when the number of parameters increased from 49 to 1001 was marginal, but the value of 49 still leads to slightly better results. Thus, it can be noticed from these results that the testing RMSE of lower parameters is maybe better than higher parameters considering the low amount of neurons, and the models with 49 parameters are sufficient to forecast the maximum temperature variable. Nevertheless, there is an adaptive amount of parameters in the model in terms of hidden layers in some cases. For example, at Tongyeong stations (Figure S4), the model having one hidden layer showed the smallest RMSE at the parameters of 1001, while three and five-hidden layer models produced the best results with 49 parameters. In another case, 49 was the best number parameters for one and five hidden layers; meanwhile, three hidden layers presented the best performance at 353 parameters at Yangpyeong station (Figure S4). Besides, it is worth noting that at the same parameter number, the values of the RMSE for the one-, three-, and five-hidden layer models were comparable in most of the stations. However, the significant differences among the three configurations of the model can be observed at the 1001 parameters such as Boryeong, Cheongju, Daewallyeong (Figure S1), Mokpo (Figure S3), Wonju, and Yangpyeong stations (Figure S4) or the 353 parameters such as Ganghwa (Figure S1), Incheon, Jangheung (Figure S2), Seosan (Figure S3), and Yangpyeong stations (Figure S4). In addition, the RMSE shows little sensitivity to changes in the number of hidden layers in some stations, especially at high parameters due to the large fluctuation of standard deviation values. For example, in Tongyeong station (Figure S4), at the same parameter number of 353 or 1001, the standard deviation values of the RMSE of three- and five-hidden layer models are considerably larger than that of the one-hidden layer model. The trend is more evident for predicting the maximum temperature as the total number of parameters increased and occured in some stations, such as Buan, Daegwallyeong, Chuncheon (Figure S1), Wonju, or Yangpyeong stations (Figure S4). It can be suspected that the performance of the model may be significantly affected when the structure of the model becomes more complex. Based on the variation model performance in terms of hidden layers and parameters, it can be concluded that both the number of parameters and hidden layers were important to the model's performance, and the selection of parameters and hidden layers needs considerable attention because of the fluctuation in error.

3.2. Effect of the Number of Hidden Layers

Figures 6–10 show the spatial distribution of the ANN performances in the test period. Accordingly, a significant decrease in error is likely to move from the eastern to the western and southern part of South Korea with 49 learnable parameters (see Figure 6).

Similar spatial distributions of the changes in RMSEs also occur in Figures 7–10 when the number of parameters is increased to 113, 169, 353, and 1001. It can be concluded that the ANN models perform better in western and southern Korea (left panels). Moreover, the visualization of the differences in the RMSE between one hidden layer and three hidden layers (middle panels) as well as between one hidden layer and five hidden layers (bottom panels) at each station is also shown in Figures 6–10 It is noticed that with the same number parameters of 49, while one-hidden layer model presented better results than the three-hidden layer model at over 60% of total stations, the five-hidden layer model performed slightly greater than the one-hidden layer model at around 79% of stations where RMSE differences greater than 0 were found (see Figure 6). However, with the increase in the number of parameters, the ANN models with one hidden layer produced better results than the three- and five-hidden layer models in almost all stations.

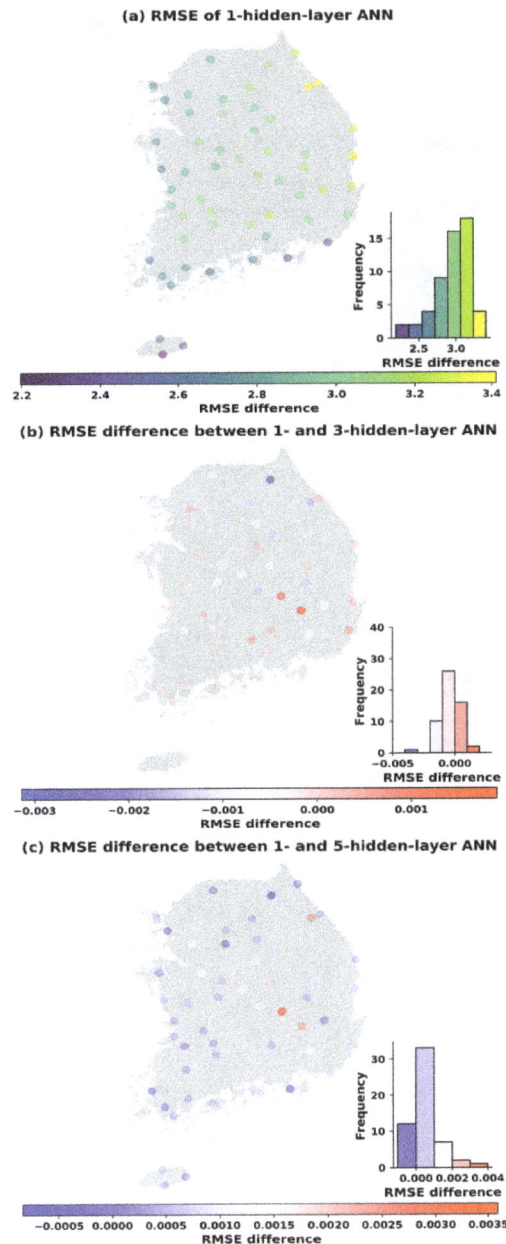

Figure 6. Panel (**a**) shows the RMSE of the test data of the 1-hidden layer model with 49 parameters; panel (**b**) shows the difference of the RMSE between the 1-hidden layer and the 3-hidden layer model (RMSE difference <0 indicates that the 1-hidden layer model performs better than the 3-hidden layer model and vice versa); and panel (**c**) shows the RMSE difference between the 1-hidden layer and the 5-hidden layer model (RMSE difference <0 indicates the that the 1-hidden layer model performs better, RMSE difference >0 shows the opposite).

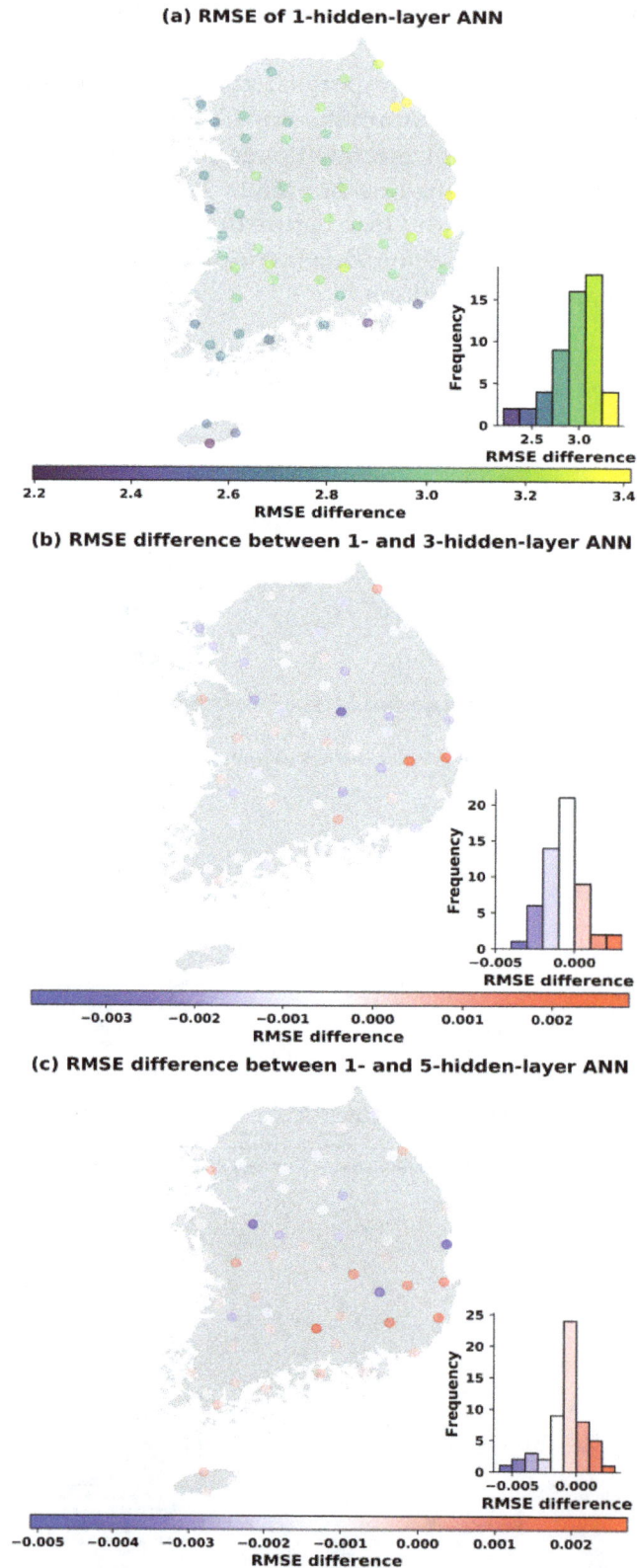

Figure 7. Panel (**a**) shows the RMSE of the test data of the 1-hidden layer model with 113 parameters; panel (**b**) shows the difference of the RMSE between the 1-hidden layer and the 3-hidden layer model (RMSE difference <0 indicates that the 1-hidden layer model performs better than the 3-hidden layer model and vice versa); and panel (**c**) shows the RMSE difference between the 1-hidden layer and the 5-hidden layer model (RMSE difference <0 indicates that the 1-hidden layer model performs better, RMSE difference >0 shows the opposite).

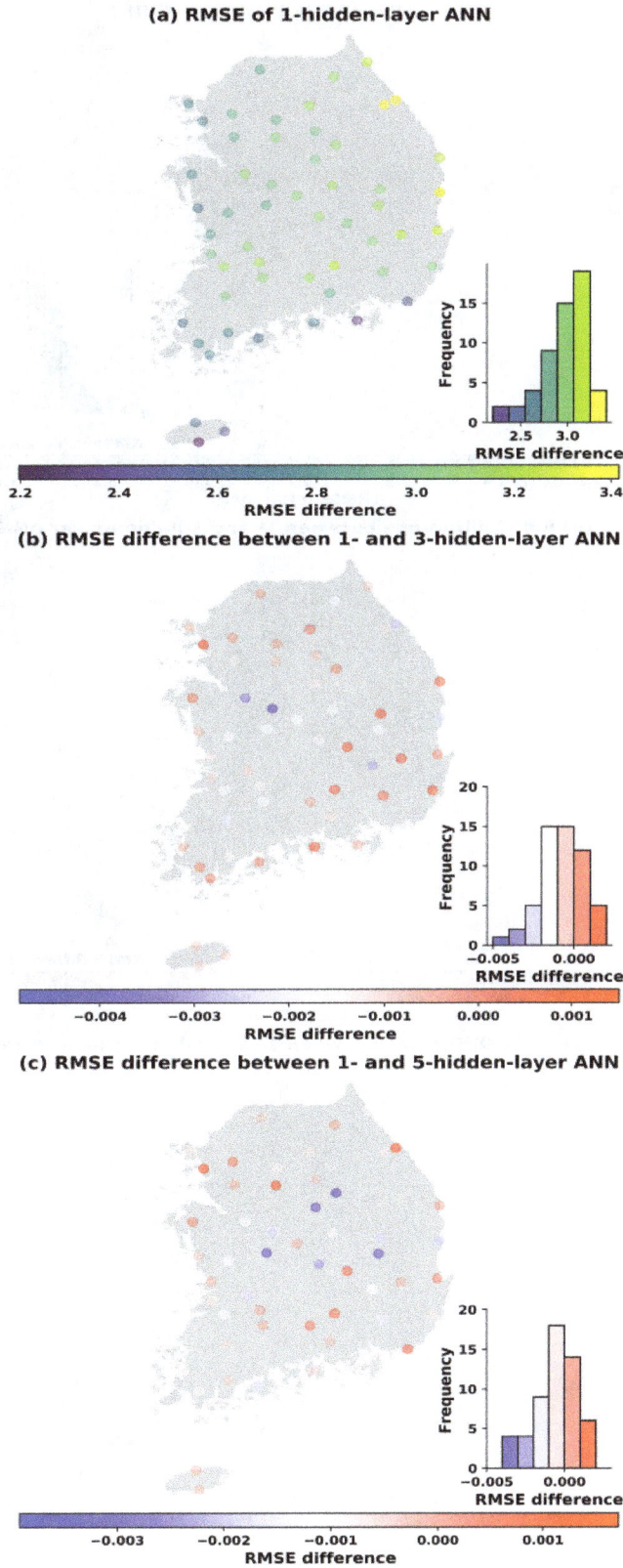

Figure 8. Panel (**a**) shows the RMSE of the test data of the 1-hidden layer model with 169 parameters; panel (**b**) shows the difference of the RMSE between the 1-hidden layer and the 3-hidden layer model (RMSE difference <0 indicates that the 1-hidden layer model performs better than the 3-hidden layer model and vice versa); and panel (**c**) shows the RMSE difference between the 1-hidden layer and the 5-hidden layer model (RMSE difference <0 indicates the that the 1-hidden layer model performs better, RMSE difference >0 shows the opposite).

(a) RMSE of 1-hidden-layer ANN

(b) RMSE difference between 1- and 3-hidden-layer ANN

(c) RMSE difference between 1- and 5-hidden-layer ANN

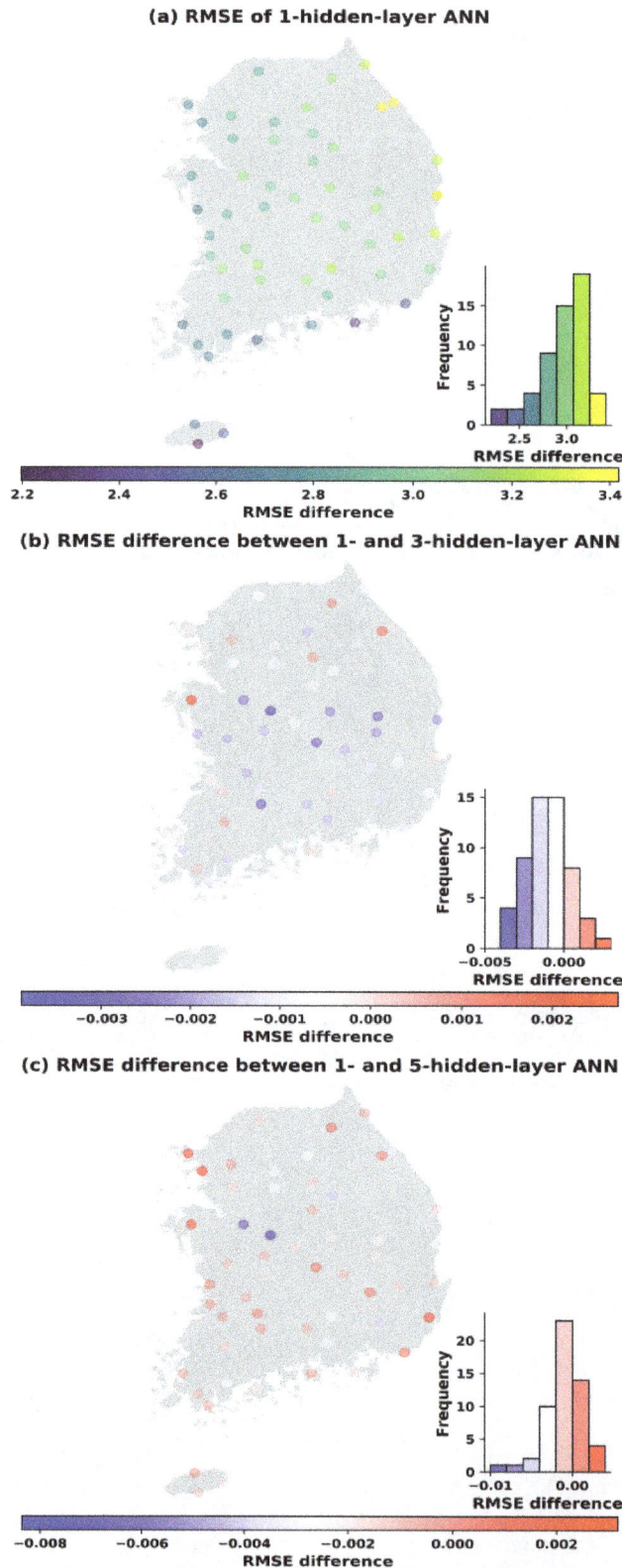

Figure 9. Panel (**a**) shows the RMSE of the test data of the 1-hidden layer model with 353 parameters; panel (**b**) shows the difference of the RMSE between the 1-hidden layer and the 3-hidden layer model (RMSE difference <0 indicates that the 1-hidden layer model performs better than the 3-hidden layer model and vice versa); and panel (**c**) shows the RMSE difference between the 1-hidden layer and the 5-hidden layer model (RMSE difference <0 indicates the that the 1-hidden layer model performs better, RMSE difference >0 shows the opposite).

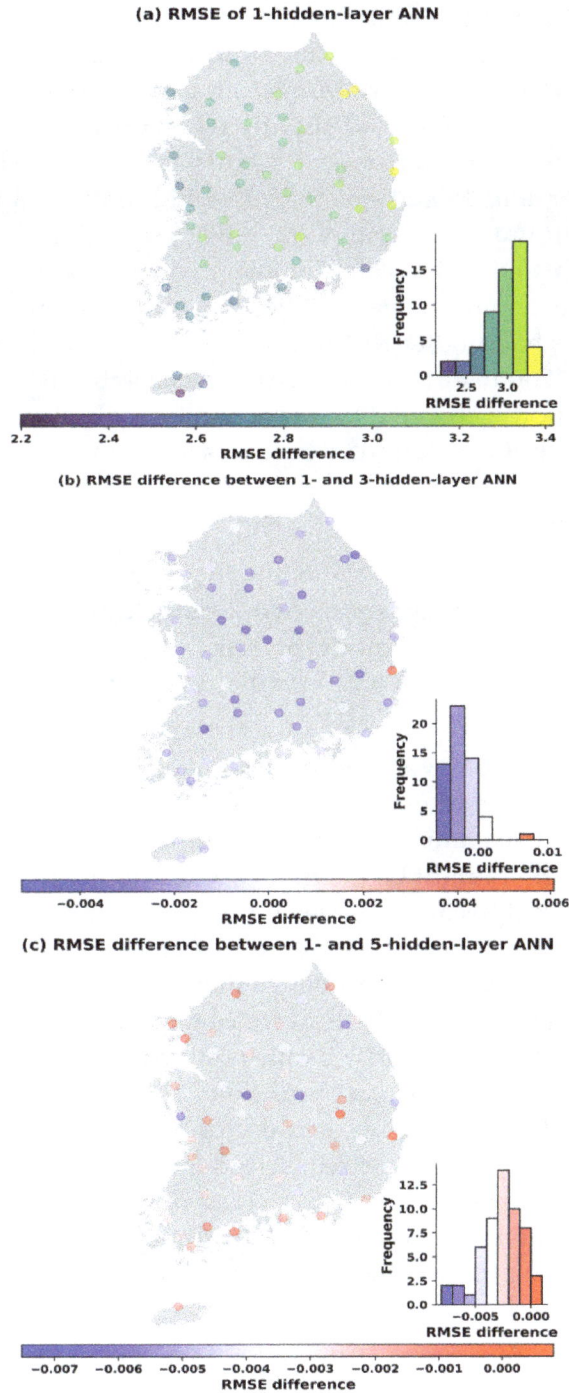

Figure 10. Panel (**a**) shows the RMSE of the test data of the 1-hidden layer model with 1001 parameters; panel (**b**) shows the difference of the RMSE between 1-hidden layer and the 3-hidden layer model (RMSE difference <0 indicates that the 1-hidden layer model performs better than the 3-hidden layer model and vice versa); and panel (**c**) shows the RMSE difference between the 1-hidden layer and the 5-hidden layer model (RMSE difference <0 indicates the that the 1-hidden layer model performs better, RMSE difference >0 shows the opposite).

For example, Figure 7 indicates that, based on the RMSE of the testing sets, the one-hidden layer model achieves a better performances at 42 and 41 stations, respectively, when the number of parameters is 113, compared to the three- and five-hidden layer models. Similarly, at the parameter number of 169, and from the histogram of RMSE differences (Figure 8), we can see that the one-hidden layer structure generally obtained a smaller error than three and five hidden layers for most sites.

However, it was noted that the multi-hidden layer models improved the performance of temperature prediction at some stations (17 stations with three hidden layers and 20 stations with five hidden layers) in comparison to one hidden layer. Moreover, the number of sites that obtained a smaller RMSE with multiple hidden layers decreased slightly when the number of parameters was increased to 353 (Figure 9). Out of 55 stations, there were 12 stations with three hidden layers having smaller errors than one hidden layer, and 18 stations with five hidden layers had better performance than one hidden layer. Finally, Figure 10 compares the ANN models in the test case in the name of the RMSE and hidden layer with 1001 trainable parameters. As the results showed, the one-hidden layer ANN generated a better result at almost every station than the three and five hidden layers. The three- and five-hidden layer ANNs had higher RMSE than one hidden layer in five and three stations out of 55 stations, respectively. The results were worth emphasizing that the deep ANN with various parameters trained for all stations generated a certain number of basins with lower performance than a single-hidden layer network, but the stations where this occurred were not always the same.

4. Summary and Conclusions

In this study, different hidden layer ANN models with various interior parameters were employed to forecast 1 day maximum temperature series over South Korea. This study aimed to explore the relationship between the size of the ANN model and its predictive capability, revealing that for future predictions of the time series of maximum temperature. In summary, the major findings of the present study are as follows.

Firstly, a deep neural network with more parameters does not perform better than a small neural network with fewer layers and neurons. The structural complexity of the ANN model can be unnecessary for unraveling the maximum temperature process. Even though the differences between these models are mostly small, it might be useful to some applications that require a small network with a lower computational cost because increasing the number of parameters may slow down the training process without substantially improving the efficiency of the network. Importantly, it can be observed that a simple network can perform better than a complex one, as also concluded in other comparisons. The authors in Lee et al. [20] showed that a large number of hidden neurons did not always lead to better performance. Similarly, in our previous study, the hybrid method of ANN and the genetic algorithm was applied to forecast multi-day ahead maximum temperature. The results demonstrated that the neural network with one hidden layer presented a better performance than the two and three hidden layers [14]. Nevertheless, too many or lesser amounts of parameters in the model can make the RMSE of the model increase, such as in the Buan station. This could be explained by an insufficient number of parameters causing difficulties in the learning data, whereas an excessive number of parameters might lead to unnecessary training time, and there is a possibility of over-fitting the training data set [21].

Secondly, although the performances of the models corresponding to different hidden layers are comparable when the number of parameters is the same, it is worth highlighting that five-hidden layer ANNs showed relatively better results compared to one and three hidden layers in the case of 49 parameters. However, when the number of parameters was large, the model with one hidden layer obtained the best solutions for forecasting problems in most stations.

Finally, the model's parameters and the degree of effectiveness of the hidden layers are relatively competitive in forecasting the maximum temperature time series due to variations in errors. Additionally, when the number of parameters is large, the significant difference of model outputs from various hidden layers can be achieved.

As future work, we are interested in investigating the effect of more parameters such as the learning rate or momentum on the system's performance. Moreover, conducting an intensive investigation on the effect of parameters on other deep learning approaches for weather forecasting, such as a recurrent neural network (RNN), long short-term memory (LSTM), and convolutional neural network (CNN) is a matter of interest for our future research. Besides, the sensitivity analysis or critical dependence of one parameter on others may be involved for further research.

Supplementary Materials:
Figure S1: RMSE of different parameters ANN corresponding to 1, 3, and 5 hidden layers for test data Andong, Boeun, Boryeong, Buan, Busan, Buyeo, Cheonan, Cheongju, Chuncheon, Chungju, Chupungryong, Daegu, Daewallyeong, Daejeon, and Ganghwa stations. The dot and the vertical lines denote the mean and the standard deviation of 5 repetitions, respectively, Figure S2: RMSE of different parameters ANN corresponding to 1, 3, and 5 hidden layers for test data at Gangneung, Goheung, Gumi, Gunsan, Gwangju, Haenam, Hapcheon, Hongcheon, Icheon, Imsil, Incheon, Inje, Jangheung, Jecheon, and Jeju stations. The dot and the vertical lines denote the mean and the standard deviation of 5 repetitions, respectively, Figure S3: RMSE of different parameters ANN corresponding to 1, 3, and 5 hidden layers for test data at Jeongeup, Jeonju, Jinju, Miryang, Mokpo, Mungyeong, Namhae, Namwon, Pohang, Sancheong, Seogwipo, Seongsan, Seosan, Seoul, and Sokcho stations. The dot and the vertical lines denote the mean and the standard deviation of 5 repetitions, respectively, Figure S4: RMSE of different parameters ANN corresponding to 1, 3, and 5 hidden layers for test data at Suwon, Tongyeong, Uiseong, Uljin, Ulsan, Wando, Wonju, Yangpyeong, Yeongchen, and Yeongdeok stations. The dot and the vertical lines denote the mean and the standard deviation of 5 repetitions, respectively.

Author Contributions: Conceptualization, formal analysis, writing—original draft, T.T.K.T.; conceptualization, resources, methodology, writing—original draft, T.L.; conceptualization, writing—original draft, writing—review and editing, J.-S.K. All authors have read and agreed to the published version of the manuscript.

Acknowledgments: The authors appreciate the journal editors and the reviewers for their useful and valuable comments for this study. We also thank the support of the State Key Laboratory of Water Resources and Hydropower Engineering Science, Wuhan University.

References

1.   Valverde Ramírez, M.C.; De Campos Velho, H.F.; Ferreira, N.J. Artificial neural network technique for rainfall forecasting applied to the São Paulo region. *J. Hydrol.* **2005**, *301*, 146–162. [CrossRef]

2.   Umakanth, N.; Satyanarayana, G.C.; Simon, B.; Rao, M.C.; Babu, N.R. Long-term analysis of thunderstorm-related parameters over Visakhapatnam and Machilipatnam, India. *Acta Geophys.* **2020**, *68*, 921–932. [CrossRef]

3.   Verma, A.P.; Swastika Chakraborty, B. Now casting of Orographic Rain Rate Using ARIMA and ANN Model. In Proceedings of the IEEE 2020, Aranasi, India, 12–14 February 2020. [CrossRef]

4.   Berbić, J.; Ocvirk, E.; Carević, D.; Lončar, G. Application of neural networks and support vector machine for significant wave height prediction. *Oceanologia* **2017**, *59*, 331–349. [CrossRef]

5.   Agrawal, K. Modelling and prediction of rainfall using artificial neural network and ARIMA techniques. *J. Ind. Geophys. Union* **2006**, *10*, 141–151.

6.   Ozoegwu, C.G. Artificial neural network forecast of monthly mean daily global solar radiation of selected locations based on time series and month number. *J. Clean. Prod.* **2019**, *216*, 1–13. [CrossRef]

7.   Young, C.-C.; Liu, W.-C.; Hsieh, W.-L. Predicting the Water Level Fluctuation in an Alpine Lake Using Physically Based, Artificial Neural Network, and Time Series Forecasting Models. *Math. Probl. Eng.* **2015**, 1–11. [CrossRef]

8.   Alotaibi, K.; Ghumman, A.R.; Haider, H.; Ghazaw, Y.M.; Shafiquzzaman, M. Future predictions of rainfall and temperature using GCM and ANN for arid regions: A case study for the Qassim region, Saudi Arabia. *Water (Switzerland)* **2018**, *10*, 1260. [CrossRef]

9.   Smith, B.A.; Mcclendon, R.W.; Hoogenboom, G. Improving Air Temperature Prediction with Artificial Neural Networks. *Int. J. Comput. Inf. Eng.* **2007**, *1*, 3159.

10.   Paras, S.; Kumar, A.; Chandra, M. A feature based neural network model for weather forecasting. *World Acad. Sci. Eng.* **2007**, *4*, 209–216.

11.   Rajendra, P.; Murthy, K.V.N.; Subbarao, A.; Boadh, R. Use of ANN models in the prediction of meteorological data. *Model. Earth Syst. Environ.* **2019**, *5*, 1051–1058. [CrossRef]

12.   Tsai, C.Y.; Lee, Y.H. The parameters effect on performance in ANN for hand gesture recognition system. *Expert Syst. Appl.* **2011**, *38*, 7980–7983. [CrossRef]

13. Hung, N.Q.; Babel, M.S.; Weesakul, S.; Tripathi, N.K. Hydrology and Earth System Sciences An artificial neural network model for rainfall forecasting in Bangkok, Thailand. *Hydrol. Earth Syst. Sci.* **2009**, *13*, 1413–1416. [CrossRef]

14. Tran, T.T.K.; Lee, T.; Shin, J.-Y.; Kim, J.-S.; Kamruzzaman, M. Deep Learning-Based Maximum Temperature Forecasting Assisted with Meta-Learning for Hyperparameter Optimization. *Atmosphere* **2020**, *11*, 487. [CrossRef]

15. Tran, T.T.K.; Lee, T. Is Deep Better in Extreme Temperature Forecasting? *J. Korean Soc. Hazard Mitig.* **2019**, *19*, 55–62. [CrossRef]

16. Sarle, W.S. Stopped Training and Other Remedies for Overfitting. Proc. 27th Symp. *Comput. Sci. Stat.* **1995**, *17*, 352–360.

17. Chollet, F. Home Keras Documentation. Available online: https://keras.io/ (accessed on 4 May 2020).

18. Abadi, M.; Agarwal, A.; Barham, P.; Brevdo, E.; Chen, Z.; Citro, C.; Corrado, G.S.; Davis, A.; Dean, J.; Devin, M.; et al. TensorFlow: Large-Scale Machine Learning on Heterogeneous Distributed Systems. *arXiv* **2016**, arXiv:1603.04467.

19. Chen, Y.; Tong, Z.; Zheng, Y.; Samuelson, H.; Norford, L. Transfer learning with deep neural networks for model predictive control of HVAC and natural ventilation in smart buildings. *J. Clean. Prod.* **2020**, *254*, 119866. [CrossRef]

20. Lee, J.; Kim, C.G.; Lee, J.E.; Kim, N.W.; Kim, H. Application of artificial neural networks to rainfall forecasting in the Geum River Basin, Korea. *Water (Switzerland)* **2018**, *10*, 1448. [CrossRef]

21. Alam, S.; Kaushik, S.C.; Garg, S.N. Assessment of diffuse solar energy under general sky condition using artificial neural network. *Appl. Energy* **2009**, *86*, 554–564. [CrossRef]

4

Impacts of Green Vegetation Fraction Derivation Methods on Regional Climate Simulations

Jose Manuel Jiménez-Gutiérrez [1], Francisco Valero [1], Sonia Jerez [2] and Juan Pedro Montávez [2,*]

[1] Departamento de Astrofísica y Ciencias de la Atmósfera, Universidad Complutense de Madrid, 28040 Madrid, Spain; jimenezg1980@gmail.com (J.M.J.-G.); valero@ucm.es (F.V.)

[2] Departamento de Física, Universidad de Murcia, 30100 Murcia, Spain; Sonia.Jerez@gmail.com

[*] Correspondence: montavez@um.es.

Abstract: The representation of vegetation in land surface models (LSM) is crucial for modeling atmospheric processes in regional climate models (RCMs). Vegetation is characterized by the green fractional vegetation cover (FVC) and/or the leaf area index (LAI) that are obtained from nearest difference vegetation index (NDVI) data. Most regional climate models use a constant FVC for each month and grid cell. In this work, three FVC datasets have been constructed using three methods: ZENG, WETZEL and GUTMAN. These datasets have been implemented in a RCM to explore, through sensitivity experiments over the Iberian Peninsula (IP), the effects of the differences among the FVC data-sets on the near surface temperature (T2m). Firstly, we noted that the selection of the NDVI database is of crucial importance, because there are important bias in mean and variability among them. The comparison between the three methods extracted from the same NDVI database, the global inventory modeling and mapping studies (GIMMS), reveals important differences reaching up to 12% in spatial average and and 35% locally. Such differences depend on the FVC magnitude and type of biome. The methods that use the frequency distribution of NDVI (ZENG and GUTMAN) are more similar, and the differences mainly depends on the land type. The comparison of the RCM experiments exhibits a not negligible effect of the FVC uncertainty on the monthly T2m values. Differences of 30% in FVC can produce bias of 1 °C in monthly T2m, although they depend on the time of the year. Therefore, the selection of a certain FVC dataset will introduce bias in T2m and will affect the annual cycle. On the other hand, fixing a FVC database, the use of synchronized FVC instead of climatological values produces differences up to 1 °C, that will modify the T2m interannual variability.

Keywords: fraction vegetation cover; regional climate model; near surface temperature

1. Introduction

Land surface features play a very important role in modulating surface-atmosphere interactions. This is an overlooked issue in climate simulations [1] and short-term weather forecasting [2]. Surface components like soil moisture, albedo, emissivity, surface roughness, vegetation type and amount are fundamental since they control the energy partition at surface [3]. Therefore, the representation of all these variables in land surface models (LSM) is crucial for modeling atmospheric processes.

Atmosphere and vegetation interact in different ways, controlling evapotranspiration, moisture availability, momentum transfer, partitioning radiation, etc. Vegetation is characterized within numerical weather prediction models (NWPM) and RCMs by FVC, LAI and the vegetation class [4]. Therefore, realistic characterization of these parameters should lead to a better reproduction of atmosphere-surface processes.

Different methodologies have been reported for obtaining FVC and LAI through satellite NDVI data [5–8]. According to Gutman and Ignatov [6] both parameters should not be used simultaneously in the same parameterization. Therefore, it is necessary to prescribe one of the indices and derive the other. Gutman and Ignatov [6] and Carlson et al. [9] argued that is preferable to derive FVC and prescribe LAI, because the exponential dependence of LAI and NDVI saturates after a certain threshold, becoming LAI insensitive to changes in NDVI. Some other authors [7,10] recommend to derive LAI (fixing FVC), because they assume that spatial and seasonal variations of NDVI are related with variations of LAI and validation of FVC is problematical because of the requirement of information at the scale of individual plant elements. Anyway, Godfrey et al. [2] argue that errors introduced by the dual specification of vegetation parameters from a single NDVI observation are likely smaller than uncertainties associated to initial conditions.

The role of vegetation parameters (FVC and LAI) is relevant for weather forecasting and climate change assessments [11]. Their impact on land surface processes has been studied in the Eta operational model [12–14] and the Weather Research and Forecasting model (WRF) [15–20]. The relevance of using realistic information of the vegetation state on RCM performance has been analyzed in several works. Meng et al. [21] and Müller et al. [22] study the impact of vegetation in concrete cases of droughts in Australia and South America. Other works investigate the contribution of near real time values of vegetation fraction to simulated precipitation. For instance, vegetation–atmosphere feedback in monsoon systems [23,24], severe convection episodes [25], or improving model performance in oasis-desert systems [26]. Other interesting studies focus on vegetation effects on regional climate simulations in complex urban areas like Los Angeles [27,28].

However, the characterization of vegetation is subjected to several sources of uncertainty. On one hand, there are several NDVI database available for obtaining FVC data. They differ in aspects such as the methods for correcting errors, type of satellite, etc. The differences among these NDVI databases reach, for example, similar values to the observed trends in the phenological phases [29]. On the other hand, several methodologies can be used to obtain FVC or LAI from NDVI data [4,6,13,30]. Crawford et al. [31] points out that differences up to 25% in LAI and FVC can easily occur among the different methodologies, similar values to the FVC interannual variability.

In addition, most of the standard configurations of NWPM and RCM use climatological values of vegetation parameters. However, the vegetation has a strong inter-annual variability [18,21,31], leading this to a non-suitable characterization of surface properties. A known limitation is that surface properties can vary at several time scales depending on climate conditions and other processes such as urbanization, forest fires or changes in crops [32].

As an example of the impact of using different approaches with FVC datasets in a LSM, Miller et al. [30] describe, using the Noah LSM, noteworthy differences in surface fluxes, comparing a five-year climatology data from the Advanced Very High Resolution Radiometer (AVHRR) and NASA's Moderate Resolution Imaging Spectrometers (MODIS) data from the year 2000. With differences of 25% in FVC in monthly-averaged values for all pixels of dry land and cropland of the same two-degree box, transpiration was modified in 30 W m^{-2}, latent heat fluxes in 10 W m^{-2}, and sensible heat fluxes in -20 W m^{-2}. Other works, where real-time satellite data have been used instead of climatological values, have found improvements in the forecasts of T2m in RCM simulations [13,33]. While other studies, indicates that the model sensitivity is not reliable with improved vegetation parameters [2,15,23], stressing the significance of minimizing errors in surface initial conditions (especially soil moisture) or canopy resistance parameterization.

The objective of this work is to assess the impact on T2m simulated by a RCM, of the uncertainty associated to the use of different NDVI data, different methodologies for obtaining FVC, as well as the temporal variability of surface properties in regional climate simulations. Our work will focus on IP, since it is characterized by a large variety of vegetation classes, with a strong seasonal and interannual climate variability that leads to important changes in the state of vegetation.

2. Methods and Data

2.1. NDVI Data

The NDVI is defined as:

$$NDVI = NIR - \frac{VIS}{NIR} + VIS, \tag{1}$$

where NIR and VIS are the amounts of near-infrared and red visible, respectively, reflected by the vegetation and captured by the satellite sensor [34]. This index was highly correlated with the photo synthetically active biomass, chlorophyll abundance and energy absorption [35] and has been widely used in studies involving land-biosphere interactions [6,13,30,31,36].

There are several global coarse-resolution satellite spectral vegetation index datasets [37]. For example, Satellite Pour l'Observation de la Terre (SPOT Vegetation) and MODIS datasets which have higher resolution and significant improvements with regard to the 1981–2004 record of the AVHRR dataset. However, the latter is an invaluable and irreplaceable archive of historical land surface information and the most used dataset for deriving NDVI due to its long record. Since 1984 several global land surface NDVI data have been derived from AVHRR, as can be the global vegetation index (GVI) dataset [38,39], the NASA Pathfinder 8-km dataset [40] and the GIMMS-NDVI[37,41]. Other datasets, such as the EFAI-NDVI [29], were derived from the NASA Pathfinder 8 km data. These datasets were corrected by taking into account artifacts from sensors, orbital drifting, atmospheric corrections and cloud screening.

Therefore, NDVI values can vary from a database to another in an remarkable way. The GIMMS-NDVI and EFAI-NDVI datasets were chosen to perform our analysis. The selection was done following criteria of NDVI data already processed, easily/free availability and temporal extension.

The EFAI-NDVI covers the period 1982–2001 and corrects the original AVHRR data to create a continuous dataset of 10-day temporal and $0.1°$ spatial resolutions with global coverage. The correction method implies a spatial interpolation of missing values and processing artifacts as well as a temporal interpolation of the NDVI series throughout a Fourier adjustment algorithm [10,29].

The GIMMS-NDVI [42] from the AVHRR dataset covers from 1982 to 2006 with a spatial resolution of 8 km. This database has been corrected for calibration, view geometry, volcanic aerosols, and other effects non related to vegetation change. In particular, NOAA-9 descending node data from September 1994 to January 1995, volcanic stratospheric aerosol correction for 1982–1984 and 1991–1994, and improved NDVI using empirical mode decomposition/reconstruction to minimize effects of the orbital drift.

2.2. Deriving FVC from NDVI Data

FVC is defined as the fraction of horizontal area associated with the photosynthetically active green vegetation that occupies a model grid cell [43]. In our experiments we derive FVC and use prescribed LAI, because of our LSM, Noah, considers this option in its configuration.

FVC calculation from NDVI data [44] can be based on linear models [6] or quadratic models [45]. These methods take as reference bare soil ($NDVI_0$) and dense green vegetation($NDVI_\infty$). Such values can be prescribed or estimated from the actual NDVI data. For instance, Montandon and Small [44] studied the impact of varying $NDVI_0$, showing how its underestimation yields a FVC overestimation.

In this study, three linear models, WETZEL, GUTMAN and ZENG have been selected as a basis to study the uncertainty in FVC calculation and its impacts on RCMs runs. These methods present different degrees of complexity, being the WETZEL/ZENG method the most simple/complex, while GUTMAN is of intermediate complexity. This latter has been extensively used in generating data for NWPMs.

The three methodologies have been applied to the GIMMS-NDVI data. Monthly FVC values have been calculated from 1982 to 2006 with a spatial resolution of $0.1°$.

2.2.1. Wetzel Method

In Chang and Wetzel [5] a two-line-segment method is presented. The slope of the linear relationship between NDVI and FVC changes when NDVI exceeds 0.547. The relation is as follows:

$$FVC = \begin{cases} 1.5(NDVI - 0.1), NDVI \leq 0.547 \\ 3.2(NDVI) - 1.08, NDVI > 0.547, \end{cases} \tag{2}$$

being that the FVC is constrained to be between 0 and 1.

The FVC data obtained by this method is used in the MM5 LSM model by Crawford et al. [31] and in the ETA model by Kurkowski et al. [13].

2.2.2. Gutman Method

This method uses the following simple linear relationship between NDVI and FVC:

$$FVC = \frac{NDVI_{val} - NDVI_0}{NDVI_\infty - NDVI_0}, \tag{3}$$

where $NDVI_0$ and $NDVI_\infty$ are the values for bare soil and dense green vegetation.

Gutman and Ignatov [6] take $NDVI_0 = 0.04$ and $NDVI_\infty = 0.52$, which correspond respectively to the minimum NDVI value of the desert cluster and the maximum NDVI value of the evergreen cluster in their study with GVI data. These values are in principle, region and season specific, since they depend on the soil and vegetation types and the vegetation chlorophyll content [46]. However, the authors take this assumption because there are many intermediate cases that make the evaluation difficult for other surface types. In this work, $NDVI_0$ and $NDVI_\infty$, have been set to 0.1 (2% percentile of bare soil) and 0.91 (98% Evergreen vegetation) using the GIMMS-NDVI database.

2.2.3. Zeng Method

This method includes procedures that involve the analysis of NDVI data as a function of biome or land cover type. This approach has been used in several works [7,15,30,43]. It follows the simple relationship between FVC and NDVI described by Gutman and Ignatov [6] (Equation (3)). The values of $NDVI_\infty$ are calculated analyzing the frequency distribution of maximum NDVI for each land cover type.

This study uses the University of Maryland Department (UMD) global land cover classification [47,48]. This land cover database has been chosen because it was created using the same NDVI data as the GIMMS-NDVI database. Figures 1a (forest types) and 2a (shrubland, cropland and grassland) show the spatial distribution as well as the relative frequency distributions of NDVI.

According to Zeng et al. [7] $NDVI_\infty$ is low sensitive to the exact percentile used for a given vegetation type. In this work the $NDVI_\infty$ is chosen getting the 98th percentile as adopted in other studies ([10,43]). For $NDVI_0$, the 5th percentile of the no-vegetation category is taken for all vegetation types. North Africa has been included to have a more suitable value for bare soil. Regarding $NDVI_{val}$, Zeng et al. [7] estimated FVC as independent of season, then $NDVI_{val}$ (Equation (3)) is the annual maximum in a given pixel in order to minimize the effects of cloud contamination. However, in our case, $NDVI_{val}$ corresponded with each of the 15-day values of the GIMMS-NDVI database in a given pixel, since we precisely wanted that FVC varies with time.

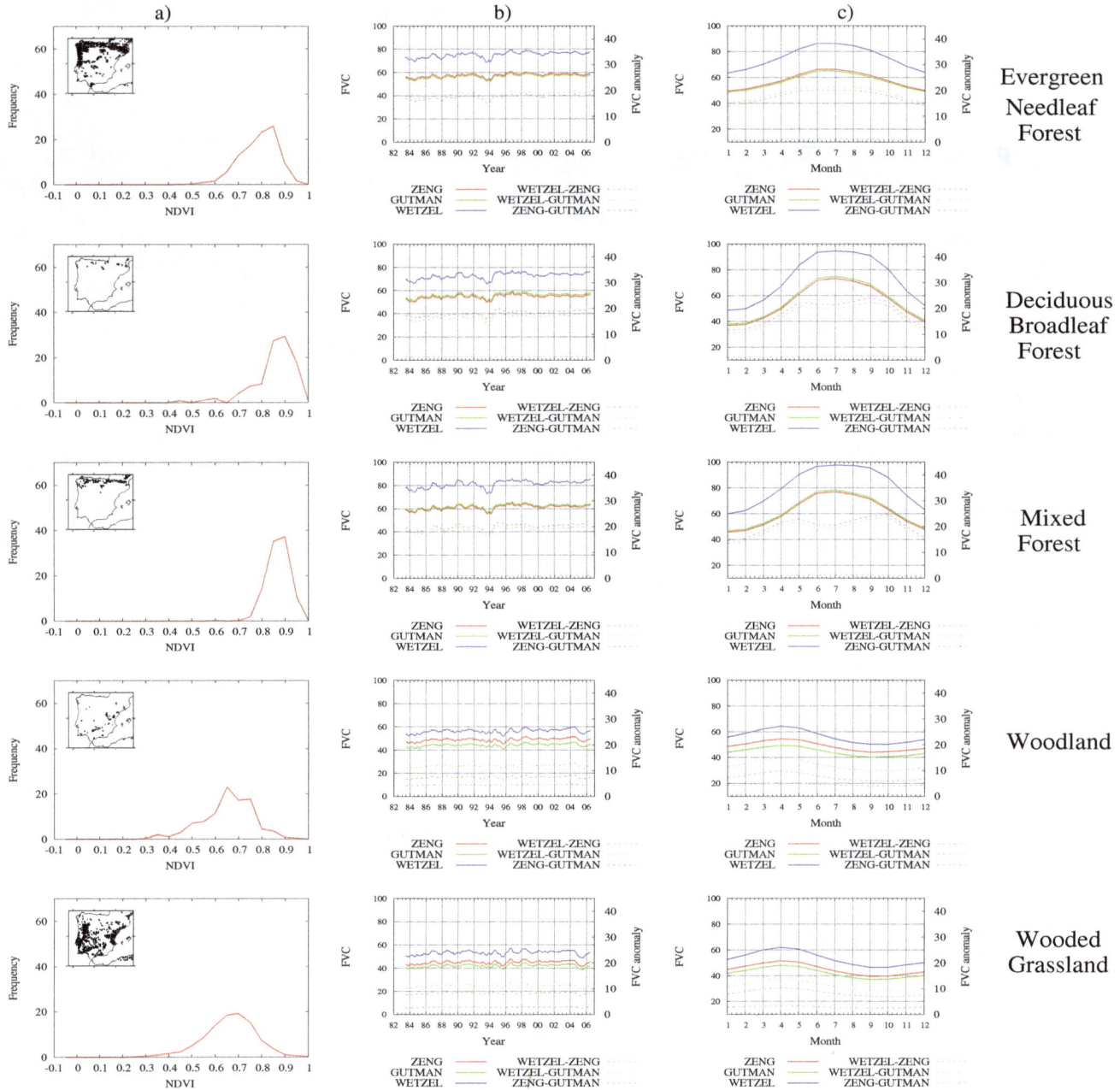

Figure 1. (a) Area covered by land type. Relative frequency distributions of NDVI in the IP by land types according to UMD global landcover. (b) FVC time series in forest land types calculated with the GUTMAN (green solid line), WETZEL (blue solid line) and ZENG (red solid line) methods (spatial average by land type). Differences between FVC methods: WETZEL-ZENG (red dashed line), WETZEL-GUTMAN (green dashed line), ZENG-GUTMAN (blue dashed line). (c) Same as b but monthly averaged values by land type for the period 1982–2006.

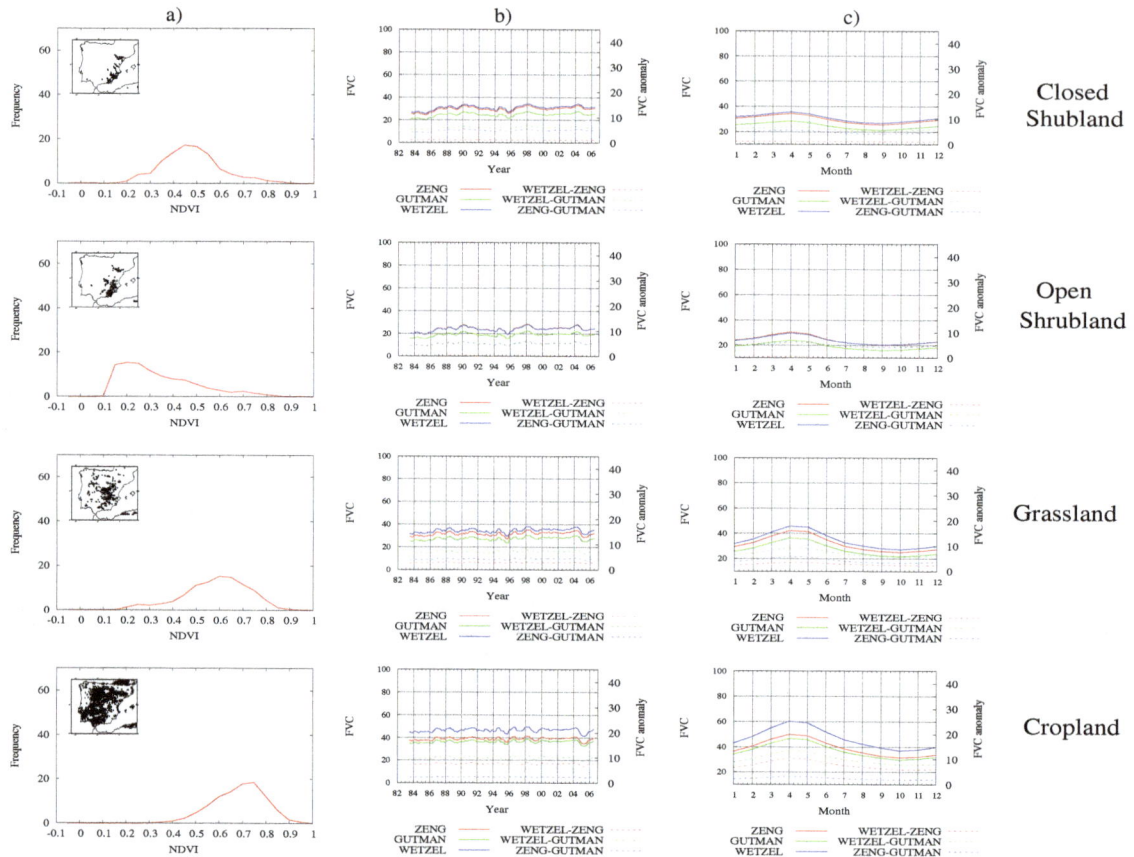

Figure 2. As Figure 1 for closed and open shrubland, grassland and croplands.

As a help for understanding differences between the methods, Figure 3 shows the relationships between FVC and NDVI for the three methods. The ZENG method (red area) provides a wider range of values of FVC depending of land type frequency distribution of NDVI. The GUTMAN method (green line) agrees with the lower correspondence between NDVI and FVC of the ZENG method, explaining the lower values of FVC. On the other hand, the WETZEL method produce higher FVC values in areas where NDVI is larger than the rupture point (0.547), especially in the north of IP.

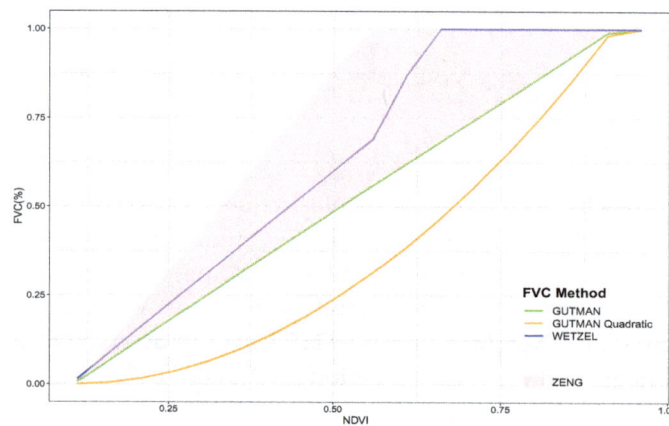

Figure 3. Relationship between fractional vegetation cover (FVC) and nearest difference vegetation index (NDVI) for the ZENG (red area), GUTMAN, WETZEL, and GUTMAN quadratic methods.

2.3. Regional Climate Model Experiments

In order to gain some insight about the sensitivity of a RCM to changes in FVC a set of simulations have been performed using a climate version of the MM5 mesoscale model [49,50]. The spatial configuration consists of two-way nested domains with 30 and 10 km horizontal resolution respectively (Figure 4) and 24 sigma levels in the vertical up to 100 mb. ERA-Interim reanalysis [51] were used to provide initial and boundary conditions, which were updated every six hours. The physics configuration consisted of Grell cumulus parameterization [52], simple ice for microphysics [53], rapid radiative transfer mode (RRTM) radiation scheme [54] and the medium-range forecast (MRF) planet boundary layer scheme [55]. The Noah LSM [56] used in this simulations was based on the coupling of the diurnally dependent Penman potential evaporation approach of Mahrt and Ek [57], the multilayer soil model of Mahrt and Pan [58], and the primitive canopy model of Pan and Mahrt [59]. It was extended by Chen and Coauthors [60] to include a complex canopy resistance approach [61,62] and a surface runoff scheme. The prognostic variables in the Noah LSM are soil moisture and temperature as well as canopy moisture and water-equivalent snow depth. Four soil layers were used with thicknesses of 10, 30, 60 and 100 cm, and an additional canopy layer for the soil model. The total soil depth was 2 m, with the root zone in the upper 1 m of soil. The lower 1-m soil layer acted as a reservoir with gravity drainage at the bottom. Ground heat flux was controlled by the usual diffusion equation for soil temperature, with heat capacity and thermal conductivity formulated as functions of the soil water content. The diffusive form of Richard's equation was used as the prognostic equation for the volumetric soil moisture content, where the hydraulic conductivity and the soil water diffusivity are also functions of the soil water content. In this LSM, vegetation type and soil texture were primary variables upon which other secondary parameters (such as minimal canopy resistance and soil hydraulic and thermal properties) were determined. Landuse-vegetation category is specified from 25-category 1 km data from USGS version 2 land cover data. The dominant vegetation type in each grid box is selected to represent the "grid level" vegetation characteristic. In Noah LSM, the FVC acts as a fundamental weighting coefficient in partitioning the total evaporation into direct evaporation from the top shallow soil layer, evaporation of precipitation intercepted by the canopy, and transpiration via canopy and roots.

The runs have been performed running the model for one year with four additional months of spin-up period. The outputs during this spin-up period, integrated only to guarantee that soil variables reach dynamic equilibrium, are ignored. The RCM used here, employing similar configurations, provided satisfactory results in climatic simulations [50,63–65].

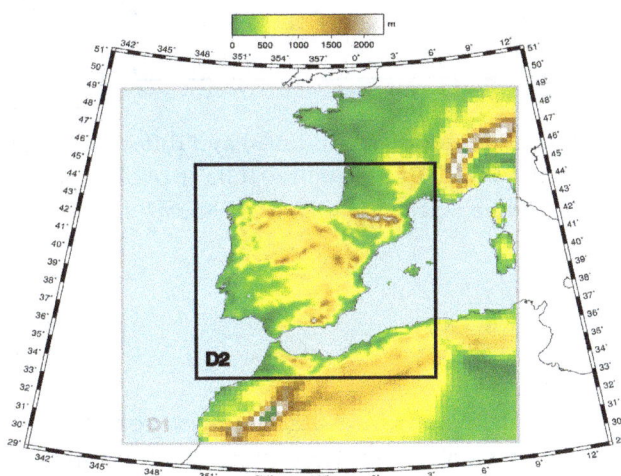

Figure 4. Regional climate model (RCM) spatial domains. Mother domain, D1 (30 km), and inner domain, D2 (10 km). Topography of the spatial domains.

Most regional models such as WRF or MM5 (Noah LSM) use a default FVC monthly climatological data calculated in a given period. A commonly used database is described in Gutman and Ignatov [6] which consist of a 5-year FVC climatology derived from AVHRR, for the period April-December of

1985, from 1986 to 1990 and the period January–March of 1991 with a spatial resolution of 0.15°. In this work, the database generated had a resolution of $0.1° \times 0.1°$.

Two set of experiments have been carried out. The first set investigated the effect of using different methodologies in constructing the FVC database used in the simulations. For this task, firstly a five-year FVC climatology has been constructed for the same period of Gutman and Ignatov [6] (April 1985–March 1991) with the ZENG, GUTMAN and WETZEL methods from the GIMMS-NDVI data. Then three runs for the year 1995 were performed using such FVC climatologies. The second set of simulations explored the role of introducing synchronous FVC data for a given year respect to the use of climatological values of this variable. For this case, two pairs of runs were carried out, choosing the ZENG method for the years 1995 (a dry year) and 1996 (a wet year). Each pair consists of a simulation using the period of the five-year FVC climatology (CLIM) described before and the other one using the FVC values for such year (YEAR). The ZENG method has been chosen, among the three options, because it presents an intermediate behaviour between WETZEL and GUTMAN (see Figure 5) as well as the fact that it has been evaluated in previous works [7].

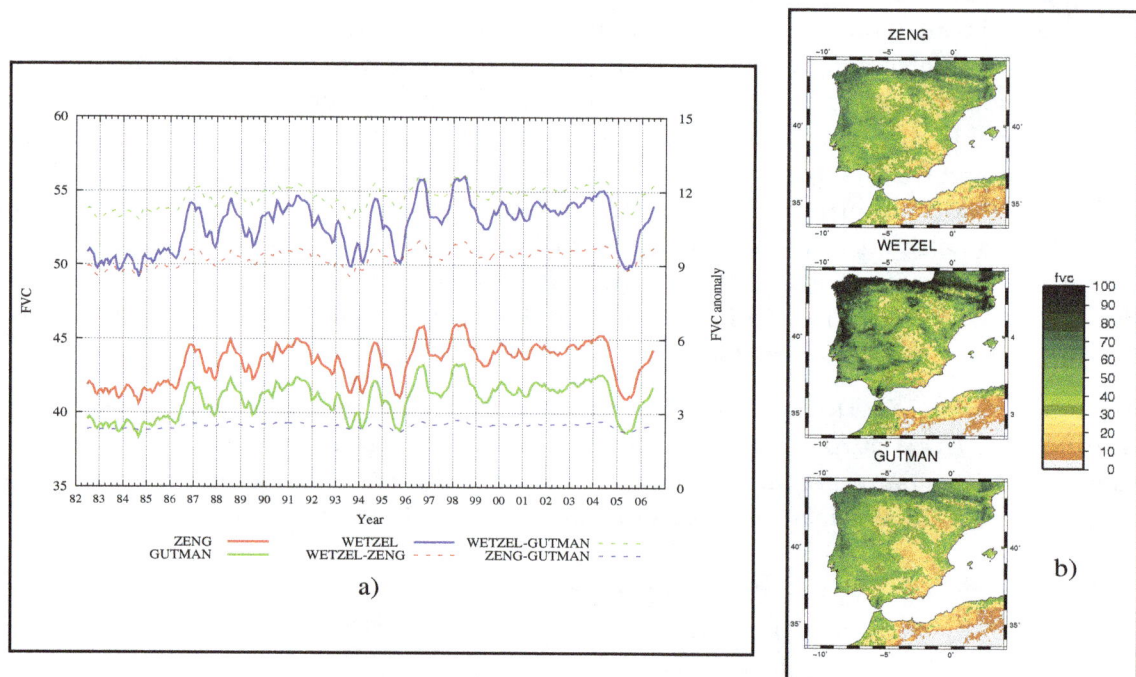

Figure 5. (**a**) FVC spatial averaged time series (solid lines) and differences between them (dashed lines) for the whole IP calculated by the WETZEL (blue line), GUTMAN (green line) and ZENG (red line) methods. (**b**) FVC temporal averaged over the whole period (1982–2006) for the IP for the ZENG, WETZEL and GUTMAN methods.

3. Results

3.1. NDVI Database Comparison

Figure 6 shows the NDVI for the two datasets in a dry year (1995) and a wet year (1996) for January and July over the IP. It is remarkable that they differed in magnitude but that spatial variability was quite similar (spatial correlation over 0.97). The differences were higher when comparing years with different precipitation regime, where NDVI values differed up to 30% in the summer (not shown).

The 12 month run-mean series of the spatially averaged NDVI over the IP for the period 1982–1998 is shown in Figure 7a. The series shows a quite different interannual variability for the whole period; the EFAI-NDVI had a greater variability and lower mean values than the GIMMS-NDVI. We attributed the disagreement (almost a periodical signal not attributable to natural variability) to the solar zenith angle correction implemented in the GIMSS-NDVI that avoids artificial trends derived from orbital drifting. At this point we concluded that the EFAI-NDVI should not be used for characterizing the vegetation properties in RCMs. Therefore, we focused our study on the GIMMS-NDVI data. In addition, GIMMS-NDVI had a longer record, because of the compatibility with AVHRR data with MODIS and SPOT vegetation satellite data, that have continued until nowadays.

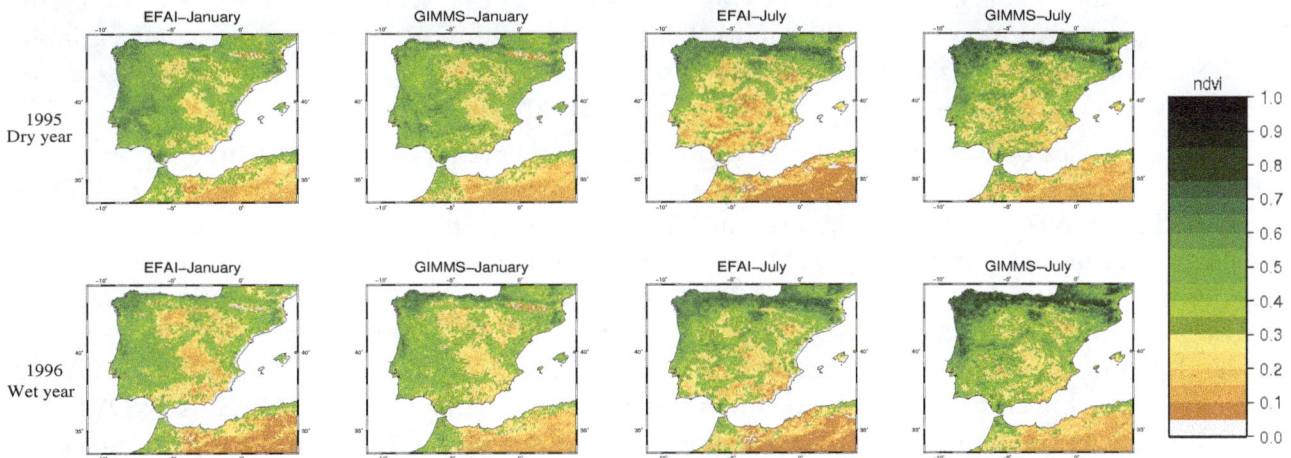

Figure 6. EFAI-NDVI and GIMMS-NDVI for January and July for a wet year (1996) and a dry year (1995) in the Iberian Peninsula (IP).

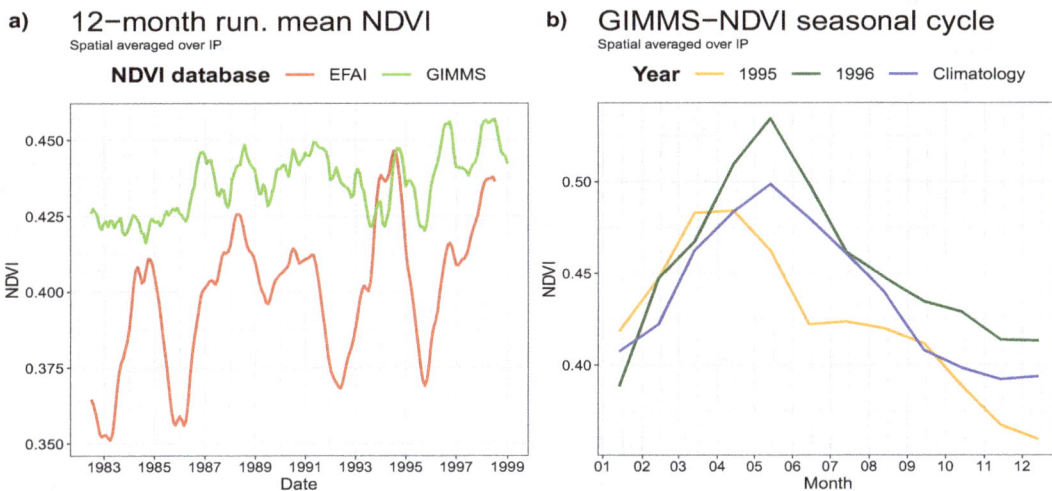

Figure 7. (a) NDVI 12 month run-mean of the spatially averaged series for the IP derived from the EFAI-NDVI (red line) and GIMMS-NDVI (green line); **(b)** GIMMS-NDVI monthly values for a wet year (1996, green line), a dry year (1995, orange line) and climatology (blue line).

Usually FVC is implemented using climatological values in NWPMs and RCMs. We compared the spatially averaged NDVI over the IP, with monthly climatological values calculated on a seven year period (1985–1991) with the monthly evolution on a specific dry (1995) and wet year (1996) (Figure 7b). The biggest NDVI differences between the dry and wet year were observed in spring, when the maximum of vegetation productivity occurred, prevailing these NDVI differences in a lesser way the rest of the year. The spatial average NDVI difference between wet and dry years was about 5% reaching values up to 15% in some areas. This elucidates that taking into account the interannual variability could be important when implementing these data in RCMs.

3.2. Analysis of FVC Retrieval Methods

FVC data were obtained for the 25 year of available NDVI data using the three methods. Figure 5a shows the temporal evolution of the FVC data spatially averaged over the IP and Figure 5b the temporal means of FVC values over the domain. Since the three methods used the same NDVI, there was a very good agreement in the temporal variation of FVC, however some important differences appear. The most remarkable difference comparing the magnitude of FVC was that WETZEL method presented FVC values 10% greater than the others. The ZENG and GUTMAN methods only differed around a 3% in magnitude. This large difference was due to the fact that when applying WETZEL method the frequency distribution of NDVI was not taken into account. Regarding the spatial variability, the ZENG and GUTMAN show very similar patterns although the ZENG method achieves a larger range of FVC values, reaching values near 10% and over 80% more frequently.

The differences between the methods depended on the land type by construction. Figures 1b and 2b depict the time series of the FVC spatially averaged by land types calculated with the GUTMAN (green solid line), WETZEL (blue solid line) and ZENG (red solid line) methods as well as the differences between them (dashed lines). The GUTMAN and ZENG methods showed very similar FVC values in forest land types (Figure 1), while the WETZEL method showed significant differences of around 20% respect to the other methods. Land types with low values of FVC, such as shrubland, cropland and grassland (Figure 2) present smaller differences. In the case of closed and open shrubland, the WETZEL method presents the closest values to the other methods. This can be explained because NDVI values below the rupture point of the WETZEL method were more frequent for this land types (Figure 2a).

Therefore, in general, places with high FVC values exhibited a good agreement between the methods that take into account the NDVI frequency distribution (GUTMAN and ZENG), showing larger differences with the WETZEL method. While in places with medium and low values of FVC, the spread among the methods was lower.

Another interesting feature was to evaluate is the differences in phenology. Figures [1] and [2] show the relative frequency distribution if NDVI (a), temporal evolution (b) and FVC annual cycle (c) for each biome and the differences between them. In the biomes where the maximum of FVC occurred in summer, forest vegetation with low hydric stress, the GUTMAN and ZENG methods had a quite similar behavior (maximum value of 60–80% in summer) while the WETZEL method gave greater values. In the rest of the biomes, the maximum of FVC occurred in spring (irrigated lands or places with summer hydric stress) and the differences between the methods were smaller than for forest land types, although there were differences that could be remarkable in cropland, grasslands and woodland areas.

3.3. Sensitivity to the FVC Estimation Method

Here we assessed the sensitivity of climate simulations to the use of different FVC data presented above. For the sake of clarity, we only present the comparisons between the ZENG and WETZEL experiments, because they show the largest FVC differences. Figure 8a depicts the FVC monthly means of ZENG for some months representative of each season (January, April, July and October) and Figure 8b the FVC differences between the ZENG and WETZEL methods. There is a clear negative difference in the FVC estimated (ZENG-WETZEL) with higher FVC values for the WETZEL data, especially in areas with the highest FVC. The biggest difference was in April, when the vegetation productivity presented the maximum rate of generation of biomass.

The effect of such differences between the ZENG and WETZEL data on the regional climate simulations for monthly mean T2m is depicted in Figure 8c,d. In general, lower FVC values led to higher temperatures, but this relationship depended on the time of the year (Figure 8d). This result was directly related to the sensible and latent heat partitioning (not shown). An increase/decrease of vegetation led to a larger/lower evapotranspiration and therefore, a higher/smaller latent heat production. This led to a reduced/intensified sensible heat manifested in a decrease/increase in T2m. Besides, there was an additional effect due to the modification of soil thermal conductivity by the presence of a vegetation canopy. Heat flux from soil was reduced when FVC increases because of heat conductivity was diminished [12,66]. This effect was of importance mostly during night when these fluxes had a dominant role and can cause higher night temperatures with reduced FVC.

It is worth noting that T2m differences were higher in April and July than in January and October with similar FVC differences (Figure 8). In January differences of 30% in FVC led to changes smaller than 0.5 °C, while in July or April can reach 1 °C in some places. This was mainly due to the greater availability of energy at these months. In addition, the greater dispersion in the quasi-linear relationship between changes in FVC and T2m (Figure 8d) for higher ΔFVC values may be relevant. Therefore, it was clear that changing the FVC database modified the climatology reproduced by a RCM, both the spatial patterns and the amplitude of the annual cycle.

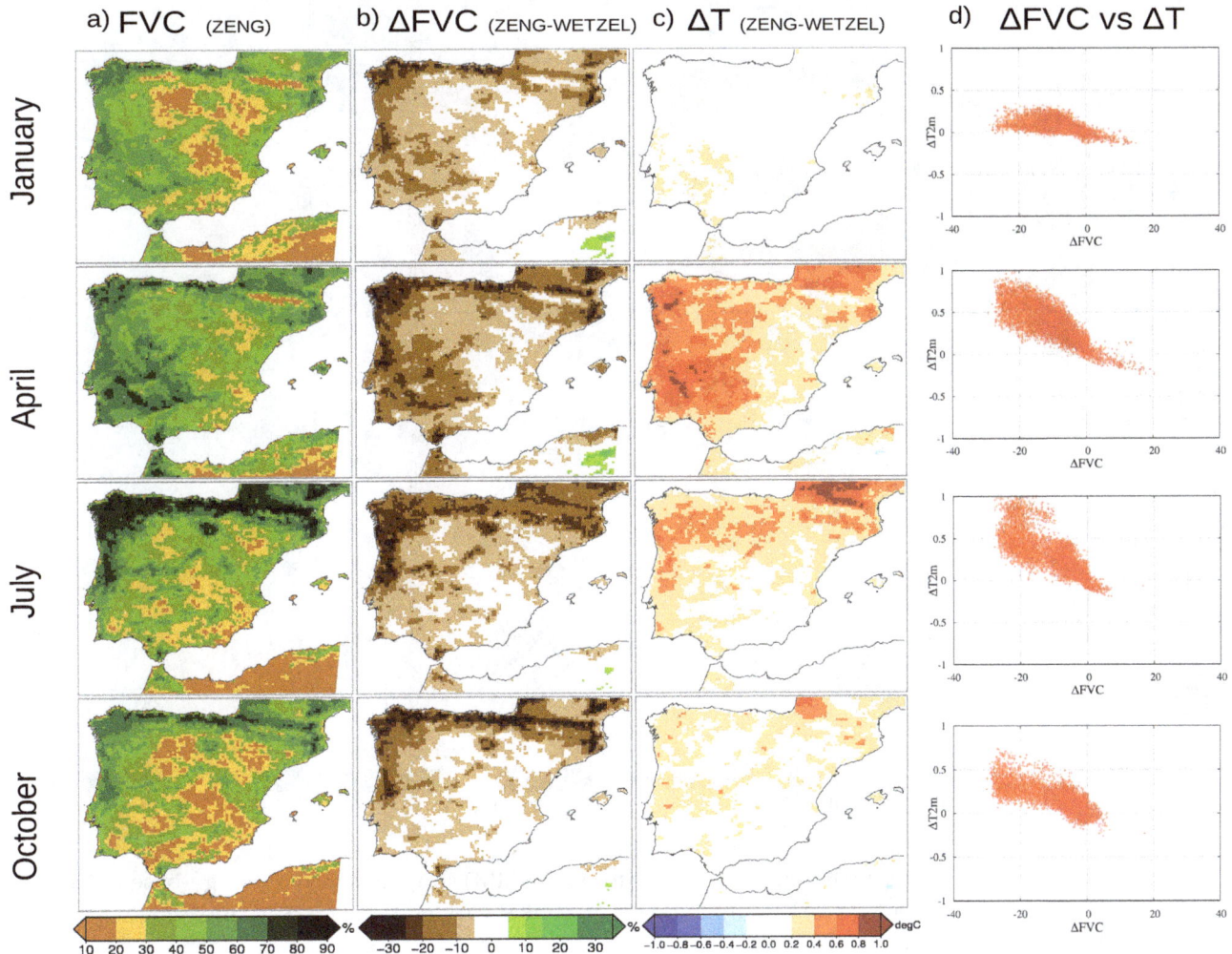

Figure 8. (**a**) Monthly means of FVC for the ZENG method. (**b**) Differences of monthly means of FVC between the ZENG and WETZEL methods. (**c**) Differences in monthly means of T2m between the ZENG and WETZEL regional climate simulations. (**d**) x–y scatter plot graph between differences of FVC (x-axis) and differences in T2m between ZENG and WETZEL methods(y-axis).

3.4. Sensitivity to FVC Interannual Variations

As mentioned above, in RCM experiments, FVC values are usually prescribed using climatologies, so they do not vary from year to year. But vegetation depended on the climate conditions as well as other factors, and it had a non negligible interannual variability. In this section we analyze the effect of varying the FVC according to the observations (real FVC) versus fixed monthly climatological/prescribed values.

Figure 9 shows the differences between the FVC 5-year climatology simulation (CLIM) and the reconstructed (YEAR) for January, April, July and October for two years, the dry 1995 and wet 1996. These years were selected because of their quite different climate conditions. Dry conditions prevailed during 1995, while 1996 was characterized for abundant precipitation that favored higher biomass production.

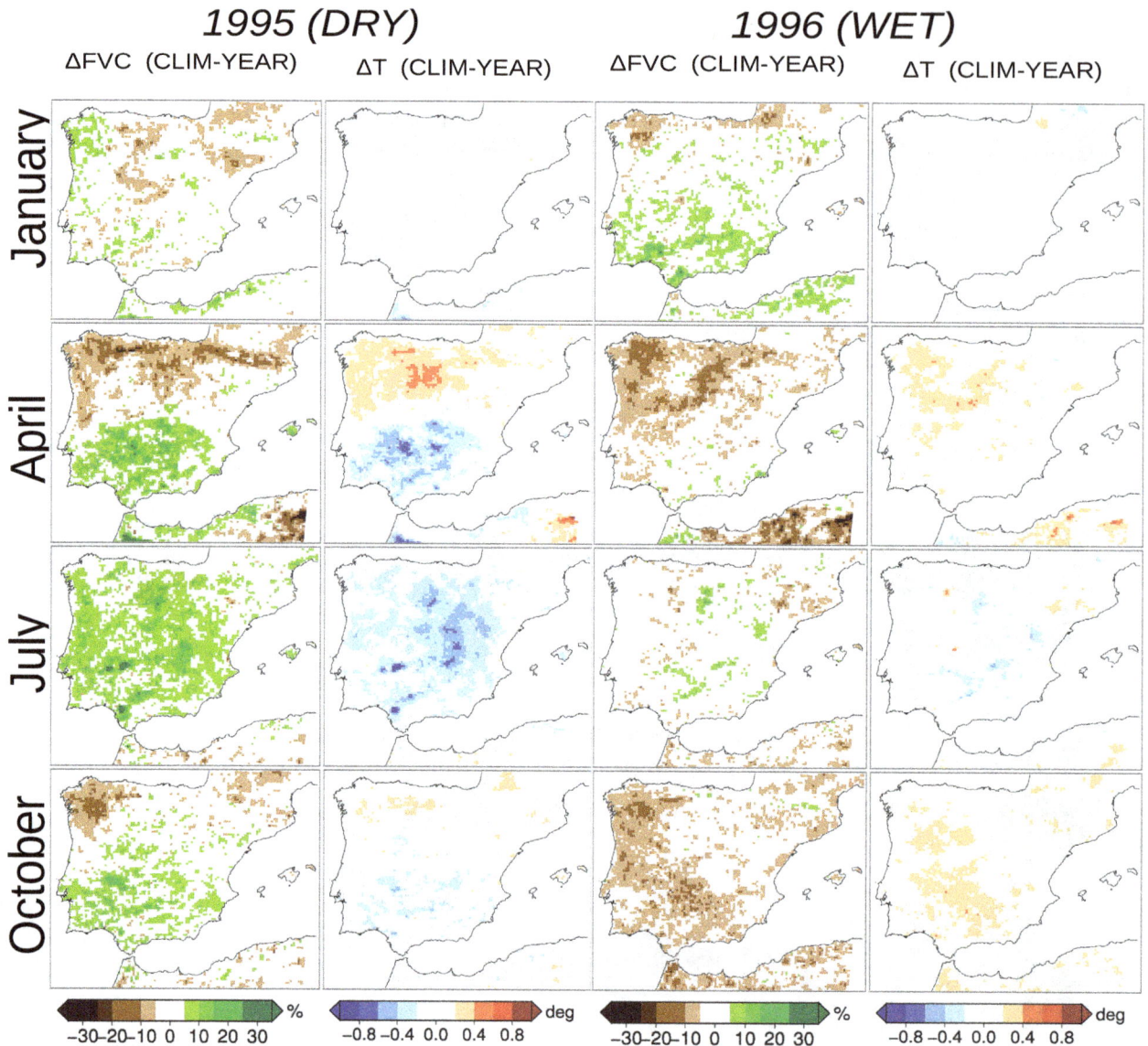

Figure 9. Differences between five-year FVC climatology (CLIM) simulation (climatology-based FVC) and FVC values for such year (YEAR) simulation (inter-annual varying FVC) for FVC and T2m in a dry year (1995) and a wet year (1996).

During the dry year, the prescribed FVC was much higher that the estimated for this year, finding large areas with ΔFVC over 10%, especially in April and July that locally reached values around 30%. The T2m temperature patterns resembled the spatial structure of ΔFVC with a clear negative bias. The largest T2m differences appeared again in July and April, with values larger than 0.4 °C in large areas that locally were over 0.8 °C.

During the wet year the largest ΔFVC appeared in April in the north–west part of the IP, and October over the South and Western Iberia. In this case differences were negative, i.e., the prescribed FVC underestimates the YEAR FVC. Interestingly, the absolute T2m differences were smaller, with slightly lower absolute ΔFVC. This is due to the fact that in July (highest sensitivity) ΔFVC is quite small, and that in April advection phenomena prevailed in the area with the largest ΔFVC.

Adding both effects at the different times suggest that interannual variability can change more that 1 °C in some areas for a given month.

4. Discussion and Conclusions

Three database of monthly FVC from 1982 to 2006 has been created for a domain covering the whole IP with a spatial resolution of 0.1° with the aim of being implemented in RCMs to improve climate simulations by providing more realistic soil physical conditions. Three different methodologies, WETZEL, GUTMAN and ZENG, were applied to the same NDVI database. The GIMMS-NDVI data was selected because it has better interannual behavior, longer record and is compatible with MODIS and SPOT vegetation data, which facilitates its temporal extension.

The comparison between the FVC databases reveals important differences between them that depend on the FVC value and the biome, being especially relevant for forest land types. Methods that use the frequency distribution of NVDI (ZENG and GUTMAN) are more similar although some differences can be found between them depending of the land type. In addition the FVC series reveal a important interannual variability, consequently prescribed FVC values can present important differences regarding the "real" FVC. These facts can be important when FVC are used in RCMs coupled to LSMs. The magnitude of the differences found causes a noteworthy impact on surface fluxes and hence modifying the regional climate.

The RCM experiments performed exhibit a not negligible effect of FVC uncertainty on the monthly climatological values. The results showed that differences of 30% of FVC, that appear in the two sensitivity experiments, can produce bias of 1° in T2m monthly values. Therefore, depending on the spatial structure of the FVC differences, the climatological patters are modified. In addition, the magnitude of the model response depends on the time of the year. This implies that the annual cycle reproduced by the model is also sensible to the FVC data used. Finally, the role of the interannual variability of the FVC series can also change the interannual variability of climatological series (T2m in our study). In the experiment showed here, the differences for some times of the year, especially spring and summer could reach 1 °C.

It is known that changes in vegetation also modifies the albedo. This was not taken into account in our experiments. Therefore, it can be expected an impact on the regional climate simulations. In fact, this impacts can be opposite to our findings since a decrease of the vegetation leads to an increase of albedo and then, less radiation is trapped by the surface and temperature could decrease. However, Crawford et al. [31] show that the impact of changing albedo on surface heat fluxes is much smaller, an order of magnitude, than changing FVC.

In this paper we just analyze the sensitivity of RCM simulations to changes in FVC. A question remaining is if the RCM simulations with these new FVC datasets can better capture the climate over the IP. The sensitivity of the model to the FVC changes is much smaller than the deviations of the model respect to the observations (not shown), therefore it is difficult to find any improvement using only one or two years of data. This is not a surprising result. For example, Gómez-Navarro et al. [67] shows that the skill score of different RCMs depends on the data-base chosen for evaluating the model. Other examples can be found in Fernández et al. [68] and Jerez et al. [64] where the authors did not find any combination of physical parameterizations that always reproduce better the observed climate. However, changing FVC with time instead of using climatological data should improve the simulation of inter-annual regional climate variability. For instance, the mean temperature differences between the dry and the wet year are better represented by the YEAR simulations (not shown).

In this work, just linear models have been tested for obtaining FVC from NDVI, because they have been more frequently used (for instance, in most of works cited along the work). Some examples are the NOAH land–surface model and the NAM Eta model. However, some authors use quadratics models. Figure 3 shows the relationship between FVC and NDVI for the linear models analyzed and a quadratic model (GUTMAN Quadratic). The quadratic methodology drives to lower values of FVC. Anyway, the FVC differences obtained with linear methods covers the range of the differences in FVC that we found.

Author Contributions: Conceptualization, J.M.J.-G. and J.P.M. Methodology, J.M.J.-G. and J.P.M. Software, J.M.J.-G. and J.P.M. and S.J. Investigation, J.M.J.-G. Formal Analysis, J.M.J.-G., J.P.M. and S.J. Data curation, J.M.J.-G., J.P.M. and S.J. Writing—original draft preparation, J.M.J.-G. and J.P.M. Writing—review and editing, J.M.J.-G., J.P.M., S.J. and F.V. Visualization, J.M.J.-G. Supervision, J.P.M. and F.V.

Acknowledgments: We acknowledge all the institutions and communities that provided free software, R community, CDO (Climate Data Operators), GMT (Generic Mapping Tools), MM5, Gnuplot, gfortran as well as the institutions supplying data (ECMWF, NASA).

References

1. Pielke, R.A. Overlooked issues in the U.S. National Climate and IPCC assesments. *Clim. Chang.* **2002**, *52*, 1–11.

2. Godfrey, C.M.; Stensrud, D.J.; Leslie, L.M. The influence of improved land surface and soil data on mesoscale model predictions. CD–ROM. In *Proceedings of the 19th Conference on Hydrology*; American Meteorological Society: Boston, MA, USA, 2002; Paper 4.7.

3. Jerez, S.; Montavez, J.; Gomez-Navarro, J.; Jimenez, P.; Jimenez-Guerrero, P.; Lorente, R.; Gonzalez-Rouco, J.F. The role of the land-surface model for climate change projections over the Iberian Peninsula. *J. Geophys. Res. Atmos.* **2012**, *117*. [CrossRef]

4. Stensrud, D.J. *Parameterization Schemes: Keys to Understanding Numerical Weather Prediction Models*; Cambridge University Press: Cambridge, UK, 2007.

5. Chang, J.T.; Wetzel, P.J. Effects of spatial variations of soil moisture and vegetation on the evolution of a prestorm environment: A numerical case study. *Mon. Weather Rev.* **1991**, *119*, 1368–1390. [CrossRef]

6. Gutman, G.; Ignatov, A. The derivation of the green vegetation fraction from NOAA/AVHRR data for use in numerical weather prediction models. *Int. J. Remote Sens.* **1998**, *19*, 1533–1543. [CrossRef]

7. Zeng, X.; Dickinson, R.E.; Walker, A.; Shaikh, M. Derivation and evaluation of global 1-km fractional vegetation cover data for land modelling. *J. Appl. Meteorol.* **2000**, *39*, 826–839. [CrossRef]

8. Li, X.; Zhang, J. Derivation of the Green Vegetation Fraction of the Whole China from 2000 to 2010 from MODIS Data. *Earth Interact.* **2016**, *20*, 1–16. [CrossRef]

9. Carlson, T.N.; Perry, E.M.; Schmugge, T.J. Remote estimates of soil moisture availability and fractional vegetation cover for agricultural fields. *Remote Sens. Environ.* **1990**, *52*, 45–70.

10. Sellers, P.J.; Los, S.O.; Tucker, C.J.; Justice, C.O.; Dazlich, D.A.; Collatz, G.J.; Randall, D.A. A revised land surface parameterization (SiB2) for atmospheric GCMs. Part II: The generation of global fields of terrestrial biophysical parameters from satellite data. *J. Clim.* **1996**, *9*, 706–737. [CrossRef]

11. Hanamean, J.M., Jr.; Pielke, R.A., Sr.; Castro, C.L.; Ojima, D.S.; Reed, B.C.; Gao, Z. Vegetation greenness impacts on maximum and minimum temperatures in northeast Colorado. *Meteorol. Appl.* **2003**, *10*, 203–215. [CrossRef]

12. Ek, M.B.; Mitchell, K.E.; Lin, Y.; Rogers, E.; Grunmann, P.; Koren, V.; Gayno, G.; Tarpley, J.D. Implementation of Noah land surface model advances in the National Centers for Environmental Prediction operational mesoscale Eta model. *J. Geophys. Res. Atmos.* **2003**, *108*, 8851. [CrossRef]

13. Kurkowski, N.P.; Stensrud, D.J.; Baldwin, M.E. Assessment of Implementing Satellite-Derived Land Cover Data in the Eta Model. *Weather Forecast.* **2003**, *18*, 404–416. [CrossRef]

14. Marshall, C.H.; Crawford, K.C.; Mitchell, K.E.; Stensrud, D.J. The impact of the land surface physics in the operational NCEP Eta model on simulating the diurnal cycle: Evaluation and testing using Oklahoma Mesonet data. *Weather Forecast.* **2003**, *18*, 748–768. [CrossRef]

15. Hong, S.; Lakshmi, V.; Small, E.; Chen, F.; Tewari, M.; Manning, K.W. Effects of vegetation and soil moisture on the simulated land surface processes from the coupled WRF/Noah model. *J. Geophys. Res. Atmos.* **2009**, *114*, D18118. [CrossRef]

16. Limei, R.; Gilliam, R.; Binkowski, F.; Xiu, A.; Pleim, J.; Band, L. Sensitivity of the Weather Research and Forecast/Community Multiscale Air Quality modeling system to MODIS LAI, FPAR, and albedo. *J. Geophys. Res. Atmos.* **2015** *120*, 8491–8511.

17. Cao, Q.; Yu, D.; Georgescu, M.; Han, Z.; Wu, J. Impacts of land use and land cover change on regional climate: A case study in the agro-pastoral transitional zone of China. *Environ. Res. Lett.* **2015**, *10*, 124025. [CrossRef]

18. Xu, L.; Pyles, R.D.; Snyder, R.H.; Monier, E.; Falk, M.; Chen, S.-H. Impact of canopy representations on regional modeling of evapotranspiration using the WRF-ACASA coupled model. *Agric. For. Meteorol.* **2017**, *247*, 79–92. [CrossRef]

19. Zhang, M.; Geping, L.; Maeyer, P.D.; Cai, P.; Kurban, A. Improved Atmospheric Modelling of the Oasis-Desert System in Central Asia Using WRF with Actual Satellite Products. *Remote Sens.* **2017**, *9*, 1273. [CrossRef]

20. Wen, J.; Lai, X.; Shi, X.; Pan, X. Numerical simulations of fractional vegetation coverage influences on the convective environment over the source region of the Yellow River. *Meteorol. Atmos. Phys.* **2013**, *120*, 1–10. [CrossRef]

21. Meng, X.H.; Evans, J.P.; McCabe, M.F. The Impact of Observed Vegetation Changes on Land–Atmosphere Feedbacks During Drought. *J. Hydrometeorol.* **2014**, *15*, 759–776. [CrossRef]

22. Müller, O.V.; Berbery, E.H.; Alcaraz-Segura, D.; Ek, M.B. Regional model simulations of the 2008 drought in southern South America using a consistent set of land surface properties. *J. Clim.* **2014**, *27*, 6754–6778. [CrossRef]

23. Matsui, T.; Lakshmi, V.; Small, E.E. The effects of satellite-derived vegetation cover variability on simulated land-atmosphere interactions in the NAMS. *J. Clim.* **2005**, *18*, 21–40. [CrossRef]

24. Notaro, M.; Chen, G.; Yu, Y.; Wang, F.; Tawfik, A. Regional Climate Modeling of Vegetation Feedbacks on the Asian–Australian Monsoon Systems. *J. Clim.* **2017**, *30*, 1553–1582. [CrossRef]

25. James, K.A.; Stensrud, D.J.; Yussouf, N. Value of real-time vegetation fraction to forecasts of severe convection in high-resolution models. *Weather Forecast.* **2009**, *24*, 187–210. [CrossRef]

26. Zhang, G.; Zhou, G.; Chen, F.; Barlage, M.; Xue, L. A Trial to Improve Surface Heat Exchange Simulation through Sensitivity Experiments over a Desert Steppe Site. *J. Hydrometeorol.* **2014**, *15*, 664–684. [CrossRef]

27. Vahmani, P.; Hogue, T. High-resolution land surface modeling utilizing remote sensing parameters and the Noah UCM: A case study in the Los Angeles Basin. *Hydrol. Earth Syst. Sci.* **2014**, *18*, 4791–4806. [CrossRef]

28. Vahmani, P.; Ban-Weiss, G. Climatic consequences of adopting drought-tolerant vegetation over Los Angeles as a response to California drought. *Geophys. Res. Lett.* **2016**, *43*, 8240–8249. [CrossRef]

29. Stockli, R.; Vidale, P.L. European plant phenology and climate as seen in a 20 year AVHRR land-surface parameter dataset. *Int. J. Remote Sens.* **2004**, *17*, 3303–3330. [CrossRef]

30. Miller, J.; Barlage, M.; Zeng, X.; Wei, H.; Mitchell, K. Sensitivity of the NCEP/Noah land surface model to the MODIS green vegetation fraction data set. *Geophys. Res. Lett.* **2006**, *33*, 237–250. [CrossRef]

31. Crawford, T.M.; Stensrud, D.J.; Mora, F.; Merchant, J.W.; Wetzel, P.J. Value of Incorporating Satellite-Derived Land Cover Data in MM5/PLACE for Simulating Surface Temperatures. *J. Hydrometeorol.* **2001**, *2*, 453–468. [CrossRef]

32. Myneni, R.B.; Keeling, C.D.; Tucker, C.J.; Asrar, G.; Nemani, R.R. Increased plant growth in the northern high latitudes from 1981 to 1991. *Nature* **1997**, *386*, 698–702. [CrossRef]

33. Refslund, J.; Dellwik, E.; Hahmann, A.; Barlage, M.; Boegh, E. Development of satellite green vegetation fraction time series for use in mesoscale modeling: Application to the European heat wave 2006. *Theor. Appl. Clim.* **2014**, *117*, 377–392. [CrossRef]

34. Pettorelli, N.; Olav, V.J.; Atle, M.; Gaillard, J.M.; Tucker, C.J.; Stenseth, N.C. Using the satellite-derived NDVI to assess ecological responses to enviromental change. *IEEE Trans. Geosci. Remote Sens.* **2005**, *20*, 503–510.

35. Myneni, R.B.; Hall, F.G.; Sellers, P.J.; Marshak, A.L. The interpretation of spectral vegetation indexes. *IEEE Trans. Geosci. Remote Sens.* **1995**, *33*, 481–486. [CrossRef]

36. Zhou, L.; Kaufmann, R.K.; Tian, Y.; Mineny, R.B.; Tucker, C.J. Relation between interannual variations in satellite measures of northern forest greenness and climate between 1982 and 1999. *J. Geophys. Res.* **2003**, *108*, ACL 3-1–ACL 3-16. [CrossRef]

37. Tucker, C.J.; Pinzon, J.E.; Brown, M.E.; Slayback, D.; Pak, E.W.; Mahoney, R.; Vermote, E.; Saleous, N.E. An Extended AVHRR 8-km NDVI Data Set Compatible with MODIS and SPOT Vegetation NDVI Data. *Int. J. Remote Sens.* **2005**, *26*, 4485–4498.

38. Tarpley, J.P.; Schneider, S.R.; Money, R.L. Global vegetation indices from NOAA-7 meteorological satellite. *J. Clim. Appl. Meteorol.* **1984**, *23*, 491–494.

39. Gutman, G.; Tarpley, D.; Ignatov, A.; Olson, S. The enhanced NOAA Global Land datasets from the

Advanced Very High Resolution Radiometer. *Int. J. Remote Sens.* **1995**, *76*, 1141–1156. [CrossRef]

40.   James, M.E.; Kalluri, S.N.V. The Pathfinder AVHRR land dataset:an improved coarse resolution dataset for terrestrial monitoring. *Int. J. Remote Sens.* **1994**, *15*, 3347–3363. [CrossRef]

41.   Pinzon, J. Using HHT to successfully uncouple seasonal and interannual components in remotely sensed data. In Proceedings of the 6th World Multiconference on Systemics, Cybernetics and Informatics, Orlando, FL, USA, 14–18 July 2002.

42.   Tucker, C.J.; Pinzon, J.E.; Brown, M.E. Global Inventory Modeling and Mapping Studies 2.0, 2004. Digital Media. Available online: http://staff.glcf.umd.edu/sns/htdocs/data/gimms/ (accessed on 8 March 2019).

43.   Gallo, K.; Tarpley, D.; Mitchell, K.; Csiszar, I.; Owen, T.; Reed, B. Monthly fractional green vegetation cover associated with land cover classes of the Conterminous USA. *Geophys. Res. Lett.* **2001**, *28*, 2089–2092. [CrossRef]

44.   Montandon, L.M.; Small, E.E. The impact of soil reflectance on the quantification of the green vegetation fraction from NDVI. *Remote Sens. Environ.* **2008**, *112*, 1835–1845. [CrossRef]

45.   Carlson, T.N.; Rypley, D.A. On the relation between NDVI, fractional vegetation cover, and leaf area index. *Remote Sens. Environ.* **1997**, *62*, 241–252. [CrossRef]

46.   Price, J.C. Estimating vegetation amount from visible and near infrared reflectances. *Remote Sens. Environ.* **1992**, *41*, 29–34. [CrossRef]

47.   Hansen, M.; DeFries, R.; Townshend, J.R.G.; Sohlberg, R. *UMD Global Land Cover Classification. 8 Kilometer. Version 1.0. 1981–1994*; Department of Geography, University of Maryland: College Park, MD, USA, 1998.

48.   Hansen, M.; DeFries, R.; Townshend, J.R.G.; Sohlberg, R. Global land cover classification at 1km resolution using a decision tree classifier. *Int. J. Remote Sens.* **2000**, *21*, 1331–1365. [CrossRef]

49.   Grell, G.; Dudhia, J.; Stauffer, D. *A Description of the Fifth-Generation Penn State/NCAR Mesoscale Model (MM5)*; NCAR Technik Note; NCAR: Boulder, CO, USA, 1994.

50.   Jerez, S.; Montavez, J.P.; Gomez-Navarro, J.J.; Jimenez-Guerrrero, P.; Jimenez, J.M.; Gonzalez-Rouco, J.F. Temperature sensitivity to the land-surface model in MM5 climate simulations over the Iberian Peninsula. *Meteorol. Z.* **2010**, *19*, 363–374. [CrossRef]

51.   Dee, D.P.; Uppala, S.M.; Simmons, A.; Berrisford, P.; Poli, P.; Kobayashi, S.; Andrae, U.; Balmaseda, M.A.; Balsamo, G.; Bauer, P.; et al. The ERA-Interim reanalysis: Configuration and performance of the data assimilation system. *Quart. J. R. Meteorol. Soc.* **2011**, *137*, 553–597. [CrossRef]

52.   Grell, G.A. Prognostic evaluation of assumptions used by cumulus parameterizations. *Mon. Weather Rev.* **1993**, *121*, 764–787. [CrossRef]

53.   Dudhia, J. Numerical study of convection observed during the winter monsoon experiment using a mesoscale two-dimensional model. *J. Atmos. Sci.* **1989**, *46*, 3077–3107. [CrossRef]

54.   Mlawer, E.; Taubman, S.; Brown, P.; Iacono, M.; Clough, S. Radiative transfer for inhomogeneous atmospheres: RRTM, a validated correlated-k model for the longwave. *J. Geophys. Res.* **1997**, *102*, 16663–16682. [CrossRef]

55.   Hong, S.; Pan, H. Nonlocal boundary layer vertical diffusion in a Medium-Range Forecast Model. *Mon. Weather Rev.* **1996**, *124*, 2322–2339. [CrossRef]

56.   Chen, F.; Dudhia, J. Coupling an advanced land-surface/hydrology model with the Penn State/NCAR MM5 modeling system. Part I: Model implementation and sensitivity. *Mon. Weather Rev.* **2001**, *129*, 569–585. [CrossRef]

57.   Mahrt, L.; Ek, M. The influence of atmospheric stability on potential evaporation. *J. Appl. Meteorol.* **1984**, *23*, 222–234. [CrossRef]

58.   Mahrt, L.; Pan, H. A two-layer model of soil hydrology. *Bound. Layer Meteorol.* **1984**, *29*, 1–20. [CrossRef]

59.   Pan, H.L.; Mahrt, L. Interaction between soil hydrology and boundary-layer development. *Bound. Layer Meteorol.* **1987**, *38*, 185–202. [CrossRef]

60.   Chen, F.; Mitchell, K.; Schaake, J.; Xue, Y.; Pan, H.L.; Koren, V.; Duan, Q.Y.; Ek, M.; Betts, A. Modeling of land surface evaporation by four schemes and comparison with FIFE observations. *J. Geophys. Res. Atmos.* **1996**, *101*, 7251–7268. [CrossRef]

61.   Noilhan, J.; Planton, S. A simple parameterization of land surface processes for meteorological models. *Mon. Weather Rev.* **1989**, *117*, 536–549. [CrossRef]

62.   Jacquemin, B.; Noilhan, J. Sensitivity study and validation of a land surface parameterization using the HAPEX-MOBILHY data set. *Bound. Layer. Meteorol.* **1990**, *52*, 93–134. [CrossRef]

63.   Gómez-Navarro, J.; Montávez, J.; Jiménez-Guerrero, P.; Jerez, S.; Lorente-Plazas, R.; González-Rouco, J.;

Zorita, E. Internal and external variability in regional simulations of the Iberian Peninsula climate over the last millennium. *Clim. Past* **2012**, *8*, 25. [CrossRef]

64. Jerez, S.; Montavez, J.P.; Jimenez-Guerrero, P.; Gomez-Navarro, J.J.; Lorente-Plazas, R.; Zorita, E. A multi-physics ensemble of present-day climate regional simulations over the Iberian Peninsula. *Clim. Dyn.* **2013**, *40*, 3023–3046. [CrossRef]

65. Lorente-Plazas, R.; Montávez, J.; Jerez, S.; Gómez-Navarro, J.; Jiménez-Guerrero, P.; Jiménez, P. A 49 year hindcast of surface winds over the Iberian Peninsula. *Int. J. Climatol.* **2015**, *35*, 3007–3023. [CrossRef]

66. Peters-Lidard, C.D.; Zion, M.S.; Wood, E.F. A soil-vegetation-atmosphere transfer scheme for modeling spatially variable water and energy balance processes. *J. Geophys. Res. Atmos.* **1997**, *102*, 4303–4324. [CrossRef]

67. Gómez-Navarro, J.; Montávez, J.; Jerez, S.; Jiménez-Guerrero, P.; Zorita, E. What is the role of the observational dataset in the evaluation and scoring of climate models? *Geophys. Res. Lett.* **2012**, *39*. [CrossRef]

68. Fernández, J.; Montávez, J.; Sáenz, J.; González-Rouco, J.; Zorita, E. Sensitivity of the MM5 mesoscale model to physical parameterizations for regional climate studies: Annual cycle. *J. Geophys. Res. Atmos.* **2007**, *112*. [CrossRef]

5

Suitable Pattern of the Natural Environment of Human Settlements in the Lower Reaches of the Yangtze River

Fan Song, Xiaohua Yang * and Feifei Wu

State Key Laboratory of Water Environment Simulation, School of Environment, Beijing Normal University, Beijing 100875, China; 201621180014@mail.bnu.edu.cn (F.S.); wufeifei_bnu@163.com (F.W.)
* Correspondence: xiaohuayang@bnu.edu.cn.

Abstract: The human settlement environment is the object on which human survival depends. In this study, six single factor suitability models and a comprehensive index model of the human settlement natural environment were established. The six single factor models included topography, hydrology, vegetation, soil, temperature and humidity, and land surface temperature. This study took 1 km × 1 km as the pixel size and relied on the ArcGIS platform to systematically and quantitatively evaluate the human settlement environment of the lower reaches of the Yangtze river. The results show that: (1) From the evaluation results of single natural elements, the topography of the study area is relatively flat, with a small number of hydraulic erosion areas. Besides, there are significant differences between the north and the south in temperature and humidity, hydrology, vegetation, and land surface temperature. (2) In 2015, the most suitable areas of human settlement environment were mainly distributed in the plains along the Yangtze river, the plain of northern Zhejiang, and the Poyang plain. The most unsuitable areas are mainly distributed in mountainous areas, such as the mountain area of southern Zhejiang and Dabie mountain area. Topography and vegetation are the dominant factors for classification. (3) From the perspective of space, the score of the human settlement natural environment in Shanghai is above the average, and the best among the other provinces is Jiangsu province, while Zhejiang, Jiangxi, and Anhui provinces have little difference. From the perspective of time, the overall level of the suitability in the lower reaches of the Yangtze river has been improved from 2005 to 2015, mainly due to the influence of temperature and humidity index and water resource index.

Keywords: human settlement suitability; human settlement natural environment index; GIS; lower reaches of Yangtze River

1. Introduction

The human settlement environment is the place where people work, live, and play. It is also the dependent object of human survival and the main transformation object of human activities [1]. The quality of the human settlement environment not only directly affects the residents' living experience, but also indirectly affects the economic and social progress of a region. The earliest research on the human settlement environment was in 1958, when the Greek scholar Doxiadis established the science of human settlement and conducted a lot of basic research on the problems of the human settlement environment [2]. The second Habitat Conference of United Nations in the 1990s put forward the "Habitat Agenda"; there were two important issues: "comfortable housing for everyone" and "sustainable development of human settlements in the process of urbanization". This indicated that sustainable human settlements should be established in a human living environment so that human beings can enjoy a harmonious, healthy, and fulfilling life with nature. In the 21st century, the concept of sustainable development has become the guiding ideology of the research on the human settlement environment. In recent years, many international scholars have done a lot of research on the sustainable development of a human settlement environment from multiple perspectives. For example, Muhammad Rusdi et al. [3] assessed the land suitability of the Banda Aceh, Indonesia region 10 years after the tsunami. The study was conducted primarily from the perspective of soil permeability, topography, and geology, so as to determine the level of suitability of the area for human habitat. Safa Mazahreh et al. [4] made a land suitability map in Jordan through an approach that integrated soil and climatic data. The analyses not only identified key factors limiting land use but also provided the type of land use for which the region was suitable. Walter Musakwa [5] and Ingrida Bagdanavičiūtė [6], respectively, established evaluation standards from the perspectives of nature-society and pollution sources, and then selected the most suitable areas for human settlement in their respective study areas, so as to provide a basis for land reform and creation of smart cities. Although their researches have their own emphases, they all make great contributions to the study of human settlement.

In China, the research on the human settlement environment started in the 1990s. Tsinghua University established the research center for human settlements in 1995. Professor Wu Liang-yong classified the human settlement environment system into five systems: natural system, human system, social system, residential system, and supporting system [7,8]. Around these five systems, scholars in different fields have conducted a lot of studies from different perspectives. Scholars in the field of architecture are more focused on the study of the residential system and support system, among the five major systems, while sociologists are more concerned with human systems and social systems [9], and environmental and ecologists are more inclined to study natural systems. There are also some scholars who study complex systems [10]. A review article by MA Ren-feng gave a detailed description of the history and current situation of research on the human settlement environment in China [11], and the evaluation method of human settlement natural environment suitability, introduced in the article, is also the main idea of this paper.

In addition to the system divisions described above, the research on the human settlement environment can also be divided from the spatial scale, which can be divided into five scales including global, regional, urban, community, and architecture [12]. Numerous classifications have also generated numerous research methods. Currently, the commonly used method is analysis in Geographic Information System. At the same time, many mathematical methods such as fuzzy comprehensive evaluation [13], factor analysis [14], and genetic algorithm [15] have been applied in this field. With the development and maturity of Geographic Information System and Remote Sensing technology, it is possible to show the advantages and disadvantages of the human settlement environment with pictures and texts. The joint analysis of GIS tool and other methods is also gradually developed [16], which is of great significance to the research on the human settlement environment suitability.

Natural system is the basic system of the human settlement environment, which directly relates to human health and body surface feeling, and indirectly constitutes the construction foundation of other systems. Among the numerous natural factors, this study selects some basic elements that are most closely related to human survival and life, including six natural elements: topography, hydrology, vegetation, soil, temperature and humidity, and land surface temperature. On the basis of the study of each individual element, the model of the human settlement natural environment index (HNEI) is established, and different weights are assigned to each element according to the actual situation of the research area. The lower reaches of the Yangtze river are one of the most developed regions in China. During a long period of economic development, a series of resource and environmental problems have emerged. The purpose of this study is as follows: 1. To reveal the spatial pattern of the human settlement natural environment suitability of five provinces in the lower reaches of the Yangtze river; 2. to explore the advantage and restriction of the human settlement environment in this region, so as to provide the corresponding research basis and decision-making support for local population spatial planning and land function planning.

2. Study Area

The study area covered the lower reaches of the Yangtze river, including five provincial-level administrative regions of Jiangxi, Anhui, Jiangsu, Zhejiang and Shanghai, with a total of 51 prefecture-level cities and one municipality directly under the central government. The region is located in longitude E113°34′–E123°25′ and latitude N24°29′–N35°20′. The total area is about 526,080 km^2. The location of the study area is shown in Figure 1. At the end of 2016, the permanent resident population of the five provinces was about 267.82 million, accounting for 19.37% of the total population of China, with a high population density. Most of the region is located in the economic radiation zone of the urban agglomeration of the Yangtze river delta. Relying on the excellent geographical advantages and economic foundation, the study area has become the region with the highest urbanization level and the most developed economy in China. Most of the region belongs to the subtropical monsoon climate zone, with obvious monsoon characteristics. In the process of rapid urbanization in this region, a lot of resources and environmental problems have been caused, resulting in great changes in the natural environment. In this study, the grid of 1 × 1 km was taken as the basic unit, which can satisfy the analysis of different spatial scales, so as to make an appropriate evaluation of the natural suitability of human settlement.

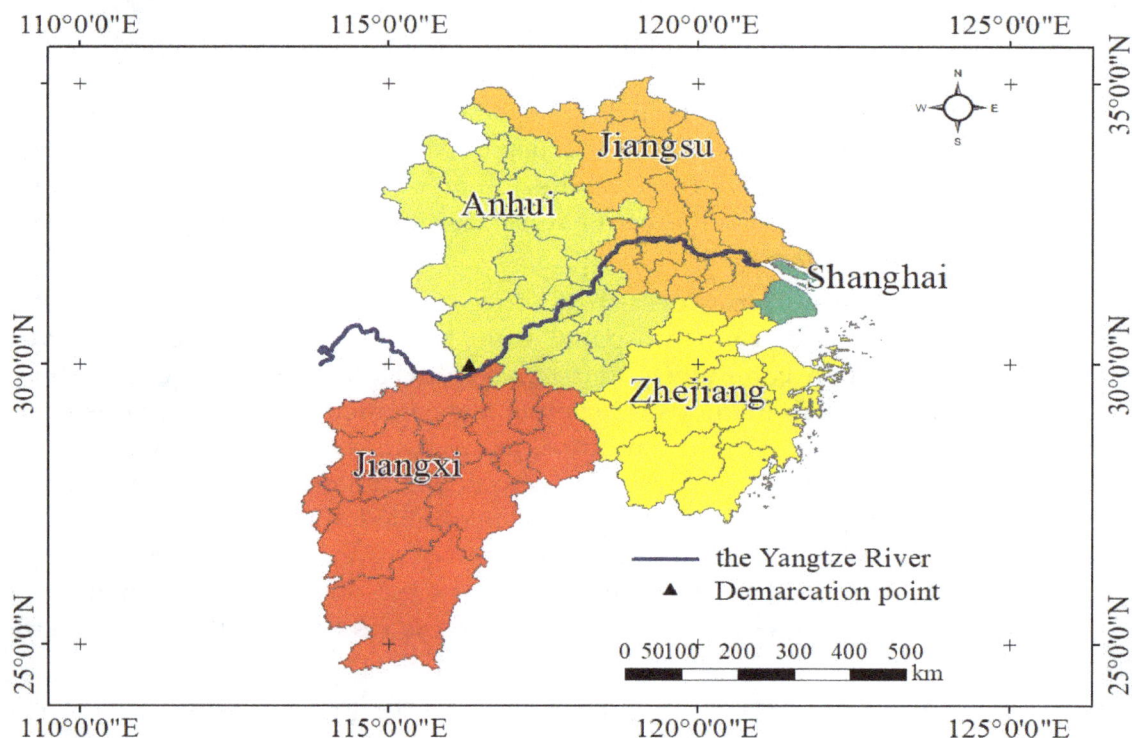

Figure 1. Location of study area.

3. Materials and Methods

3.1. Data Sources

The six indexes selected in this study were: relief degree of land surface (RDLS), water resource index (WRI), land cover index (LCI), soil suitability index (SSI), temperature–humidity index (THI), and land surface temperature index (LSTI). The data used for calculation included: digital elevation model data, land use data, annual precipitation spatial interpolation data, normalized difference vegetation index data [17], soil erosion spatial distribution data [18], annual data for natural indicators of weather stations, and MODIS land surface temperature data. The data types, year of data, and data source are shown in the Table 1.

Table 1. Details of the data.

The Name of the Data	Data Types	Year of Data	Data Source
Digital elevation model data	Raster data, 90 × 90 m	2015	The resource and environmental data cloud platform of the Chinese academy of sciences (http://www.resdc.cn/)
Land use data	Raster data, 1 × 1 km	2005,2010,2015	The resource and environmental data cloud platform of the Chinese academy of sciences (http://www.resdc.cn/)
Annual precipitation spatial interpolation data	Raster data, 1 × 1 km	2004–2006,2009–2011,2014–2016	The resource and environmental data cloud platform of the Chinese academy of sciences (http://www.resdc.cn/)
Normalized difference vegetation index data	Raster data, 1 × 1 km	2005,2010,2015	The resource and environmental data cloud platform of the Chinese academy of sciences (http://www.resdc.cn/)
Soil erosion spatial distribution data	Raster data, 1 × 1 km	2015	The resource and environmental data cloud platform of the Chinese academy of sciences (http://www.resdc.cn/)
Annual data for natural indicators of weather stations	Point data	2004–2006,2009–2011,2014–2016	Chinese academy of sciences (http://www.resdc.cn/)
MODIS land surface temperature data	Raster data, 1 × 1 km	2005,2010,2015	United States Atmosphere Archive & Distribution System, Distributed Active Archive Center (https://ladsweb.modaps.eosdis.nasa.gov/).

3.2. Methods of Single Factor Model

3.2.1. Relief Degree of Land Surface

Relief degree of land surface refers to the difference between the highest and lowest elevations in an area. It reflects the macroscopic changes of the region's terrain, which is of great reference value for the site selection of human settlements [19]. The key to calculate RDLS is to determine the size of the reference area, which can well express the topographic characteristics of the study area. Referring to the research results of Feng Zhi-ming [20], Wang Yong-li [21], and other scholars, the formula of RDLS established in this paper is as follows:

$$RDLS = ALT/1000 + \{[\max(H) - \min(H)] \times [1 - P(A)/A]\}/500 \tag{1}$$

where, ALT refers to the average altitude within the selected reference area; Max(H) and min(H) refer to the highest and lowest elevations within the reference area; P(A) refers to the flat area within the reference area, whose slope is less than 5°; A refers to the total area of the reference area; and 500 refers to China's benchmark mountain height.

The neighborhood analysis method based on raster data was used to calculate the size of the reference area [22]. The neighborhood analysis tool of ArcGIS was used to analyze the altitude difference within the grid, from 3×3 pixels to 69×69 pixels. According to the calculation results, the scatter diagram of the corresponding relationship between the grid size and the altitude difference was established, and the logarithmic function of the curve equation was determined. The size of the reference area grid can be determined according to the change of curve growth rate. In this study, the determined optimal grid size was 19×19 (2.92 km^2). According to Formula (1), the RDLS of the lower reaches of the Yangtze river can be obtained. The result in 2015 is shown in Figure 3a.

3.2.2. Water Resource Index

Water is the source of life, and the basic need of human settlement construction and development. Although the lower reaches of the Yangtze river are relatively rich in water resources, they are unevenly distributed. In order to express this difference, referred to the environmental assessment standards of the ministry of environmental protection of the People's Republic of China, and relevant research results [23,24], we selected two indexes of "rainfall" and "distance from water source" to build the water resource index model. The formula of WRI is as follows:

$$WRI = \alpha P + \beta D \tag{2}$$

where P is the normalized rainfall; D is the normalized distance from the water source; A and B are the weights of two factors, which are respectively 0.8 and 0.2 in this study. In order to reduce the contingency of data, the average rainfall in 2004–2006, 2009–2011, and 2014–2016 was used as the values in 2005, 2010, and 2015, respectively. The required water source location was obtained by extracting the "water area" type in the land use data and the "Euclidean Distance" in ArcGIS was used to calculate the "distance from water source". In this paper, the distance within 10 km from the water source was calculated and reversely normalized, and the value of the remaining areas was assigned to zero. The calculation result in 2015 is shown in Figure 3b.

3.2.3. Land Cover Index

Land is an important material for human survival and development. Land coverage and utilization not only affect the stability of natural ecosystem, but also continuously affect the development of social economy [25]. In recent years, with the rapid development of urbanization, inappropriate land use has become more and more serious. Inappropriate land use has become a problem of universal concern, among which vegetation cover has been a hot topic. Vegetation is an important intermediate link connecting the atmospheric environment, soil environment, and water environment, which can regulate the climate and have a significant impact on the human settlement environment [26]. In this paper, the land cover index model was constructed by referring to the ecological environment evaluation standard of the State Environmental Protection Administration [27] and previous research results [24]:

$$LCI = LT_i \times NDVI \tag{3}$$

where, LT_i refers to the weight of each land use type, and $NDVI$ refers to the normalized difference vegetation index.

According to the research purpose, relevant literatures [28,29] and technical specifications, weight was assigned to each land use type, and the results are shown in Table 2. Then, the land cover index can be obtained according to Formula (3). The calculation result in 2015 is shown in Figure 3c.

Table 2. Weight of each land use type.

Land-Use Type	Weight	Classification	Weight
Arable land	0.25	Paddy field	0.65
		Dry field	0.35
Woodland	0.18	Forest land	0.34
		Shrub land	0.26
		Sparse wood	0.2
		Other woodland	0.2
Grass land	0.15	Grassland with high coverage	0.4
		Grassland with moderate coverage	0.33
		Grassland with low coverage	0.27
Water area	0.17	Canal	0.31
		Lake	0.29
		Reservoir and pit-pond	0.2
		Mudflat	0.1
		Bottomland	0.1
Construction land	0.2	Urban and town land	0.44
		Village land	0.35
		Other construction land	0.21
Unused land	0.05	Wetland	0.48
		Bare land	0.28
		Bare rock	0.24

3.2.4. Soil Suitability Index

Soil erosion is a serious environmental problem, which not only damages land resources and affects normal industrial and agricultural behaviors, but also may aggravate drought and waterlogging. It is one of the important evaluation indexes of the human settlement environment [30]. China is one of the countries with serious soil erosion, which has affected the economic and social development. It can be known from the existing studies that soil erosion exists in some areas of the study area, especially in Jiangxi province [31,32]. Therefore, soil erosion was taken as an evaluation index in this study. We used soil erosion data for reclassification to obtain soil suitability index of five provinces in the lower reaches of the Yangtze river, as shown in Figure 3d.

3.2.5. Temperature–Humidity Index

Among the many climatic elements, air temperature directly affects the exchange of heat and water between human body and the outside world, so the air temperature directly affects the comfort degree of human body. Due to the difference in humidity between the north and the south, there will be a difference in apparent temperature [33]. Therefore, this study adopted the widely used temperature–humidity index (THI) to express this effect [34,35], so as to reflect the temperature suitability through the comprehensive effect of temperature and humidity. The calculation formula is as follows:

$$THI = T - 0.55(1 - f)(T - 58)$$
$$T = 1.8t + 32 \tag{4}$$

where t is Celsius, T is Fahrenheit, and f is relative humidity. In order to reduce the contingency of data, the average temperature and relative humidity data in 2004–2006, 2009–2011, and 2014–2016 was used as the values in 2005, 2010, and 2015, respectively. Firstly, the location of weather stations in the study area and the corresponding temperature and humidity information were input in ArcGIS. The locations of weather stations are shown in Figure 2. After data processing, the kriging interpolating model was used to carry out spatial interpolation processing in five provinces, and the spatial distribution data of various climatic elements were obtained. We mainly used the "Spatial Analyst tools-interpolation-Kriging" function in ArcGIS. Finally, according to Formula (4), the 2015 result is shown in Figure 3e.

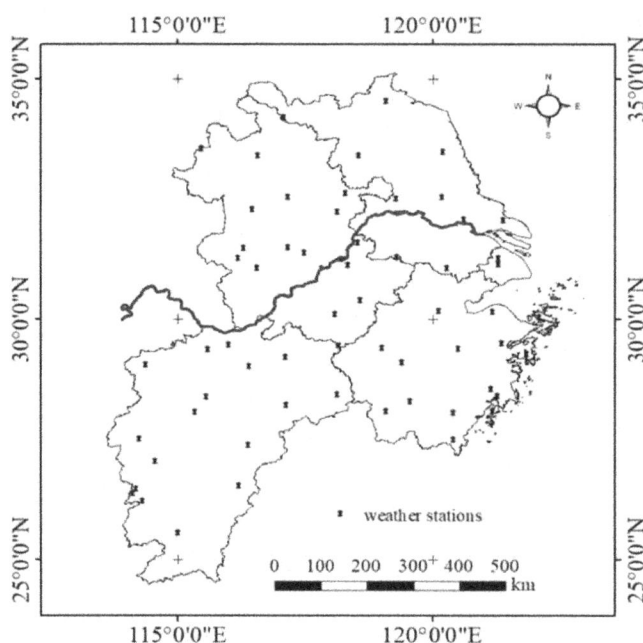

Figure 2. Locations of weather stations.

3.2.6. Land Surface Temperature Index

The rapid development of urbanization has changed the urban climate, and one of the remarkable characteristics is the urban heat island effect. On the one hand, the urban heat island effect directly affects the energy utilization, hydrological environment, and air quality. On the other hand, the urban heat island effect also indirectly affects the quality of urban living environment and the health status of residents [36]. Land surface temperature is a key parameter of urban surface energy balance and a very important parameter for monitoring urban heat island effect.

The classification method of land surface temperature was to segment the temperature density on the basis of the original land surface temperature data [37]. The surface temperature in the region was divided into five grades, which were: low temperature ($T < T_{mean} - 1.5T_{std}$); sub-low temperature ($T_{mean} - 1.5T_{std} < T < T_{mean} - 0.5T_{std}$); medium temperature ($T_{mean} - 0.5T_{std} < T < T_{mean} + 0.5T_{std}$); sub-high temperature ($T_{mean} + 0.5T_{std} < T < T_{mean} + 1.5T_{std}$); high temperature ($T > T_{mean} + 1.5T_{std}$). Where T_{mean} is the average value of land surface temperature and T_{std} is the standard deviation of land surface temperature. The result of land surface temperature index in 2015 is shown in Figure 3f.

3.3. Methods of Human Settlement Natural Environment Model

3.3.1. Scoring Criteria

After the completion of the single factor evaluation, it is necessary to carry out the graph superposition to complete the evaluation of human settlement natural environment. Therefore, each single factor model should be graded according to the actual situation of the study area and existing studies [38]. The single factor evaluation results in 2015 were divided into five grades and assigned with scoring values, as shown in Table 3.

Table 3. Scoring criteria of the single factors.

	RDLS	WRI	LCI
I (90 grades)	[−0.02, 0.15]	(0.67, 0.96]	(0.1209, 0.1495]
II (70 grades)	(0.15, 0.44]	(0.54, 0.67]	(0.0807, 0.1209]
III (50 grades)	(0.44, 1.03]	(0.41, 0.54]	(0.0562, 0.0807]
IV (30 grades)	(1.03, 1.68]	(0.25, 0.41]	(0.0323, 0.0562]
V (10 grades)	(1.68, 3.30]	[0, 0.25]	[−0.004, 0.0323]
	SSI	THI	LSTI
I (90 grades)	mired	(61, 63]	low temperature
II (70 grades)	mild	(59, 61], (63, 66]	sub-low temperature
III (50 grades)	moderate	(57, 59]	medium temperature
IV (30 grades)	strong	(55, 57]	sub-high temperature
V (10 grades)	very strong and violent	[46, 55]	high temperature

Note: RDLS refers to Relief Degree of Land Surface; WRI refers to Water Resource Index; LCI refers to Land Cover Index; SSI refers to Soil Suitability Index; THI refers to Temperature–Humidity Index; LSTI refers to Land Surface Temperature Index.

3.3.2. Establishment of Human Settlement Natural Environment Index

On the basis of single factor evaluation, a human settlement natural environment index (HNEI) model was constructed to evaluate the natural suitability of human settlement in five provinces of the lower reaches of the Yangtze river. The formula is as follows:

$$HNEI = \alpha RDLS + \beta WRI + \chi LCI + \delta SSI + \gamma THI + \varepsilon LSTI + 0.2GDPI \tag{5}$$

where $\alpha\beta\chi\delta\gamma\varepsilon$ are the corresponding weights of each index respectively, which were obtained by referring to existing studies [4,28,38] through the analytic hierarchy process. The results were as follows: $\alpha = 0.2324$, $\beta = 0.0794$, $\chi = 0.3579$, $\delta = 0.0435$, $\gamma = 0.1697$, $\varepsilon = 0.1172$.

0.2GDPI (Gross Domestic Product Index) is the economic correction coefficient. Since selected natural indexes such as LCI and LSTI are affected largely by population distribution, the economic coefficient was added to eliminate such influence. At the same time, economically developed areas have relatively complete residential infrastructure, which is also one of the necessary conditions for human settlement. The classification of GDPI is shown in Table 4.

Table 4. The classification of Gross Domestic Product Index (GDPI).

Grades	Classification of GDP (unit: Yuan per km^2)
10	[0, 424]
20	(425, 558]
30	(558, 983]
40	(983, 2337]
50	(2337, 6654]
60	(6654, 20,415]
70	(20,415, 64,282]
80	(64,282, 204,114]
90	(204,114, 649,851]
100	(649,851, 2,070,711]

4. Results and Discussion

4.1. Results of Single Natural Factor

(1) RDLS: As can be seen from Figure 3a, the RDLS in most areas was low, with the maximum of 3.3 and the minimum of −0.03. From the perspective of province, there were some high-RDLS areas in Anhui, Zhejiang, and Jiangxi provinces, which were distributed in the Dabie Mountain area, mountainous area of southern Anhui, mountain area of southern Zhejiang, and the Mufu Mountain, Jiuling Mountain in northwest Jiangxi province.

(2) WRI: As can be seen from Figure 3b, the WRI showed a trend of gradual decrease from south to north, with high suitability areas concentrated in central and northern Jiangxi, southern Anhui, and most areas of southwest Zhejiang. The low suitability areas were mainly concentrated in the central and northern Anhui and most areas of Jiangsu province.

(3) LCI: As shown in Figure 3c, the LCI had the following characteristics: 1. In terms of spatial distribution, a medium-high-low pattern was formed from north to south. The dominant vegetation types in the three regions were dry field, paddy field, and woodland, respectively. 2. In terms of provinces, Jiangsu province had the best suitability. Except for the uninhabitable lake areas, the scores of the other regions were all above the average. The LCI of Anhui from north to south also presented a medium-high-low distribution. Because the built-up areas in Shanghai were too concentrated, the LCI of the northern Shanghai was poor. The most suitable areas of Jiangxi and Zhejiang were the areas of basins and plains.

(4) SSI: As can be seen from Figure 3d, soil erosion in the lower reaches of the Yangtze river was only caused by hydraulic erosion without wind erosion and freeze–thaw erosion, which was related to the geographical location of the region. The numerous tributaries along the Yangtze river and the abundant rainfall contributed to this phenomenon.

(5) THI: According to Figure 3e and the reference standard, the THI of the lower reaches of the Yangtze river was suitable for human habitation. From the perspective of distribution, the THI basically increased with the increase of latitude. The overall temperature–humidity index of Jiangxi province was the highest, while that of Anhui province and Jiangsu province was lower. In addition, the THI of mountainous areas was obviously lower than that of plain areas at the same latitude.

(6) LSTI: According to the actual situation of the research area, we believed that the lower the land surface temperature is, the better the habitability is. Figure 3f shows that Shanghai had the highest land surface temperature among the five provinces, because Shanghai had a high population density and a high proportion of built-up area. Due to the existence of a large number of lakes, such as Tai Lake and Hongze Lake, Jiangsu province had reserved a large area of low temperature area, while the high temperature area was mostly concentrated in the urban agglomeration in southern Jiangsu. There was a close relationship between land surface temperature and topography in Zhejiang and Anhui provinces. High temperature areas were mainly concentrated in plain and basin areas; low

temperatures areas were concentrated in hilly and mountainous areas. Jiangxi province is located in the south of the five provinces, the sun is the strongest, so the land surface temperature was relatively high.

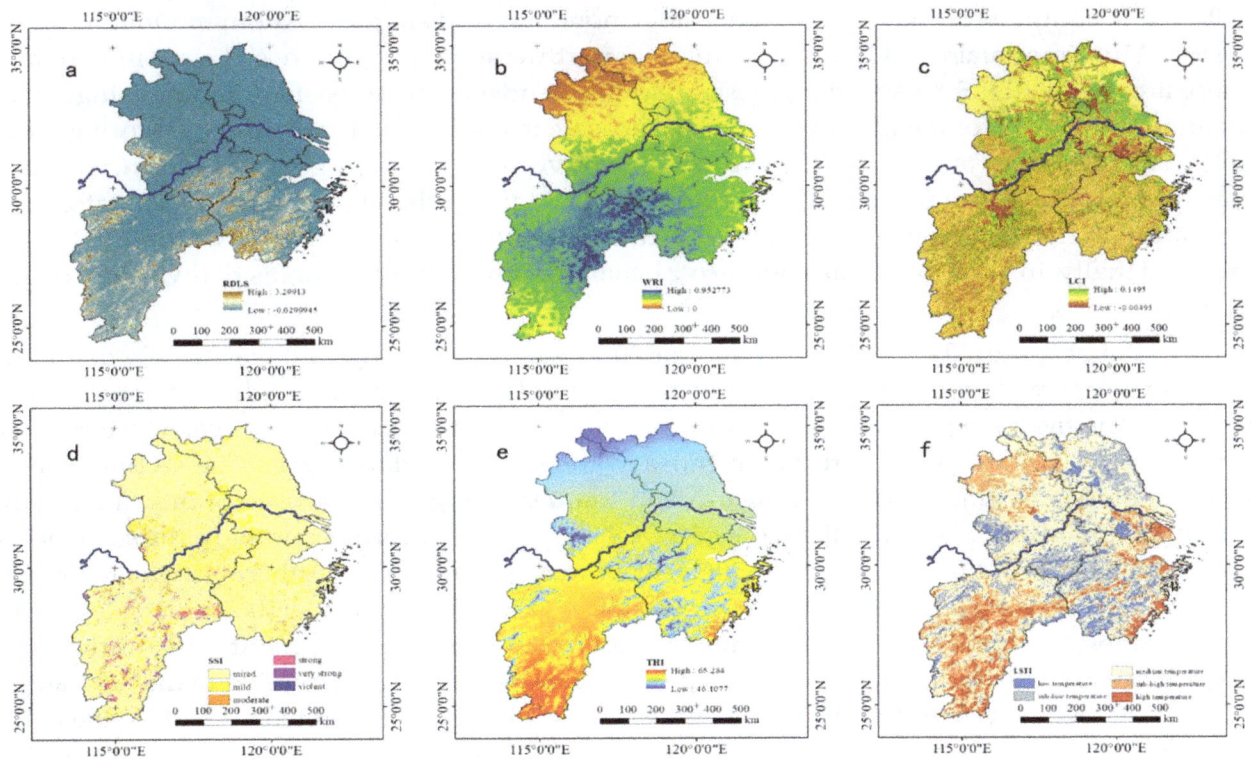

Figure 3. Suitability evaluation of single natural elements. (**a**) RDLS; (**b**) WRI; (**c**) LCI; (**d**) SSI; (**e**) THI; (**f**) LSTI.

4.2. Results of Human Settlement Natural Environment

The distribution diagram of Human Settlement Natural Environment Index (HNEI) obtained after superposition calculation is shown in Figure 4a. It can be seen from the figure that the range of HNEI was 22.2–99.3. The results are expressed in five levels respectively: very unsuitable ($22 \leq$ HNEI ≤ 48), unsuitable ($48 <$ HNEI ≤ 57), medium suitable ($57 <$ HNEI ≤ 66), suitable ($66 <$ HNEI ≤ 77), and very suitable ($77 <$ HNEI ≤ 100).

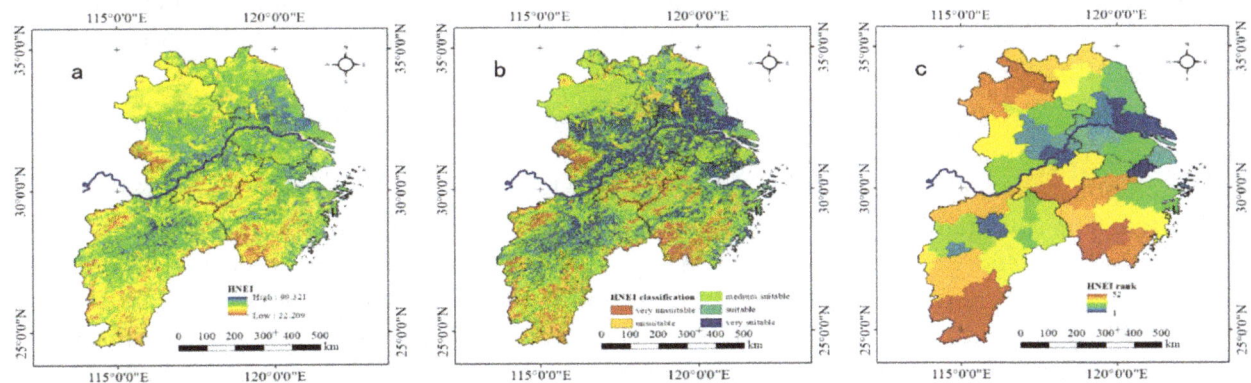

Figure 4. Evaluation results of Human Settlement Natural Environment Index (HNEI). (**a**) HNEI; (**b**) HNEI classification; (**c**) HNEI rank.

As can be seen from Figure 4b, the regions with high HNEI were mainly distributed along the Yangtze river, the northern Zhejiang plain, and the Poyang lake plain. The specific analysis is as follows:

(1) Very unsuitable area: This area was the most unsuitable area for human habitation in the study area, which was not suitable for development and had a fragile ecological environment. This area accounted for 7.27% of the total area. As can be seen from the figure, this part of the region was mainly distributed in southern Zhejiang mountain, Dabie mountain, and southern Anhui mountain. Besides, there was Mufu mountain and Jiuling mountain in northwest Jiangxi province. The dominant factor in these areas was RDLS. In addition, these areas also tended to be consistent in other single factor evaluation results. As for the LCI, most of these areas were covered by forest, and the scoring criteria was IV (30 grades). In terms of LSTI, due to the high vegetation coverage, most scoring criteria in these areas were II (70 grades) or above. In terms of THI, due to the high terrain, the somatosensory temperature in these areas was relatively cold, and most of the scoring criteria were III (50 grades) or below. Finally, in addition to the southern Zhejiang mountains, other areas had varying degrees of erosion.

(2) Unsuitable area: This area accounted for 20.14% of the total area and was widely distributed. The reasons for the formation of such areas were complex and can be roughly divided into five categories: The first category was the surrounding area of the very unsuitable area. The terrain and other factors in this area were better than those in the very unsuitable area. The second category was distributed in mountainous and hilly areas in southern Jiangxi. The terrain of this category was between mountain and plain, and the score of RDLS increased, but correspondingly, the score of LSTI decreased. The WRI of this area differed widely, there was less rainfall in southern Jiangxi and more in eastern Jiangxi. The biggest difference from the first category was in the THI. Such a category had lower latitudes and higher temperatures, making it more suitable for settlement. The third category was the western Zhejiang hilly region and the eastern Zhejiang hilly region. Similar to the first category, the dominant factors were RDLS and LCI. In general, this category was limited by THI, so the natural ecological conditions were slightly worse than those of the second category. However, the GDP of this area was higher than that of the second category; there was a strong ability to construct and transform the residential environment. After correction, they were all in the unsuitable area. The fourth category was located in the lake area of northern Anhui and northern Jiangsu. The water area itself was uninhabitable and the temperature was low, so it was in the unsuitable area. The fifth category was the northern region of the study area. The limiting factor in this category was the THI, and the temperature was relatively cold.

(3) Very suitable area: This area had the most harmonious combination of various natural factors, and was the most suitable place for human production and life. This category accounted for 21.19% of the total area, mostly distributed in the plain areas along the Yangtze river, the plain areas in northern Zhejiang, and the Poyang lake plain. There were two dominant factors in this area. The first one was LCI, with scores of 70 or above. The vegetation coverage in these regions was basically paddy fields, which were suitable for living. The second was RDLS. These areas were all flat and suitable for large-scale group living.

(4) Suitable area: This category was also suitable for human habitation, accounting for 16.91% of the total area. This area was relatively complex, which can be roughly divided into three categories: The first category was the area around the Yangtze river basin, which was distributed around the very suitable area. The dominant factor in this region was the LCI, and most of the vegetation was paddy field. The second category was parts of Nantong city, Yancheng city, and part of northern Jiangsu. The dominant factor of this region was still the LCI, and most of them were dryland regions with relatively good hydrothermal conditions. In addition, the correlation degree between this category and the LSTI was relatively high, and most of them were located in the region with medium temperature. The third category was the area around Poyang lake plain in Jiangxi province and the basin area in the south of Poyang lake. The dominant factors in this part were WRI and RDLS, and most of them were in flat areas and areas with abundant water resources.

(5) Medium suitable area: This category covered the largest area among the five categories, accounting for 34.56% of the total area, with a large distribution in each province. Generally, there

were the following distribution areas: The first category was the part of northern Anhui and northern Zhejiang province, the dominant factor was the land cover index, and the land use type was mostly dry land. The limiting factors were the THI and WRI. The temperature in most regions was low and the amount of water resources was small; the score of the two indexes were mostly below 30 points. The second category covered most of Shanghai and some parts of southern Jiangsu province. The dominant factor of this category was LCI, and the limiting factor was LSTI. The main land use type was construction land, and the natural conditions of this region were more suitable for living. However, it was located in the Yangtze river delta region, and the built-up area had a large population density, so the LSTI was higher. In fact, it was no longer suitable for adding too many people, so it was classified in the medium suitable area. The third category was Tai lake, Poyang lake, and other lake areas, which were similar to the lake in unsuitable areas. The fourth category was the area between the suitable areas and the unsuitable areas. If such areas can follow the local development rules and make rational development and utilization in the development process, they are highly likely to become suitable areas. On the contrary, if the ecological environment is damaged or the climate changes greatly, they are also likely to become unsuitable areas.

The ranking of HNEI in 2015 is shown in Figure 4c. Similar to Figure 4b, cities with a high index were mainly distributed along the Yangtze river, the northern Zhejiang plain, and the Poyang lake plain. Shanghai's HNEI ranked twelfth. Among the other four provinces, Jiangsu province had the best HNEI. There were six cities in the top 15: Taizhou, Nantong, Yangzhou, Zhenjiang, Changzhou, and Nanjing. The overall scores of the other three provinces were similar. The top 15 cities in Zhejiang province are Jiaxing and Zhoushan. Jiangxi province had Nanchang, Xinyu. Anhui province had Wuhu, Hefei, Ma'anshan, and Tongling.

4.3. Time Evolution Analysis of HNEI

In order to explore the evolution of HNEI in the study area, WRI, LCI, THI, and LSTI in 2010 and 2005 were recalculated using relevant data to obtain the HNEI of 2010 and 2005. The results were graded using the same criteria as in 2015, as shown in Figure 5 and Table 5.

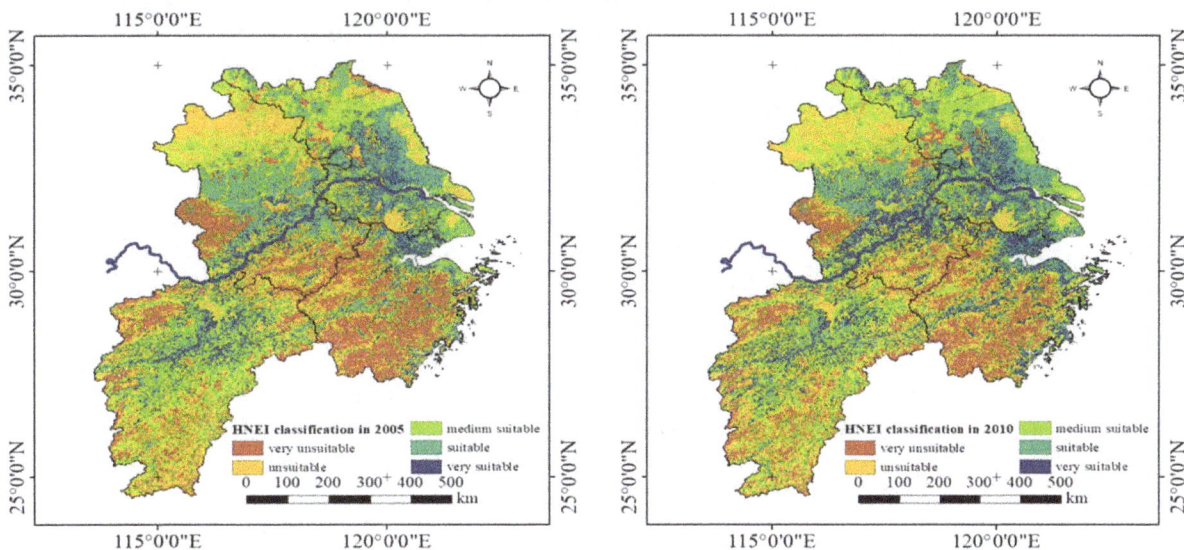

Figure 5. Results of natural suitability of human settlements in 2005 and 2010. HNEI classification in 2005; HNEI classification in 2010.

According to the results in Figure 5 and the result of 2015, the natural suitability of human settlement in the lower reaches of the Yangtze river showed little change, and the high suitability areas were still concentrated along the Yangtze river, the northern Zhejiang plain, and the Poyang lake plain. However, the overall level of HENI was slightly improved; it can be seen from the results in

Table 5 that the area of very unsuitable was significantly reduced, while the area of very suitable was significantly increased. It was found that the leading factors are THI and WRI, and the improvement of these two indexes increased the overall score, making the natural suitability of human settlement somewhat improved compared with that of ten years ago.

Table 5. The change of area proportion of each district.

	Very Unsuitable	Unsuitable	Medium Suitable	Suitable	Very Suitable
2015	7.20%	20.14%	34.56%	16.91%	21.19%
2010	14.30%	30.82%	22.29%	18.31%	14.28%
2005	17.80%	31.93%	20.04%	20.59%	9.64%

5. Conclusions

(1) In this paper, six natural factors, including relief degree of land surface (RDLS), water resource index (WRI), land cover index (LCI), soil suitability index (SSI), temperature-humidity index (THI), and land surface temperature index (LSTI), were selected to establish the human settlement natural environment index (HNEI) model. With ArcGIS as a tool, the natural suitability of human settlement in five provinces in the lower reaches of the Yangtze river was quantitatively evaluated and analyzed.

(2) The human settlement environment includes five major systems: nature, human, society, residence, and support. There is a strong correlation between each system. The main research object of this study is the natural system, and the GDP Index was used to correct the repeated interference of human factors on various elements. The method is feasible to restore the human ecological system objectively and truly from the angle of nature.

(3) In 2015, the most suitable areas accounted for 22.06% of the total area. The areas with the highest suitability index were mainly distributed in the plains along the Yangtze river, the northern Zhejiang plain, and the Poyang lake plain. The dominant factors were RDLS and LCI, and the vegetation cover was basically paddy field, which is suitable for living. Suitable areas accounted for 16.85% of the total area. Some areas were distributed around very suitable areas. In addition, there were parts of paddy fields and dry lands in Jiangsu province, and areas with better hydrothermal conditions in Jiangxi province. The most unsuitable areas accounted for 7.27% of the total area, and were mainly distributed in southern Zhejiang mountain, Dabie mountain, southern Anhui mountain, Mufu mountain, and Jiuling mountain. Relief degree of land surface (RDLS) was the dominant factor, and soil erosion areas were also mainly distributed in these areas. The unsuitable areas accounted for 20.49% of the total area. Some areas were distributed around the very unsuitable areas, while others were mainly distributed in southern Jiangxi (the dominant factors were RDLS and THI), western and eastern Zhejiang (the dominant factor was RDLS), and northern Anhui (the dominant factor was THI). The area of medium suitable area accounted for 33.33%, which was widely distributed in each province. Whether these areas will develop into suitable or unsuitable areas in the future depends, to a great extent, on urban development policies and means, which are the key areas that should be paid attention to.

(4) From the perspective of space, the human settlement natural environment index of Shanghai in 2015 was above the average. The best among the rest of the provinces was Jiangsu province, with six cities ranking in the top 15. Zhejiang province, Jiangxi province, and Anhui province had two, two, and four cities, respectively. From the perspective of time, the spatial distribution of the natural suitability of human settlements did not change much from 2005 to 2015, but the overall suitability level was improved for a certain extent, mainly due to the improvement of the temperature–humidity index (THI) and the water resource index (WRI).

Author Contributions: Conceptualization, F.S. and X.H.Y.; methodology, F.S. and F.F.W.; data curation, F.S.; writing, F.S.; visualization, F.S.; project administration, X.H.Y.

References

1. Aynur, M.; Shi, P.; Zhao, G.J.; Yan, F.; Xue, G.L. Evaluations of living environment suitability of Hotan Prefecture in Xinjiang based on GIS. *Arid Land Geogr.* **2012**, *35*, 847–855. (In Chinese)

2. Xie, X.Y.; Zeng, X.; Li, J. Evaluation of Nature Suitability for Human Settlement in Chongqing. *Resour. Environ. Yangtze Basin* **2014**, *23*, 1351–1359. (In Chinese)

3. Muhammad, R.; Ruhizal, R.; Mohd, S.; Ahamad, S. Land evaluation suitability for settlement based on soil permeability, topography and geology ten years after tsunami in Banda Aceh, Indonesia. *Egypt. J. Remot Sens. Space Sci.* **2015**, *18*, 207–215.

4. Safa, M.; Majed, B.; Doaa, A.H. GIS approach for assessment of land suitability for different land use alternatives in semi arid environment in Jordan: Case study (Al Gadeer Alabyad-Mafraq). *Inf. Proc. Agric.* **2019**, *6*, 91–108.

5. Walter, M.R.; Tshesane, M.; Kangethe, M. The strategically located land index support system for human settlements land reform in South Africa. *Cities* **2017**, *60*, 91–101.

6. Bagdanavičiūtė, I.; Valiūnas, J. GIS-based land suitability analysis integrating multi-criteria evaluation for the allocation of potential pollution sources. *Environ. Earth Sci.* **2013**, *68*, 1797–1812. [CrossRef]

7. Wu, L.Y. Search for the Theory of Science of Human Settlement. *Planners* **2001**, *17*, 5–8. (In Chinese)

8. Wu, L.Y. *Introduction to Sciences of Human Settlements*; China Architecture & Building Press: Beijing, China, 2001. (In Chinese)

9. Wei, W.; Shi, P.J.; Feng, H.C.; Wang, X.F. Study on the Suitability Evaluation of the Human Settlements Environment in Arid Inland River Basin—A Case Study on the Shiyang River Basin. *J. Nat. Resour.* **2012**, *27*, 1940–1950. (In Chinese)

10. Wang, Y.; Jin, C.; Lu, M.Q.; Lu, Y.Q. Assessing the suitability of regional human settlements environment from a different preferences perspective: A case study of Zhejiang Province, China. *Habitat Int.* **2017**, *70*, 1–12. [CrossRef]

11. Ma, R.F.; Wang, T.F.; Zhang, W.Z.; Yu, J.H.; Wang, D.; Chen, L.; Jiang, Y.P.; Feng, G.Q. Overview and progress of Chinese geographical human settlement research. *J. Geogr. Sci.* **2016**, *26*, 1159–1175. [CrossRef]

12. Deng, M.L.; Zhang, B.; Gao, Y.R.; Yu, B. Assessment of Human Settlements in Prefecture-level Cities of Sichuan Province. *Sichuan Environ.* **2009**, *28*, 42–47. (In Chinese)

13. Xia, Y.; Lin, A.W.; Zhu, H.J. Evolvement of Spatial Pattern of Urban Human Settlement Environment Suitability in Yangtze River Delta. *Ecol. Econ.* **2017**, *33*, 112–117. (In Chinese)

14. Li, X.; Li, X.X.; Wang, T.; Li, L. Human Settlement Assessment in Yellow River Valley Based on Factor Analysis. *Environ. Sci. Technol.* **2010**, *33*, 189–193. (In Chinese)

15. Li, M.; Li, X.M. Application Research on Quality Evaluation of Urban Human Settlements Based on The BP Neural Network Improved by GA. *Econ. Geogr.* **2007**, *32*, 99–103. (In Chinese)

16. Zhang, X.R.; Fang, C.L.; Wang, Z.B.; Ma, H.T. Urban Construction Land Suitability Evaluation Based on Improved Multi-criteria Evaluation Based on GIS (MCE-GIS): Case of New Hefei City, China. *Chin. Geogr. Sci.* **2013**, *23*, 740–753. [CrossRef]

17. Xu, X.L. China Annual Normalized Difference Vegetation Index Spatial Distribution Data. Resource and Environment Data Cloud Platform, Chinese Academy of Sciences. Available online: http://www.resdc.cn/DOI/doi.aspx?DOIid=49 (accessed on 2 November 2018). (In Chinese).

18. Xu, X.L. Spatial Distribution Data of Soil Erosion in China. Resource and Environment Data Cloud Platform, Chinese Academy of Sciences. Available online: http://www.resdc.cn/DOI/doi.aspx?DOIid=47 (accessed on 2 November 2018). (In Chinese).

19. Cheng, S.J.; Zhu, Z.L.; Bai, L.B. GIS-based Assessment on Ecological Suitability for Human Settlement—A Case Study in the Central Arid Zone in Ningxia. *Arid Zone Res.* **2015**, *32*, 176–183. (In Chinese)

20. Feng, Z.M.; Tang, Y.; Yang, Y.Z.; Zhang, D. Relief degree of land surface and its influence on population distribution in China. *J. Geogr. Sci.* **2008**, *18*, 237–246. [CrossRef]

21. Wang, Y.L.; Qi, P.C.; Li, D.; Ma, X.L. Relief and Suitability Evaluation of the Human Settlements in Shaanxi Province. *J. Northwest Normal Univ.* **2013**, *49*, 96–101. (In Chinese)

22. Lang, L.L.; Cheng, W.M.; Zhu, Q.J.; Long, E. A Comparative Analysis of the Multi-criteria DEM Extracted Relief–Taking Fujian Low Mountainous Region as an Example. *Geo-Inf. Sci.* **2007**, *1*, 135–136. (In Chinese)

23. Hao, H.M.; Ren, Z.Y. Evaluation of Nature Suitability for Human Settlement in Shaanxi Province Based on Grid Data. *Acta Geogr. Sin.* **2009**, *64*, 498–506. (In Chinese)

24. Feng, Z.M.; Yang, Y.Z.; Zhang, D.; Tang, Y. Natural environment suitability for human settlements in China based on GIS. *J. Geogr. Sci.* **2009**, *19*, 437–446. [CrossRef]

25. Hao, H.M.; Ren, Z.Y.; Xue, L.; Jiang, Y.F. Quantitative Study on Response of Ecosystem to Land Use/Cover Changes in Yulin Area Based on 3S Technology. *Prog. Geogr.* **2007**, *96*, 106–130. (In Chinese)

26. Peng, S.Z.; Yu, K.L.; Li, Z.; Wen, Z.M.; Zhang, C. Integrating potential natural vegetation and habitat suitability into revegetation programs for sustainable ecosystems under future climate change. *Agric. For. Meteorol.* **2019**, *269*, 270–284. [CrossRef]

27. State Environmental Protection Administration. *Environmental Protection Industry Standard of the People's Republic of China: Technical Criterion for Ecosystem Status Evaluation*; China Environmental Science Press: Beijing, China, 2006. (In Chinese)

28. Deng, S.B.; Zhang, Q.N. GIS-Based Evaluation of Natural Suitability of Human Settlement Environment in Guangdong Province. *Acta Scientiarum Nat. Univ. Sunyatseni* **2014**, *53*, 127–134. (In Chinese)

29. Li, Y.C.; Liu, C.X.; Zhang, H.; Gao, X. Evaluation on the human settlements environment suitability in the Three Gorges Reservoir Area of Chongqing based on RS and GIS. *J. Geogr. Sci.* **2011**, *21*, 346–358. [CrossRef]

30. Qiao, H.J.; Ding, M.J.; Li, L.H.; Chen, Z.P.; Hou, H.Y. Design and Achievement of Soil Erosion Assessment System of Jiangxi Province. *Soil Water Conserv. China* **2014**, *46*, 63–69. (In Chinese)

31. Teng, H.F.; Hu, J.; Zhou, Y.; Zhou, L.Q.; Shi, Z. Modelling and mapping soil erosion potential in China. *J. Integr. Agric.* **2019**, *18*, 251–264. [CrossRef]

32. Zhao, J.L.; Yang, Z.Q.; Gerard, G. Soil and water conservation measures reduce soil and water losses in China but not down to background levels: Evidence from erosion plot data. *Geoderma* **2019**, *337*, 729–741. [CrossRef]

33. Yu, S.; Dai, W.Y. Fujian Province Tour Climate Evaluation. *J. Fujian Normal Univ.* **2005**, 103–106. (In Chinese)

34. Aynur, M.; Wahap, H.; Gulgena, H.; Rebiyam, M. Assessment of residential environment suitability for southern Xinjiang based on GIS. *J. Arid Land Resour. Environ.* **2012**, *26*, 11–17. (In Chinese)

35. Halik, W.; Mamat, A.; Dang, J.H.; Deng, B.S.; Tiyip, T. Suitability analysis of human settlement environment within the Tarim Basin in Northwestern China. *Quatern. Int.* **2013**, *311*, 175–180. [CrossRef]

36. Bokaie, M.; Zarkesh, M.K.; Arasteh, P.D.; Hosseini, A. Assessment of Urban Heat Island based on the relationship between land surface temperature and Land Use/ Land Cover in Tehran. *Sustain. Cities Soc.* **2016**, *23*, 94–104. [CrossRef]

37. Qiao, Z.; Sun, Z.Y.; Sun, X.H.; Xu, X.L.; Yang, J. Prediction and analysis of urban thermal environment risk and its spatio-temporal pattern. *Acta Econ. Sin.* **2019**, *39*, 1–10. (In Chinese)

38. Zhu, B.Y.; Li, G.Z.; Liu, C.Y.; Liu, J.F.; Xu, C.H. Evaluation of the natural suitability of human settlements environment in Jilin Province based on RS and GIS. *Remot Sens. Land Resour.* **2013**, *25*, 138–142. (In Chinese)

6

Summertime Urban Mixing Layer Height over Sofia, Bulgaria

Ventsislav Danchovski

Department of Meteorology and Geophysics, Faculty of Physics, Sofia University, Sofia 1164, Bulgaria; danchovski@phys.uni-sofia.bg.

Abstract: Mixing layer height (MLH) is a crucial parameter for air quality modelling that is still not routinely measured. Common methods for MLH determination use atmospheric profiles recorded by radiosonde but this process suffers from coarse temporal resolution since the balloon is usually launched only twice a day. Recently, cheap ceilometers are gaining popularity in the retrieval of MLH diurnal evolution based on aerosol profiles. This study presents a comparison between proprietary (Jenoptik) and freely available (STRAT) algorithms to retrieve MLH diurnal cycle over an urban area. The comparison was conducted in the summer season when MLH is above the full overlapping height of the ceilometer in order to minimize negative impact of the biaxial LiDAR's drawback. Moreover, fogs or very low clouds which can deteriorate the ceilometer retrieval accuracy are very unlikely to be present in summer. The MLHs determined from the ceilometer were verified against those measured from the radiosonde, which were estimated using the parcel, lapse rate, and Richardson methods (the Richardson method was used as a reference in this study). We found that the STRAT and Jenoptik methods gave lower MLH values than radiosonde with an underestimation of about 150 m and 650 m, respectively. Additionally, STRAT showed some potential in tracking the MLH diurnal evolution, especially during the day. A daily MLH maximum of about 2000 m was found in the late afternoon (18–19 LT). The Jenoptik algorithm showed comparable results to the STRAT algorithm during the night (although both methods sometimes misleadingly reported residual or advected layers as the mixing layer (ML)). During the morning transition the Jenoptik algorithm outperformed STRAT, which suffers from abrupt changes in MLH due to integrated layer attribution. However, daytime performance of Jenoptik was worse, especially in the afternoon when the algorithm often cannot estimate any MLH (in the period 13–16 LT the method reports MLHs in only 15–30% of all cases). This makes day-to-day tracing of MLH diurnal evolution virtually impracticable. This problem is possibly due to its early version (JO-CloVis 8.80, 2009) and issues with real-time processing of a single profile combined with the low signal-to-noise ratio of the ceilometer. Both LiDAR-based algorithms have trouble in the evening transition since they rely on aerosol signature which is more affected by the mixing processes in the past hours than the current turbulent mixing.

Keywords: mixing layer; urban area; ceilometer; radiosonde

PACS: 01.30.-y; 01.30.Ww; 01.30.Xx

1. Introduction

The effect of air quality on human health is a serious problem, especially in densely populated areas. Hence, a lot of effort is being made to better understand the processes controlling pollution levels, particularly in numerical modelling. Key input parameters of these models are meteorological variables, which are needed to be identified in order to calculate the production, diffusion, transport and scavenging of atmospheric pollutants. These harmful substances are dispersed vertically within the mixing layer (ML) due to its inherent turbulence. According to Seibert et al. [1], ML is "...the layer adjacent to the ground over which pollutants or any constituents emitted within this layer or entrained into it become vertically dispersed by convection or mechanical turbulence within a time scale of about an hour". However, one should bear in mind that there are situations when time-scales of the dominant processes (such as diabatic processes like radiative cooling in the evening transition, or unsteadiness of pressure gradients, or intermittent turbulence due to breaking gravity waves, just to name a few) are much longer so that the ML is unsteady [2]. Obviously, near-ground pollution levels will depend on the mixing layer height (MLH) since it constrains the dispersion volume. Thus, the MLH is vitally important to be identified especially in urban areas where pollution sources and inhabitants are much greater [3–10]. Moreover, urban MLH can be characterized by enormous temporal and spatial variability due to inhomogeneity in surface roughness and heating in cities [11]. Therefore, MLH is worthwhile to be continuously monitored and also compared with parametrizations in numerical weather and/or pollution prediction models [12–14].

Despite its importance, MLH is not a part of routine measurements. Furthermore, because it is associated with the spatial distribution of turbulence, we need turbulence profiles to determine MLH. Consequently, TKE (turbulent kinetic energy)-based criteria (MLH is marked by the level where TKE drops below a predefined threshold) are often used in numerical models with turbulence closure of order 1.5 or higher for MLH determination [1,15]. Moreover, profiles of the TKE and its dissipation rate can be measured by remote sensing instruments [16,17]. Therefore, Doppler LiDARs [18] and sodars [19,20] can serve as "turbulence profilers" but the former are quite expensive and the latter have limited vertical range. Fortunately, vertical profiles of non-reactive scalar meteorological variables should be nearly constant with height within a well-mixed boundary layer [21], so we can detect the MLH by looking for abrupt changes in the uniformly distributed profiles of these tracers [22,23].

Regardless of the wide variety of remote sensing methods, the most used instrument for MLH detection is the radiosonde, which is still used as a reference. The derivation of the MLH from the radiosonde profiles of the atmospheric temperature, humidity, and wind dates back to 1960s [24]. Moreover, these radiosonde-based methods are still used independently [25] or as a reference for validating MLH measurements from remote sensing instruments [26]. However, the radiosonde also has some drawbacks as it measures atmospheric properties along its flight, which is slant instead of vertical, due to horizontal wind. Therefore, the radiosonde profiles do not coincide with rising thermal or vertical profiles derived from the remote sensing instruments. Additionally, the radiosondes' main limitation is their coarse temporal resolution since they are usually launched no more than twice a day.

The necessity of continuous MLH monitoring can be met by operating ground-based remote sensing instruments. A comprehensive review of existing techniques for MLH determination through ground-based remote sensing instruments, along with their advantages and limitations, can be found in Wiegner et al. [27] and Emeis et al. [28]. It is worthwhile to note that individual disadvantages of each instrument in the MLH diurnal cycle determination can be overcome if apparatuses are used together [29].

One should note that relying on ground-based remote sensing instruments in MLH estimation cannot provide good spatial representativeness, especially over areas with non-homogeneous land-use and/or complex topography. Fortunately, this lack of information can be filled if space-based remote sensing instruments are used. Among them are the Cloud-Aerosol LiDAR with Orthogonal Polarization (CALIOP) [30] and the Moderate Resolution Imaging Spectroradiometer (MODIS) [31], which are the most used for determination of the atmospheric boundary layer height over continents and oceans [32–35]. The radio occultation method based on global position system signals can provide vertical profiles of the refractivity index that can be used in MLH retrieval [36,37].

In recent years, laser-based remote sensing instruments, especially automatic LiDARs and ceilometers (ALC), have become more affordable and widely used in the field of atmospheric research, particularly for MLH determination [38–45]. We should also note the considerable efforts made in the COST Action ES 1303 TOPROF [46] which provides standards for calibrated profiles of the aerosols, winds, temperature, and humidity to fill the observational gap in the lower troposphere. These quality controlled observations are delivered in near real-time through the EUMETNET Composite Observing System (EUCOS) network E-PROFILE [47] to the national weather services in order to improve numerical weather prediction.

Different LiDAR-based methods for MLH retrieval from the range corrected signal are summarized in Haeffelin et al. [48]. These methods are the basis of many proprietary [49,50] and in-house [51–54] algorithms explicitly designed for MLH retrieval from ceilometers' data. Possible performance improvement of these LiDAR-based techniques can be achieved by monitoring diurnal variations in Radon-222 [55] or it can be used alone to evaluate MLH [56].

The objectives of the present study are to evaluate the performance of a proprietary algorithm as well as a popular, freely available algorithm in detecting of MLH from ceilometer data over an urban area. Both methods are evaluated against MLHs retrieved using radiosonde profiles as a reference. The structure and evolution of the mixing layer over Sofia in summertime is also discussed. To highlight the advantages and disadvantages of both algorithms, an analysis was performed in the summer when MLH is high enough to minimize the negative effects due to incomplete overlapping in the near-field range of the ceilometer. The paper is organized as follow: measurement sites and specifications of the instruments used for observations, as well as details about the collected data and methods applied to determine the MLH, are in Section 2. Inter-comparison of the three radiosonde-based methods is in Section 3.1. Verification of the MLH retrieved from the ceilometer compared to the MLH measured from the radiosonde is in Section 3.2. Diurnal evolution of the MLH over Sofia derived from the ceilometer data by both the proprietary and the freely available algorithm, as well as a discussion of their main benefits and drawbacks and suggestions for performance improvement, are in Section 3.3. Statistical analysis of the MLH diurnal cycle is discussed in Section 3.4. The article ends with summary of findings.

2. Data and Methodology

Sofia is the largest and the most densely populated city in Bulgaria with roughly 1,400,000 inhabitants. The city is located in a valley that is almost fully encircled by mountains; therefore, the micro- and meso-scale processes, as well as the ML dynamics, are heavily influenced by both complex orography and urban territory. To perform our analysis of urban MLH we used 3 months of intensive measurements, from 1 June until 31 August 2015. The data used in this study were obtained from a continuously operating ceilometer, Jenoptik CHM 15k (in 2014, the company G. Lufft Mess- und Regeltechnik GmbH acquired the product segment of ceilometers from Jenoptik and now the ceilometer is known as Lufft CHM 15k), and a balloon sounding launched on a daily basis at 12:00 UTC (14:00 LT).

The CHM15k (firmware version 0.63) is operated by the Department of Meteorology and Geophysics, Sofia University. The ceilometer is situated in the city centre on the territory of the University Astronomical Observatory in the park "Borisova gradina" (Figure 1). The CHM15k is an eye-safe biaxial LiDAR system equipped with an Nd: YAG solid-state near-infra-red laser operating at 1064 nm. It emits pulses with an energy of 8 μJ and a repetition frequency of 5–7 kHz. The ceilometer provides data with a vertical resolution of 15 m, the maximum height of the signal is 15,000 m and the temporal resolution is set up to 60 s. As CHM15k is a biaxial LiDAR it suffers from incomplete overlap in the near range since only a small portion of the laser beam gets into the receiver field of view. According to the manufacturer, the overlapping is ~1% at 15th bin (225 m), ~10% at 24th bin (360 m), ~50% at 40th bin (600 m), ~90% at 57th bin (855 m), ~99% at 78th bin (1170 m) and full overlap is achieved at 120th bin (1800 m). Further details about the instrument can be found in [57].

The MLH was retrieved from the ceilometer profiles by supposing that aerosol concentration is rapidly adapted to the thermal stratification of the ML and that aerosol loading above the city is not dominated by advection. The manufacturer's software includes a proprietary algorithm for automatically deriving the MLH every minute (software JO-CloVis version 8.80) [58]. Because the Jenoptik algorithm is proprietary not much is known about it. Haeffelin et al. [48] reported that the algorithm uses vertical derivatives and wavelet transforms on the range-corrected signal to identify local minima which are used as MLH, however, it is not specified which version is referred to and there is no information on how signal-to-noise ratio is enhanced in the pre-processing. A freely available Structure of the Atmosphere (STRAT) algorithm [52], which is designed for the retrieval of aerosol vertical profiles in the atmospheric boundary layer and free troposphere, is used for comparison. In contrast to the Jenoptik algorithm that is based on vertical gradients of backscatter signal in a single profile, STRAT uses both temporal and vertical gradients (G_t and G_v) by using Sobel 2-D derivation operators. The global gradient is calculated as $G = \sqrt{G_t{}^2 + G_v{}^2}$. The edges in backscatter are kept if G is greater than predefined thresholds. Additionally, edges in low signal-to-noise ratio zones are rejected. Finally, minimum (450 m due to overlapping) and maximum (here, 3000 m during the day and 1500 m at night are used) allowed heights are applied and three global gradients—the strongest, the second strongest, and the lowest-height—are reported as MLH candidates [48]. The MLH is then determined as the lowest-height candidate at night, during day quality control based on relative change in backscatter around each candidate is performed and the first existing one in the line strongest, second strongest and lowest is selected as MLH [18].

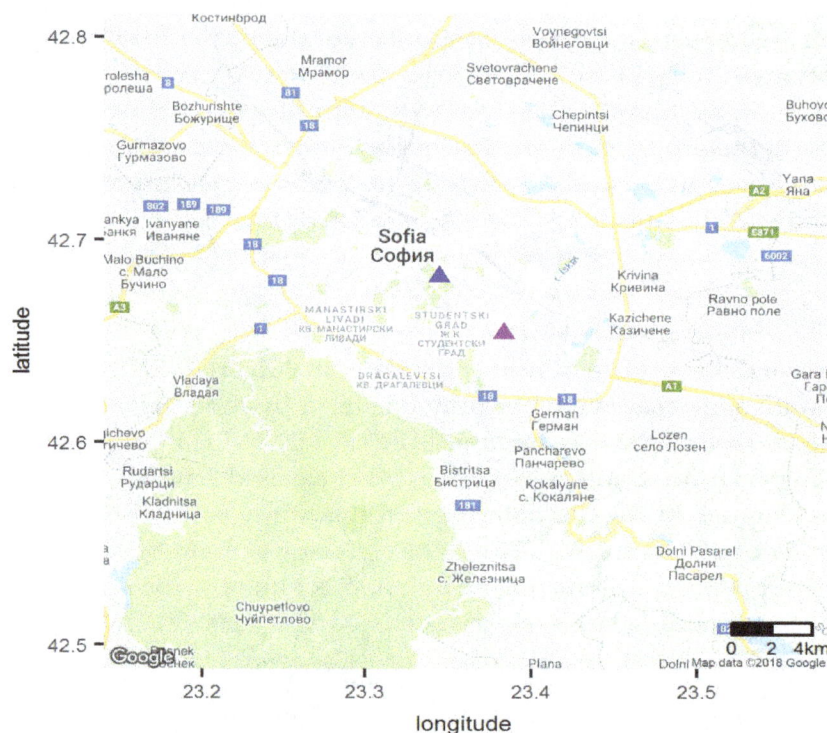

Figure 1. The locations of the ceilometer and the radiosonde indicated by a blue and a purple triangle in the Sofia valley. (Source of the map is Google LLC).

The atmospheric sounding system is the Vaisala's MW41 which is located in the Central Aerological Observatory on the territory of the National Institute of Meteorology and Hydrology, which is about 4.4 km south-east from the ceilometer (Figure 1). In this study, low resolution radiosonde data are used, which are freely available at Integrated Global Radiosonde Archive (IGRA) [59]. Archived radiosonde data consist of atmospheric parameters recorded at mandatory and significant levels which were used to restore the atmospheric profiles by linear interpolation.

Following de Haij et al. [60], three different MLH detecting algorithms were applied to the radiosonde data. The Bulk Richardson (Ri) method is based on the Richardson number which is the ratio of thermally and mechanically driven turbulence. According to this method, MLH is the level where the bulk Richardson number exceeds predefined threshold values [61–63]. In this study, the commonly recommended value of 0.21 was used. It is worth noting that the Ri method is suitable for both convective and stable conditions. In the parcel method [24,64], the MLH is determined by extending dry-adiabatically surface temperature to its intersection point with temperature profile. However, this method provides reliable results only for unstable convective boundary layer as it neglects wind shear effects on vertical mixing. The last method for MLH determination from radiosonde data is the lapse rate method [21,65]. It is based on threshold values of vertical gradients of potential temperature (θ) and relative humidity (RH). Adhering to de Haij et al. [60], negative gradients of RH and a gradient of θ >2 K/km were used as the basis for this study. As the selected critical value of potential temperature gradient is more or less subjectively chosen, the performance of lapse rate values of 0.5, 1, 1.5, 2.5, 3, 3.5, and 4 K/km was also tested.

As main synoptic-scale systems are associated with the suppression or stimulation of parcel ascending, it is interesting to examine their role on mixing layer height [66]. Therefore, the difference (Δp) of surface layer atmospheric pressure (p) and its smoothed value (p_{smooth}, which is obtained by low pass filter with cut-off 6 days) is calculated by Equation (1). Then Δp is standardized by Equation (2), i.e., the Δp is rescaled to have a mean of zero (subtraction of the mean value $\overline{\Delta p}$) and a standard deviation of one (division by the standard deviation $\sigma_{\Delta p}$).

$$\Delta p = p - p_{smooth} \tag{1}$$

$$\Delta p_{std} = \frac{\Delta p - \overline{\Delta p}}{\sigma_{\Delta p}} \tag{2}$$

Finally, the atmospheric pressure (atm.press) is classified as "Low" if Δp_{std} is smaller than -0.5 while it is marked as "High" if Δp_{std} is higher than 0.5. If the Δp_{std} values are greater than -0.5 but less than 0.5, atmospheric pressure is marked as "Normal".

3. Results and Discussions

3.1. Inter-Comparison of Radiosonde-Based MLH Retrieval Methods

The three aforementioned radiosonde-based algorithms—Richardson, parcel and lapse-rate—were applied on the dataset for a total of 92 days (for 28 days the atmospheric pressure was "High", for 43 "Normal", and for 21 "Low"). MLH values were successfully estimated at 92, 92, and 81 days, respectively. The estimated MLHs were then compared against one another on Figure 2. The perfect correlation between the Richardson and parcel method indicates that in summer at 14:00 LT (12:00 UTC) the urban mixing layer over Sofia is dominated by thermally driven turbulence. It is a fairly expected result since the study period took place in summer and radiosonde launching occurred in the early afternoon. The box-plot shows that slightly higher MLH values are related to prevailing low atmospheric pressure and that when atmospheric pressure is marked as normal or high, MLHs are slightly decreased; however, the observed difference is not statistically significant (Wilcoxon–Mann–Whitney test with a significance level of 5% was performed).

The lapse rate method shows worse alignment with the Richardson and parcel methods, therefore, we tested how a threshold value of vertical gradient θ influences concurrence with the other two approaches. We found that lapse rate values of 1, 1.5 and 2 K/km perform similarly and Pearson correlation coefficients with respect to the Ri method are about 0.89. However, the correlation diminishes if smaller or higher threshold values are used. It is also worth mentioning that a negative vertical gradient of the relative humidity is not changed because it agrees with the mixing layer conception (the Earth's surface is the water vapour source and free atmosphere is low in humidity, so the humidity gradient should be negative at MLH). Keeping in mind that the Ri method incorporates both mechanical and buoyancy production of turbulent mixing we choose it as a reference in the following analysis. Performance of the parcel and lapse rate methods using the set of critical values mentioned above were evaluated against the Richardson method and is summarised in Table 1.

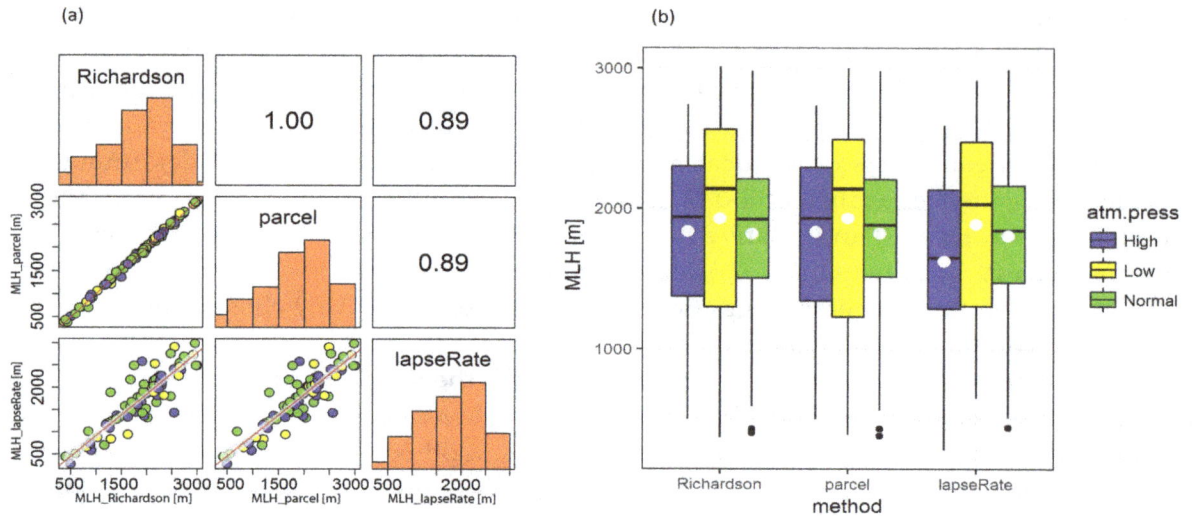

Figure 2. Inter-comparison of the three radiosonde-based MLH methods. The correlation matrix (**a**) shows correlation coefficients in the upper-right triangle, the diagonal shows a histogram of each method, and the lower-left triangle shows scatter-plots and linear regression lines with corresponding 95% confidence intervals. The box and whisker plot (in the style of Tukey) is on plot (**b**). The box lines correspond to the 25, 50 and 75 percentiles. The lower and upper whiskers represent the lowest values still within 1.5 IQR (inter-quantile range) of the lower quartile, and the highest values still within 1.5 IQR of the upper quartile. The data beyond the end of the whiskers signify outliers and are plotted as black dots. White dots indicate mean values. In both figures, atmospheric pressure is color-coded as "High" (blue), "Low" (yellow), "Normal" (green).

Table 1. Skill scores (MD—mean deviation; RMSD—root-mean-square deviation, r—Pearson correlation coefficient, slope—linear regression slope, intercept—linear regression intercept) of parcel and lapse rate methods compared against the Richardson method as a reference in the MLH determination.

Method	MD [m]	RMSD [m]	r	Slope	Intercept [m]
parcel	1	37	1.00	0.99	11
lapse rate 0.5 K/km	−215	396	0.85	0.82	520
lapse rate 1.0 K/km	−200	351	0.89	0.88	402
lapse rate 1.5 K/km	−194	342	0.89	0.89	388
lapse rate 2.0 K/km	−153	330	0.89	0.85	417
lapse rate 2.5 K/km	−87	403	0.81	0.75	554
lapse rate 3.0 K/km	−40	431	0.78	0.70	601
lapse rate 3.5 K/km	39	493	0.75	0.61	719
lapse rate 4.0 K/km	106	543	0.73	0.57	773

3.2. Inter-Comparison of MLHs Derived from Ceilometer and Radiosonde Data

MLHs calculated from radiosonde data are often used for reference since they are based on the thermodynamic structure of the lowest atmosphere that directly reflects changes in the surface forcing. However, since routine balloon launching usually occurs only twice a day so it does not allow for MLH diurnal evolution to be tracked. Low-cost ceilometers that provide backscatter power profiles are a tempting alternative because they operate continuously.

To evaluate the overall performance of the ceilometer-based methods in the MLH determination, the calculated values are compared against the Richardson method estimates from the radiosonde data. Since the radiosonde in Sofia is launched once a day at 12:00 UTC (14:00 LT), the ceilometer-retrieved MLHs from within a 20 min timespan are averaged and used in the comparison. After this procedure the size of the STRAT's datasets at "High", "Normal" and "Low" atmospheric pressure is reduced to 18 (64%), 24 (56%) and 10 (48%) days, respectively. The Jenoptik algorithm successfully estimates

MLHs in 13 (46%), 17 (40%) and 11 (52%) days at "High", "Normal" and "Low" atmospheric pressure, respectively. In other words, both ceilometer-based algorithms cannot estimate MLHs in about half of the days with "Low" atmospheric pressure. The percentage of the Jenoptik-retrieved MLHs becomes even lower at "Normal" and "High" pressure, while the performance of STRAT is slightly increased. The left and right panels of Figure 3 show a correlation matrix and box and whiskers plots of the MLH determined by STRAT, Jenoptik and Ri methods at different atmospheric pressures. It is evident that both LiDAR-based algorithms tend to underestimate MLH compared to radiosonde (Richardson). We should bear in mind that MLH estimation from the ceilometer and the radiosonde data rely on different tracers, which may contribute to the observed discrepancy. When optically thick clouds or rain are presented the backscatter signal can be strong enough to saturate the ceilometers receiver so the cloud base or somewhere under the cloud within the rain column is reported as the MLH. To prove the hypothesis, the data was spited to rainy (if nonzero ceilometer's precipitation index is registered from 11:20 to 11:40 UTC) and dry cases. The analysis showed that the difference between Ri and STRAT, and between Ri and Jenoptik are statistically non-significant (t-test with a significant level of 0.05 is performed) in rainy days. In the rest of the days the Ri method reports about 180 m (750 m) higher MLH than STRAT (Jenoptik) and the results are statistically significant. Additionally the role of low clouds was tested. In days with low clouds (if cloud base height is < 1500 m) the difference between Ri and STRAT and between Ri and Jenoptik are evaluated as a statistically non-significant. In the rest of the days radiosonde estimates are 212 m and 493 m higher than STRAT and Jenoptik respectively and t-test showed that both results are statistically significant. The observed underestimation of the MLH by the ceilometer could be attributed to the difference in land surface type. The ceilometer is situated in the park (while the radiosonde is in a built-up area) so one can expect that some of the solar energy is consumed during evapotranspiration; therefore the rest of the energy that would produce the thermally driven turbulence, and thus MLH raising, is reduced. To prove this, the number of consecutive days with no precipitation (ceilometer's precipitation index is used for the classification) are used to split the data into categories. It was found that the difference between Ri and STRAT methods (the Jenoptik is not included since it shows significant deviation from the Richardson method, see Figure 3) increases with the number of consecutive droughty days, which is supposed to be a result of the lack of available water for evaporation in a built-up zone. Additionally, if the drought period is long, MLH becomes higher and the ceilometer-based method experiences difficulties that are supposed to be a result of diminished backscatter due to the increased volume for aerosol dispersion (Figure 4).

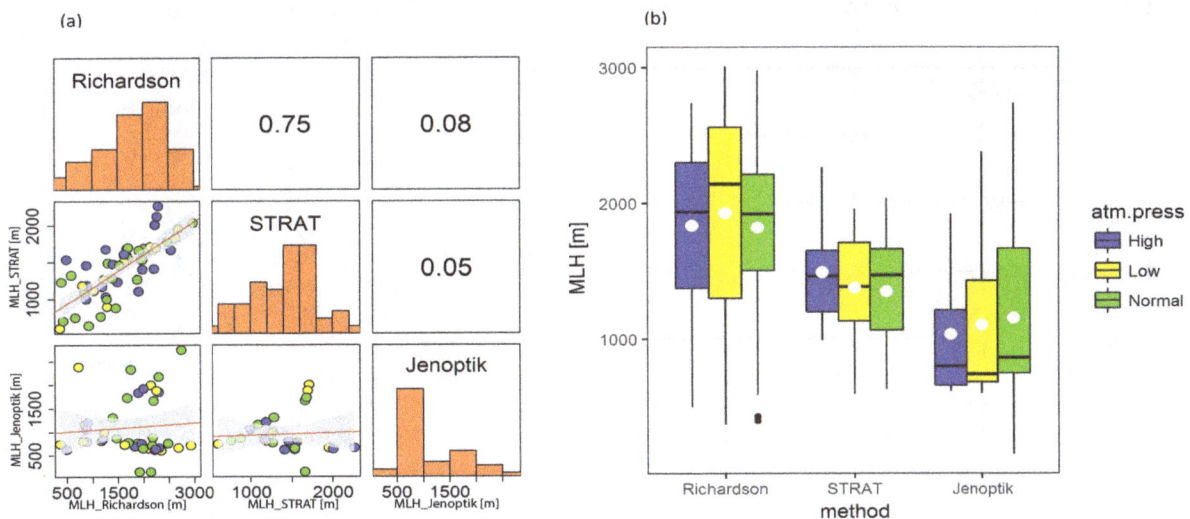

Figure 3. A correlation matrix (**a**) and Tukey's box and whiskers plot (**b**) of radiosonde-(Richardson) and LiDAR-based (STRAT and Jenoptik) algorithms for MLH detection. Conventions are the same as in Figure 2.

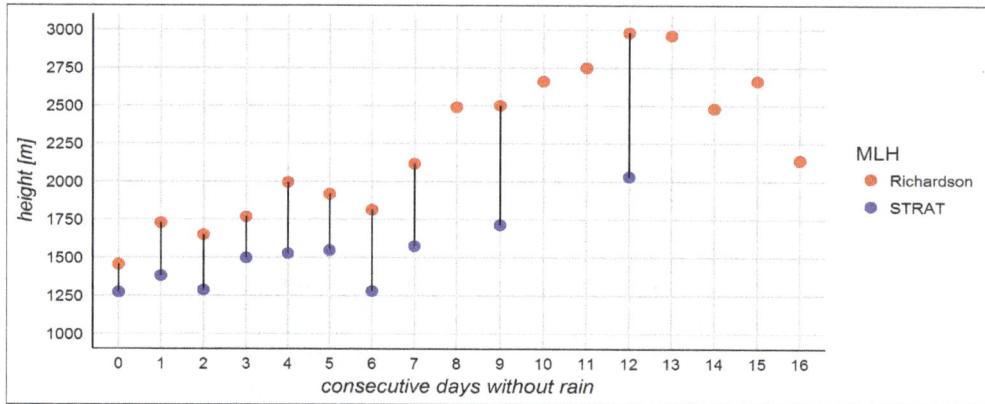

Figure 4. The dependence of drought duration (in number of dry days) on the mean MLH determined by Richardson and STRAT methods.

Skill scores of LiDAR-based algorithms against the Ri method are listed in Table 2. It can be seen that the average underestimation of MLH by STRAT and Jenoptik is ~160 m and ~660 m respectively. The STRAT-estimated values of MLH are reasonably comparable with those retrieved from radiosonde profiles, but Jenoptik's performance is quite unpromising and needs further clarification.

Table 2. Skill scores (MD—mean deviation; RMSD—root-mean-square deviation, r—Pearson correlation coefficient, slope—linear regression slope, intercept—linear regression intercept) of aerosol-based algorithms (Jenoptik and STRAT) compared against the Richardson method as a reference in the MLH determination.

Method	MD [m]	RMSD [m]	r	Slope	Intercept [m]
Jenoptik	−665	1086	0.08	0.0.7	970
STRAT	−162	467	0.75	0.45	701

3.3. Diurnal Evolution of the MLH Determined by the Ceilometer—A Case Study

To elucidate the above-mentioned ceilometer's capacity to track the MLH diurnal cycle, a case study is first considered. In Figure 5 diurnal evolution of the range-corrected ceilometer signal (PR^2) on July 24 is presented along with MLHs determined according to STRAT and Jenoptik algorithms. Radiosonde-derived MLH by the Richardson method is also plotted for comparison.

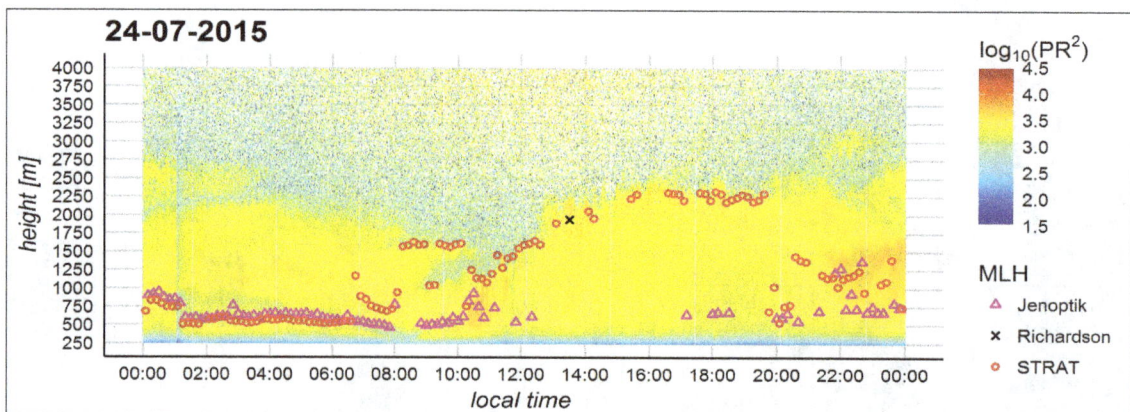

Figure 5. Time-height cross section of the ceilometer's range-corrected backscatter power (PR^2 in arbitrary units) on 24 July 2015. The MLH retrieved from ceilometer's data by Jenoptik and STRAT algorithms are marked by magenta triangles and red circles, respectively (for clarity, the Jenoptik MLHs are plotted with the same temporal resolution as STRAT—10 min). Radiosonde-based MLH according to the Ri method is presented by black "x" marks.

As observed, the range corrected ceilometer's signal reveals some characteristic features in the MLH diurnal evolution. The backscatter power within the first 500–700 m is high in the early hours of the night which can be associated with mechanically mixed aerosols within the nocturnal boundary layer. As seen, the layer was shrinking and at ∼8:00 LT (less than 2 h after sunrise) it had disappeared. One may expect a new convective layer to be identifiable at that moment but we should keep in mind that the ceilometer has virtually zero overlapping in the first ∼200 m (overlapping is <1%) so that the first signs of the rising thermals are visible at ∼9:00 LT. Above the nocturnal layer, there is a zone with decreased signal that is capped by a high backscatter layer, which most likely outlines aerosol burden air in the residual layer, or it is a result of advection at that elevation. The ceilometer's signal also depicts the daytime evolution of the MLH with its typical growth due to the solar heating of the surface. After sunrise, thermals start forming and rising due to positive buoyancy. These updrafts produce turbulent mixing so that the diminished vertical backscatter within ML in the afternoon results from an increased volume for aerosol dispersion. An enhanced signal close to the ML top in the afternoon can be attributed to hygroscopic growth of aerosols due to increased relative humidity. As can be seen, MLH reached its maximum (∼2250 m) at ∼16:00 LT and remains almost constant until ∼19:00 LT. In the evening, the thermals cease to form (in the absence of cold air advection), allowing turbulence to decay in the formerly well mixed layer. A new nocturnal layer starts forming and overhead air associated with the new residual layer becomes decoupled from the mechanical source of turbulence on the ground. However, the evening transition period is non-stationary as heat fluxes decrease over a few hours after sunset so that the aerosol vertical distribution does not respond to surface forcing within an hour [2]. At that part of the day the ceilometer profiles are mostly a result of the turbulence dynamics in the recent periods, therefore, they do not reveal the present ML but its history. That problem is inherent to all remote sensing instruments which use aerosol backscatter to trace the ML but can be overcome if a "true turbulence profiler" is used. As observed, the MLH determined by the Ri method is ∼1940 m which corresponds very well to the aerosol distribution depicted by the ceilometer backscatter signal at the moment of balloon launching (at 13:30 LT that day).

It is also noticeable that the STRAT algorithm plausibly represents the diurnal evolution of the MLH. However, in the time interval from sunrise (6:09 LT) to approximately 10:30 LT, which corresponds to the morning transition period, STRAT misleadingly reports an overhead backscatter gradient (associated with the residual layer) as MLH instead of the one closest to the ground. Similar behaviour is found across all days and seems to be due to the layer attribution technique implemented in STRAT. According to Haeffelin et al. [48] the algorithm reports the strongest, second strongest, and the lowest gradients in backscatter and then, depending on the local time, it constructs a diurnal evolution of its "best estimate" (used here as MLH) which is the lowest gradient during the night and the strongest gradient during the day. Thus, the STRAT method reports abrupt changes in MLH around sunrise and sunset instead of smooth transitions from the nocturnal to convective boundary layer, and vice versa. Possible improvement of layer attributions and representations of the MLH diurnal evolution can be achieved through the use of statistical analysis [67] or graph theory [68].

It can be seen that the overall consistency of the MLHs reported by the Jenoptik algorithm with the observed aerosol distribution and evolution is relatively poorer than the consistency of the STRAT's MLHs. However, Jenoptik outperforms STRAT in the morning transition, although neither method can track the MLH from 8 to 9 LT when the MLH is in the zone of incomplete overlapping. The performance of the Jenoptik method during daytime is much worse and it cannot represent the MLH evolution. The method cannot report MLH from 13 to 17 LT and it significantly underestimates ML depth around noon and in the late afternoon. It should be noted that STRAT also locates these aerosol gradients at intermediate levels (Figure 6) but reports them as the lowest and the second strongest candidates, which are then successfully filtered out by the attribution procedure in the algorithm. This worsened performance of the Jenoptik method is likely to be a result of the immaturity of the outdated version of the algorithm used. Additionally, the Jenoptik method operates in real-time so it is likely to use only the current backscatter profile without taking into account previous measurements. Therefore, the signal-to-noise ratio (SNR) will be lower, which may result in poor performance compared to STRAT. Consequently, the poorer daytime performance of the Jenoptik can be attributed to reduced backscatter signal within the increased depth of the MLH (and enlarged volume for aerosol dispersion) and augmented background signal due to the higher sun elevation angle. Data shows the STRAT method also has similar troubles with backscatter gradient detection from ~13:00 LT to ~15:00 LT when only a few MLHs are reported. However, the process of smoothing incorporated within the algorithm enhances the SNR, enabling the MLH evolution to be tracked against the Jenoptik algorithm.

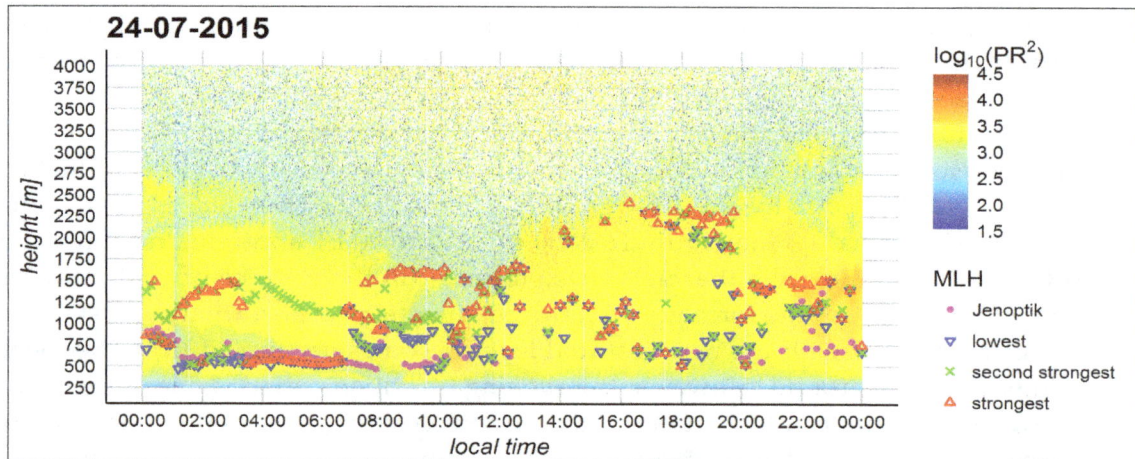

Figure 6. Time-height cross section of the ceilometer's range-corrected backscatter power (PR^2 in arbitrary units) on 24 July 2015. The MLH calculated by the Jenoptik algorithm (magenta dots) and STRAT's candidates (the strongest gradient—red triangles, the second strongest gradient—green "x" marks, and the lowest gradient—blue upside down triangles) are also shown.

3.4. Diurnal Evolution of the MLH Determined by Ceilometer—A Statistical Analysis

To compare the performance of both LiDAR-based algorithms we first make the datasets comparable. Since the Jenoptik algorithm has a 1-min resolution but STRAT's temporal resolution is 10 min, Jenoptik-derived MLHs are averaged in 10-minute intervals. The availability of STRAT- and Jenoptik-derived MLHs after applying the described procedure is presented in Figure 7. As seen, MLH data availability of both methods show similar patterns related to the diurnal cycle. The STRAT algorithm manages to estimate MLHs in about 70–95% of the cases but in the afternoon its availability drops to 50–70% with minimum of ~45% at 14 LT. In contrast, the Jenoptik method provides MLHs in about 60–85% of the cases but in the afternoon it hardly reaches even 35–40% with a minimum of ~15% at 15 LT. The observed diurnal pattern in MLH availability in both LiDAR-based methods is closely related to decreased SNR due to reduced aerosol concentration (due to increased volume for aerosol dispersion) and increased background signal (due to higher solar radiation) in the afternoon. Neither of the two applied algorithms show a clear atmospheric pressure dependency.

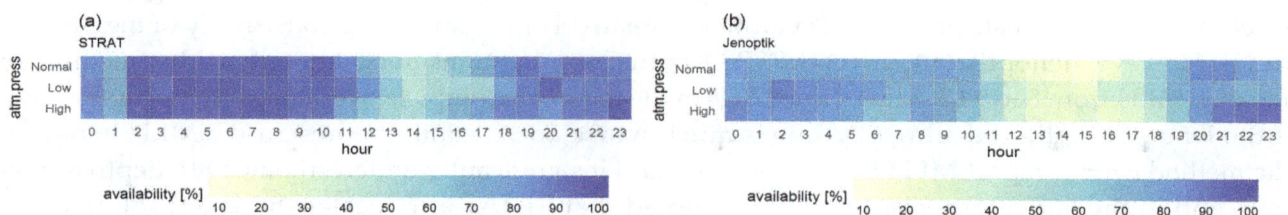

Figure 7. Diurnal evolution of the availability of MLH determined by STRAT (**a**) and Jenoptik (**b**) algorithms at "Normal", "Low" and "High" atmospheric pressure in summer of 2015.

The aforementioned features of both LiDAR-based techniques are also visible if all data for the MLH daytime progress are summarised and presented as box-plots (Figure 8). From midnight to

7 LT STRAT and Jenoptik algorithms provide comparable MLHs and most of the estimated values are in the range of 500–1000 m. However, the Jenoptik also shows several quite large values marked as outliers (most of them in the range 1500–3500 m) which are a result of improper selection of high aerosol layers that cannot be related to the near-ground turbulence. When atmospheric pressure is "Normal" the Jenoptik algorithm also reports a few quite low MLH in the ceilometer's incomplete overlapping zone which should be treated as incorrect values (most likely they are result of multiple scattering). The morning transition is marked by STRAT as an abrupt jump that is a result of its layer attribution criterion, while the Jenoptik represents the transition less steeply. Daytime performance of both algorithms is, thus, easily distinguishable. The MLHs retrieved by Jenoptik are often in the first 1 km and rarely reach 2 km. As was noted, the algorithm tends to report mid-level gradients that are also marked by STRAT lowest-height and/or second strongest gradients. However, in STRAT these mid-level gradients in the ML are successfully filtered out by the successive layer attribution. The daily maximum of MLH (sometimes more than 2000 m) is registered in the late afternoon (~18:00 LT), a few hours before sunset and, more importantly, during peak car traffic, which can help against excessive concentrations of air pollution. The evening transition is hard to be correctly traced by the LiDAR's backscatter profile as the aerosol signature is more related to turbulent mixing in the past than the current state. Therefore, although showing different behaviours, it is difficult to designate one of the two methods as more reliable. It is worth noting that there are a large number of outliers in the retrieved MLHs by the Jenoptik algorithm; most of them are related to high aerosol layers due to advection or residual layers at night. As seen, both methods report lower daytime MLHs in "Low" atmospheric pressure, especially Jenoptik algorithm, whose estimates do not reach 1 km in 50% of cases.

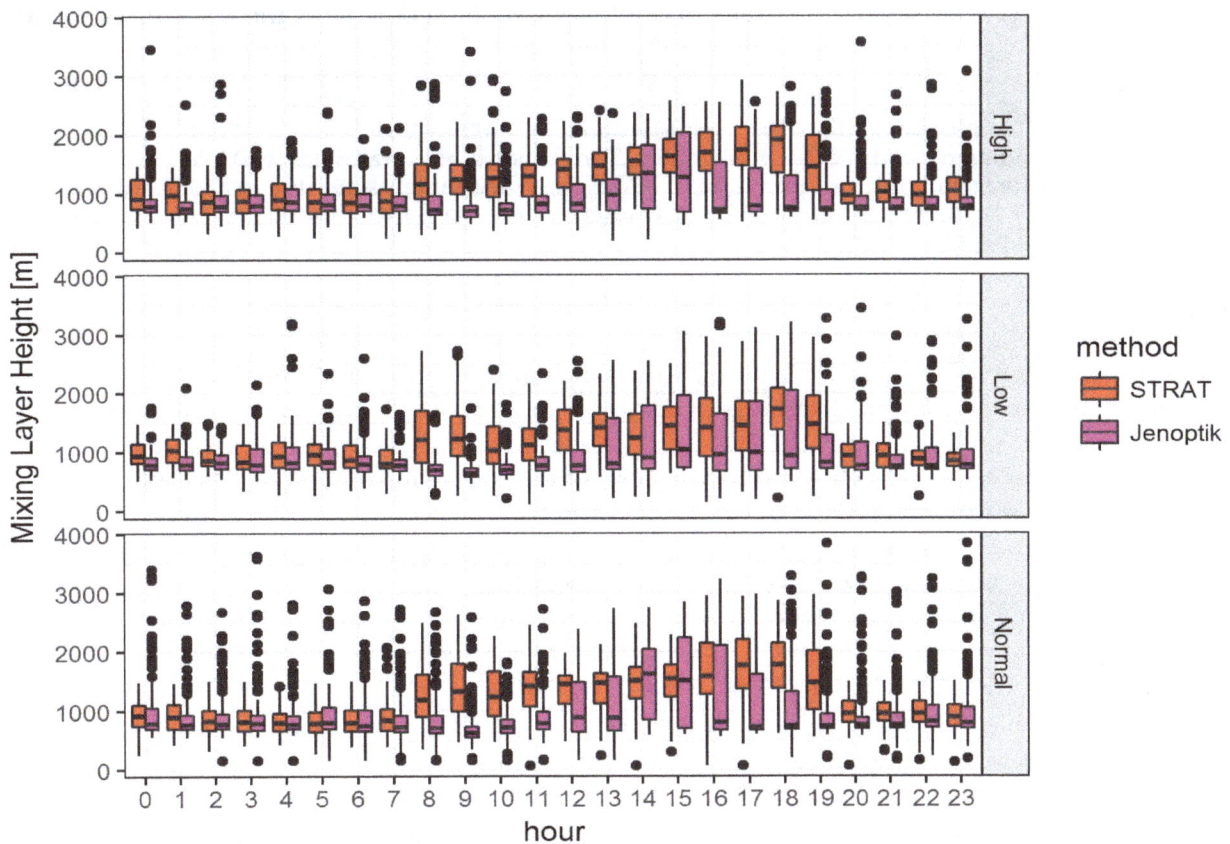

Figure 8. Diurnal cycle of the MLH over Sofia determined by STRAT (red) and Jenoptik (magenta) algorithms as a box and whiskers plot (in Tukey's style) at "High", "Low" and "Normal" atmospheric pressure in summer of 2015.

4. Conclusions

In this paper, MLHs derived by different algorithms over three summer months from radiosonde and ceilometer data were analysed and compared. It was shown that the Richardson and parcel methods produce identical MLHs which indicates that ML is primarily thermally driven while the lapse-rate method underestimates the MLH. It was also found that threshold values for potential temperature higher than 2 K/km or smaller than 1K/km deteriorate the agreement of lapse rate with the Richardson and parcel methods. Based on performed comparison against the Richardson method, it was shown that the ceilometer tends to underestimate the MLH. The observed discrepancy was mainly attributed to different land surface types where the instruments are situated and the distance between the sites. Additionally, the proprietary algorithm has difficulty with the low SNR of the ceilometer and frequently cannot report any MLH. In contrast, STRAT handles this better in that it incorporates SNR enhancement. It was shown that the ceilometer-derived aerosol profiles provide consistent with expected MLH information, which can be used to trace the urban MLH dynamics during day. However, the Jenoptik algorithm has difficulty (low availability of reported MLH due to reduced SNR when ML is high) primarily due to the early version of the software used in this work. The hampered tracking of the MLH by the proprietary algorithm may also be a result of the real-time operation on a single profile without making use of the previously collected data. The primary issues of both LiDAR-based techniques were identified as layer attribution, particularly at night and during transition periods when high aerosol layers were mistakenly used by the algorithms. It was underlined that incomplete overlapping of the ceilometer impacts the detection of low MLH at night. Based on the performed statistical analysis it was shown that the STRAT algorithm reconstructs expected MLH dynamics during the day, with maximums in the late afternoon. On the other hand, the Jenoptik method rarely reports MLH values in daytime, which embarrasses the tracking of the MLH diurnal evolution.

Acknowledgments: This study would not be possible without TOPROF—European COST action ES1303 and the advices and recommendations of all TOPROF members. The author is also grateful to NOAA's National Centers for Environmental Information for providing the IGRA. Acknowledgements are due to all contributors to the **R** project. The author would also like to thank anonymous readers whose valuable comments and corrections significantly improved paper quality.

References

1.   Seibert, P.; Beyrich, F.; Gryning, S.E.; Joffre, S.; Rasmussen, A.; Tercier, P. Review and intercomparison of operational methods for the determination of the mixing height. *Atmos. Environ.* **2000**, *34*, 1001–1027. [CrossRef]

2.   Momen, M.; Bou-Zeid, E. Analytical reduced models for the non-stationary diabatic atmospheric boundary layer. *Bound.-Layer Meteorol.* **2017**, *164*, 383–399. [CrossRef]

3.   Schäfer, K.; Emeis, S.; Hoffmann, H.; Jahn, C. Influence of mixing layer height upon air pollution in urban and sub-urban areas. *Meteorol. Z.* **2006**, *15*, 647–658. [CrossRef]

4.   Schäfer, K.; Wagner, P.; Emeis, S.; Jahn, C.; Muenkel, C.; Suppan, P. Mixing layer height and air pollution levels in urban area. In *Proceedings of SPIE—Remote Sensing of Clouds and the Atmosphere XVII*; SPIE: Bellingham, WA, USA, 2012; Volume 8534, p. 853409. [CrossRef]

5.   Yuan, J.; Bu, L.; Huang, X.; Gao, H.; Sa, R. Particulate Characteristics during a Haze Episode Based on Two Ceilometers with Different Wavelengths. *Atmosphere* **2016**, *7*, 20. [CrossRef]

6.   Zang, Z.; Wang, W.; Cheng, X.; Yang, B.; Pan, X.; You, W. Effects of Boundary Layer Height on the Model of Ground-Level PM2.5 Concentrations from AOD: Comparison of Stable and Convective Boundary Layer Heights from Different Methods. *Atmosphere* **2017**, *8*, 104. [CrossRef]

7.   Geiß, A.; Wiegner, M.; Bonn, B.; Schäfer, K.; Forkel, R.; Schneidemesser, E.v.; Münkel, C.; Chan, K.L.; Nothard, R. Mixing layer height as an indicator for urban air quality? *Atmos. Meas. Tech.* **2017**, *10*, 2969–2988. [CrossRef]

8. Mues, A.; Rupakheti, M.; Münkel, C.; Lauer, A.; Bozem, H.; Hoor, P.; Butler, T.; Lawrence, M.G. Investigation of the mixing layer height derived from ceilometer measurements in the Kathmandu Valley and implications for local air quality. *Atmos. Chem. Phys.* **2017**, *17*, 8157–8176. [CrossRef]

9. Zeng, S.; Zhang, Y. The Effect of Meteorological Elements on Continuing Heavy Air Pollution: A Case Study in the Chengdu Area during the 2014 Spring Festival. *Atmosphere* **2017**, *8*, 71. [CrossRef]

10. Kotthaus, S.; Halios, C.H.; Barlow, J.F.; Grimmond, C. Volume for pollution dispersion: London's atmospheric boundary layer during ClearfLo observed with two ground-based lidar types. *Atmos. Environ.* **2018**, *190*, 401–414. [CrossRef]

11. Li, X.X.; Britter, R.E.; Norford, L.K.; Koh, T.Y.; Entekhabi, D. Flow and pollutant transport in urban street canyons of different aspect ratios with ground heating: Large-eddy simulation. *Bound.-Layer Meteorol.* **2012**, *142*, 289–304. [CrossRef]

12. Baklanov, A. The mixing height in urban areas—A review. *Mix. Heights Invers. Urban Areas COST Action* **2002**, *715*, 9–28.

13. Schäfer, K.; Emeis, S.; Jahn, C.; Münkel, C.; Schrader, S.; Höß, M. New results from continuous mixing layer height monitoring in urban atmosphere. In *Proceedings of SPIE—Remote Sensing of Clouds and the Atmosphere XIII*; SPIE: Bellingham, WA, USA, 2008; Volume 7107, p. 71070A. [CrossRef]

14. Vishnu, R.; Kumar, Y.B.; Sinha, P.R.; Rao, T.N.; Samuel, E.J.J.; Kumar, P. Comparison of mixing layer heights determined using LiDAR, radiosonde, and numerical weather prediction model at a rural site in southern India. *Int. J. Remote Sens.* **2017**, *38*, 6366–6385. [CrossRef]

15. Banks, R.F.; Tiana-Alsina, J.; Rocadenbosch, F.; Baldasano, J.M. Performance evaluation of the boundary-layer height from lidar and the Weather Research and Forecasting model at an urban coastal site in the north-east Iberian Peninsula. *Bound.-Layer Meteorol.* **2015**, *157*, 265–292. [CrossRef]

16. Kumer, V.M.; Reuder, J.; Dorninger, M.; Zauner, R.; Grubišić, V. Turbulent kinetic energy estimates from profiling wind LiDAR measurements and their potential for wind energy applications. *Renew. Energy* **2016**, *99*, 898–910. [CrossRef]

17. O'Connor, E.J.; Illingworth, A.J.; Brooks, I.M.; Westbrook, C.D.; Hogan, R.J.; Davies, F.; Brooks, B.J. A method for estimating the turbulent kinetic energy dissipation rate from a vertically pointing Doppler lidar, and independent evaluation from balloon-borne in situ measurements. *J. Atmos. Ocean. Technol.* **2010**, *27*, 1652–1664. [CrossRef]

18. Schween, J.; Hirsikko, A.; Löhnert, U.; Crewell, S. Mixing-layer height retrieval with ceilometer and Doppler lidar: From case studies to long-term assessment. *Atmos. Meas. Tech.* **2014**, *7*, 3685–3704. [CrossRef]

19. Lokoshchenko, M.A. Long-term sodar observations in Moscow and a new approach to potential mixing determination by radiosonde data. *J. Atmos. Ocean. Technol.* **2002**, *19*, 1151–1162. [CrossRef]

20. Emeis, S.; Türk, M. Frequency distributions of the mixing height over an urban area from SODAR data. *Meteorol. Z.* **2004**, *13*, 361–367. [CrossRef]

21. Stull, R.B. *An Introduction to Boundary Layer Meteorology*; Springer Science & Business Media: Berlin, Germany, 1988. [CrossRef]

22. Sicard, M.; Perez, C.; Comeren, A.; Baldasano, J.M.; Rocadenbosch, F. Determination of the mixing layer height from regular lidar measurements in the Barcelona area. In *Remote Sensing of Clouds and the Atmosphere VIII*; SPIE: Bellingham, WA, USA, 2004; Volume 5235, pp. 505–517. [CrossRef]

23. Cimini, D.; De Angelis, F.; Dupont, J.C.; Pal, S.; Haeffelin, M. Mixing layer height retrievals by multichannel microwave radiometer observations. *Atmos. Meas. Tech.* **2012**, *6*, 2941–2951. [CrossRef]

24. Holzworth, G.C. Estimates of mean maximum mixing depths in the contiguous United States. *Mon. Weather Rev.* **1964**, *92*, 235–242. [CrossRef]

25. Wang, X.; Wang, K. Estimation of atmospheric mixing layer height from radiosonde data. *Atmos. Meas. Tech.* **2014**, *7*, 1701–1709. [CrossRef]

26. Hennemuth, B.; Lammert, A. Determination of the atmospheric boundary layer height from radiosonde and lidar backscatter. *Bound.-Layer Meteorol.* **2006**, *120*, 181–200. [CrossRef]

27. Wiegner, M.; Emeis, S.; Freudenthaler, V.; Heese, B.; Junkermann, W.; Münkel, C.; Schäfer, K.; Seefeldner, M.; Vogt, S. Mixing layer height over Munich, Germany: Variability and comparisons of different methodologies. *J. Geophys. Res. Atmos.* **2006**, *111*. [CrossRef]

28. Emeis, S.; Schäfer, K.; Münkel, C. Surface-based remote sensing of the mixing-layer height—A review. *Meteorol. Z.* **2008**, *17*, 621–630. [CrossRef] [PubMed]

29. Beyrich, F.; Görsdorf, U. Composing the diurnal cycle of mixing height from simultaneous sodar and wind profiler measurements. *Bound.-Layer Meteorol.* **1995**, *76*, 387–394. [CrossRef]

30. Winker, D.M.; Vaughan, M.A.; Omar, A.; Hu, Y.; Powell, K.A.; Liu, Z.; Hunt, W.H.; Young, S.A. Overview of the CALIPSO mission and CALIOP data processing algorithms. *J. Atmos. Ocean. Technol.* **2009**, *26*, 2310–2323. [CrossRef]

31. Kaufman, Y.J.; Tanré, D.; Remer, L.A.; Vermote, E.; Chu, A.; Holben, B. Operational remote sensing of tropospheric aerosol over land from EOS moderate resolution imaging spectroradiometer. *J. Geophys. Res. Atmos.* **1997**, *102*, 17051–17067. [CrossRef]

32. Luo, T.; Yuan, R.; Wang, Z. Lidar-based remote sensing of atmospheric boundary layer height over land and ocean. *Atmos. Meas. Tech.* **2014**, *7*, 173–182. [CrossRef]

33. Zhang, W.; Guo, J.; Miao, Y.; Liu, H.; Zhang, Y.; Li, Z.; Zhai, P. Planetary boundary layer height from CALIOP compared to radiosonde over China. *Atmos. Chem. Phys.* **2016**, *16*, 9951–9963. [CrossRef]

34. Liu, Z.; Mortier, A.; Li, Z.; Hou, W.; Goloub, P.; Lv, Y.; Chen, X.; Li, D.; Li, K.; Xie, Y. Improving Daytime Planetary Boundary Layer Height Determination from CALIOP: Validation Based on Ground-Based Lidar Station. *Adv. Meteorol.* **2017**, *2017*, 5759074. [CrossRef]

35. Feng, X.; Wu, B.; Yan, N. A method for deriving the boundary layer mixing height from modis atmospheric profile data. *Atmosphere* **2015**, *6*, 1346–1361. [CrossRef]

36. Sokolovskiy, S.; Kuo, Y.H.; Rocken, C.; Schreiner, W.; Hunt, D.; Anthes, R. Monitoring the atmospheric boundary layer by GPS radio occultation signals recorded in the open-loop mode. *Geophys. Res. Lett.* **2006**, *33*. [CrossRef]

37. Guo, P.; Kuo, Y.H.; Sokolovskiy, S.; Lenschow, D. Estimating atmospheric boundary layer depth using COSMIC radio occultation data. *J. Atmos. Sci.* **2011**, *68*, 1703–1713. [CrossRef]

38. García-Franco, J.; Stremme, W.; Bezanilla, A.; Ruiz-Angulo, A.; Grutter, M. Variability of the Mixed-Layer Height Over Mexico City. *Bound.-Layer Meteorol.* **2018**, *167*, 493–507. [CrossRef]

39. Knepp, T.N.; Szykman, J.J.; Long, R.; Duvall, R.M.; Krug, J.; Beaver, M.; Cavender, K.; Kronmiller, K.; Wheeler, M.; Delgado, R.; et al. Assessment of mixed-layer height estimation from single-wavelength ceilometer profiles. *Atmos. Meas. Tech.* **2017**, *10*, 3963. [CrossRef] [PubMed]

40. Peng, J.; Grimmond, C.S.B.; Fu, X.; Chang, Y.; Zhang, G.; Guo, J.; Tang, C.; Gao, J.; Xu, X.; Tan, J. Ceilometer-Based Analysis of Shanghai's Boundary Layer Height (under Rain-and Fog-Free Conditions). *J. Atmos. Ocean. Technol.* **2017**, *34*, 749–764. [CrossRef]

41. Nemuc, A.; Nicolae, D.; Talianu, C.; Carstea, E.; Radu, C. Dynamic of the lower troposphere from multiwavelength LIDAR measurements. *Roman. Rep. Phys.* **2009**, *61*, 313–323.

42. Ungureanu, I.; Stefan, S.; Nicolae, D. Investigation of the cloud cover and planetary boundary layer (PBL) characteristics using ceilometer CL-31. *Roman. Rep. Phys.* **2010**, *62*, 396–404.

43. Wang, W.; Gong, W.; Mao, F.; Pan, Z. An improved iterative fitting method to estimate nocturnal residual layer height. *Atmosphere* **2016**, *7*, 106. [CrossRef]

44. Li, H.; Yang, Y.; Hu, X.M.; Huang, Z.; Wang, G.; Zhang, B. Application of Convective Condensation Level Limiter in Convective Boundary Layer Height Retrieval Based on Lidar Data. *Atmosphere* **2017**, *8*, 79. [CrossRef]

45. Caicedo, V.; Rappenglück, B.; Lefer, B.; Morris, G.; Toledo, D.; Delgado, R. Comparison of aerosol lidar retrieval methods for boundary layer height detection using ceilometer aerosol backscatter data. *Atmos. Meas. Tech.* **2017**, *10*, 1609–1622. [CrossRef]

46. Illingworth, A. TOPROF (COST Action ES1303)—Towards Operational Ground Based Profiling with Ceilometers, Doppler Lidars and Microwave Radiometers for Improving Weather Forecasts. Available online: http://www.toprof.imaa.cnr.it/ (accessed on 30 November 2018).

47. EUMETNET Composite Observing System—E-PROFILE. Available online: http://eumetnet.eu/activities/observations-programme/current-activities/e-profile/ (accessed on 30 November 2018).

48. Haeffelin, M.; Angelini, F.; Morille, Y.; Martucci, G.; Frey, S.; Gobbi, G.; Lolli, S.; O'dowd, C.; Sauvage, L.; Xueref-Rémy, I.; et al. Evaluation of mixing-height retrievals from automatic profiling lidars and ceilometers in view of future integrated networks in Europe. *Bound.-Layer Meteorol.* **2012** *143*, 49–75. [CrossRef]

49. Münkel, C.; Eresmaa, N.; Räsänen, J.; Karppinen, A. Retrieval of mixing height and dust concentration with lidar ceilometer. *Bound.-Layer Meteorol.* **2007**, *124*, 117–128. [CrossRef]

50. Uzan, L.; Egert, S.; Alpert, P. Ceilometer evaluation of the eastern Mediterranean summer boundary layer height—First study of two Israeli sites. *Atmos. Meas. Tech.* **2016**, *9*, 4387–4398. [CrossRef]

51. Stachlewska, I.; Piądłowski, M.; Migacz, S.; Szkop, A.; Zielińska, A.; Swaczyna, P. Ceilometer observations of the boundary layer over Warsaw, Poland. *Acta Geophys.* **2012**, *60*, 1386–1412. [CrossRef]

52. Morille, Y.; Haeffelin, M.; Drobinski, P.; Pelon, J. STRAT: An automated algorithm to retrieve the vertical structure of the atmosphere from single-channel lidar data. *J. Atmos. Ocean. Technol.* **2007**, *24*, 761–775. [CrossRef]

53. Poltera, Y.; Martucci, G.; Collaud Coen, M.; Hervo, M.; Emmenegger, L.; Henne, S.; Brunner, D.; Haefele, A. PathfinderTURB: An automatic boundary layer algorithm. Development, validation and application to study the impact on in-situ measurements at the Jungfraujoch. *Atmos. Chem. Phys. Discuss.* **2017**. [CrossRef]

54. Kotthaus, S.; Grimmond, C.S.B. Atmospheric boundary-layer characteristics from ceilometer measurements. Part 1: A new method to track mixed layer height and classify clouds. *Q. J. R. Meteorol. Soc.* **2018**, *144*, 1525–1538. [CrossRef]

55. Griffiths, A.; Parkes, S.; Chambers, S.; McCabe, M.; Williams, A. Improved mixing height monitoring through a combination of lidar and radon measurements. *Atmos. Meas. Tech.* **2013**, *6*, 207–218. [CrossRef]

56. Galeriu, D.; Melintescu, A.; Stochioiu, A.; Nicolae, D.; Balin, I. Radon, as a tracer for mixing height dynamics—An overview and RADO perspectives. *Roman. Rep. Phys.* **2011**, *63*, 115–127.

57. Heese, B.; Flentje, H.; Althausen, D.; Ansmann, A.; Frey, S. Ceilometer lidar comparison: Backscatter coefficient retrieval and signal-to-noise ratio determination. *Atmos. Meas. Tech.* **2010**, *3*, 1763–1770. [CrossRef]

58. Jenoptik. *Cloud Height Meter CHM 15k—User Manual*; JENOPTIK Laser, Optical Systems GmbH: Jena, Germany, 2009.

59. Durre, I.; Vose, R.S.; Wuertz, D.B. Overview of the integrated global radiosonde archive. *J. Clim.* **2006**, *19*, 53–68. [CrossRef]

60. De Haij, M.; Wauben, W.; Baltink, H.K. *Continuous Mixing Layer Height Determination Using the LD-40 Ceilometer: A Feasibility Study*; Royal Netherlands Meteorological Institute (KNMI): De Bilt, The Netherlands, 2007.

61. Vogelezang, D.; Holtslag, A. Evaluation and model impacts of alternative boundary-layer height formulations. *Bound.-Layer Meteorol.* **1996**, *81*, 245–269. [CrossRef]

62. Menut, L.; Flamant, C.; Pelon, J.; Flamant, P.H. Urban boundary-layer height determination from lidar measurements over the Paris area. *Appl. Opt.* **1999**, *38*, 945–954. [CrossRef] [PubMed]

63. Sicard, M.; Pérez, C.; Rocadenbosch, F.; Baldasano, J.; García-Vizcaino, D. Mixed-layer depth determination in the Barcelona coastal area from regular lidar measurements: Methods, results and limitations. *Bound.-Layer Meteorol.* **2006**, *119*, 135–157. [CrossRef]

64. Holzworth, G.C. Mixing depths, wind speeds and air pollution potential for selected locations in the United States. *J. Appl. Meteorol.* **1967**, *6*, 1039–1044. [CrossRef]

65. Garrett, A. Comparison of Observed Mixed-Layer Depths to Model Estimates Using Observed Temperatures and Winds, and MOS Forecasts. *J. Appl. Meteorol.* **1981**, *20*, 1277–1283. [CrossRef]

66. Dang, R.; Li, H.; Liu, Z.; Yang, Y. Statistical analysis of relationship between daytime Lidar-derived planetary boundary layer height and relevant atmospheric variables in the semiarid region in Northwest China. *Adv. Meteorol.* **2016**, *2016*. [CrossRef]

67. Lotteraner, C.; Piringer, M. Mixing-height time series from operational ceilometer aerosol-layer heights. *Bound.-Layer Meteorol.* **2016**, *161*, 265–287. [CrossRef]

68. de Bruine, M.; Apituley, A.; Donovan, D.P.; Klein Baltink, H.; de Haij, M.J. Pathfinder: Applying graph theory to consistent tracking of daytime mixed layer height with backscatter lidar. *Atmos. Meas. Tech.* **2017**, *10*, 1893–1909. [CrossRef]

7

Investigating the Effect of Different Meteorological Conditions on MAX-DOAS Observations of NO_2 and CHOCHO in Hefei, China

Zeeshan Javed [1], Cheng Liu [1,2,3,4,*], Kalim Ullah [5,*], Wei Tan [3], Chengzhi Xing [1] and Haoran Liu [1]

[1] School of Earth and Space Sciences, University of Science and Technology of China, Hefei 230026, China
[2] Key Lab of Environmental Optics & Technology, Anhui Institute of Optics and Fine Mechanics, Chinese Academy of Sciences, Hefei 230031, China
[3] Center for Excellence in Regional Atmospheric Environment, Institute of Urban Environment, Chinese Academy of Sciences, Xiamen 361021, China
[4] Anhui Province Key Laboratory of Polar Environment and Global Change, USTC, Hefei 230026, China
[5] Department of Meteorology, COMSATS University Islamabad, Islamabad 44000, Pakistan
* Correspondence: chliu81@ustc.edu.cn (C.L.); kalim_ullah@comsats.edu.pk (K.U.);

Abstract: In this work, a ground-based remote sensing instrument was used for observation of the trace gases NO_2 and CHOCHO in Hefei, China. Excessive development and rapid economic growth over the years have resulted in the compromising of air quality in this city, with haze being the most prominent environmental problem. This is first study covering observation of CHOCHO in Hefei ($31.783°$ N, $117.201°$ E). The observation period of this study, i.e., July 2018 to December 2018, is divided into three different categories: (1) clear days, (2) haze days, and (3) severe haze days. The quality of the differential optical absorption spectroscopy (DOAS) fit for both CHOCHO and NO_2 was low during severe haze days due to a reduced signal to noise ratio. NO_2 and CHOCHO showed positive correlations with PM2.5, producing R values of 0.95 and 0.98, respectively. NO_2 showed strong negative correlations with visibility and air temperature, obtaining R values of 0.97 and 0.98, respectively. CHOCHO also exhibited strong negative correlations with temperature and visibility, displaying R values of 0.83 and 0.91, respectively. The average concentration of NO_2, CHOCHO, and PM2.5 during haze days was larger compared to that of clear days. Diurnal variation of both CHOCHO and NO_2 showed a significant decreasing trend in the afternoons during clear days due to photolysis, while during haze days these two gases started to accumulate as their residence time increases in the absence of photolysis. There was no prominent weekly cycle for both trace gases.

Keywords: MAX-DOAS; CHOCHO; NO_2; meteorological conditions; haze days

1. Introduction

China has been ranked near the bottom of the Global Environmental Sustainability Index due to its persistent and extensive air pollution [1]. Widespread air pollution in China can be attributed to a variety of factors: high output from production and manufacturing, an extensive rise in the number of automobiles, a massive economic boom, and high population growth. Poor air quality results in the degradation of human health as well as negative impacts on terrestrial ecosystems and the built environment.

Nitrogen dioxide (NO_2) and glyoxal (CHOCHO) have significant roles in varying the chemistry of the troposphere [2]. Glyoxal may originate from natural as well as anthropogenic activities [3–5]. The molecule is the smallest alpha-dicarbonyl with the highest predominance in the troposphere [6]. Glyoxal is generated as an intermediate product in most volatile organic compound (VOC) oxidation cycles [5,7,8] and acts as an indicator for secondary organic aerosol (SOA) formation in the atmosphere [4,7,9]. In addition, CHOCHO is also produced as an oxidation product for alkyne, isoprene, and various aromatic hydrocarbons. The concentrations of CHOCHO are not directly impacted by vehicular discharges [5] because they are only in fractions.

The formation of tropospheric ozone and the destruction of stratospheric ozone are catalyzed by nitrogen dioxide (NO_2) along with other species in the troposphere [10]. The oxidation of nitrogen dioxide forms nitric acid under favorable circumstances in the atmosphere [11]. It increases the risk and incidence of infections related to the respiratory tract [12]. Combustion of fossil fuels in urban settlements has been regarded as the main source of nitrogen dioxide [13]. Additionally, nitrogen dioxide may act as a precursor for the formation of aerosols under certain meteorological conditions. Nitrate aerosols constitute a significant proportion of fine particulates in the urban environment.

In the lower atmosphere, a substantial driving factor towards pollutant distribution in terms of chemical behavior and residence time is believed to be the meteorological condition of the locality. There exist several pieces of evidence in the literature that highlight the effect of various meteorological factors on the distribution of trace gases and aerosols in the atmosphere [14]. The overall tropospheric profile including gaseous pollutants and a variety of particles along with their meteorological parameters present a dynamic and multidimensional picture of the atmosphere, aiding towards a better understanding of pollutant sources, features, and sinks, as well as their dependence on weather conditions.

Multi-axis differential optical absorption spectroscopy (MAX-DOAS) is a type of passive spectroscopy system that has been in use to observe tropospheric trace gases over the past decade. There have been some reported studies on monitoring air quality in China using MAX-DOAS instruments. Long-term observations have resulted in the demonstration of monthly, weekly, and diurnal cycles of SO_2, NO_2, and HCHO in China, and these long-term based observations have been published in Ma et al. [15], Wang et al. [16], and Hendrick et al. [17]. Concentrations of tropospheric nitrogen dioxide were obtained using MAX-DOAS mounted at four different locations in Shanghai in 2010 during the Shanghai World Expo [18]. A MAX-DOAS system was deployed in the Eastern part of China to monitor NO_2, HCHO, and SO_2 in the Yangtze River Delta (YRD) region. Shanghai, Hefei, and Nanjing were designated for long-term monitoring. Vertical profiles and vertical column densities (VCDs) of tropospheric trace gases were monitored. Minimum values of SO_2 and NO_2 were observed at noon, whereas maxima for HCHO occurred during noontime. The broad-spectrum concentration of the pollutants was observed to gradually decrease from Shanghai to Hefei [19]. The effect of haze and non-haze conditions on the retrieval of CHOCHO from MAX-DOAS observations in Beijing has been investigated [20].

The city of Hefei, which supports a population of 683 people/km^2, is the capital city of Anhui Province. Excessive development and rapid economic growth over the years have resulted in the compromising of air quality in this city, with haze being the most prominent environmental problem. The prevalence of haze has dramatically impacted public health and has cast negative impressions on transportation networks and production systems. The city usually witnesses hazy days during the autumn and winter months of the year. This can be understood by taking into account the crop burning practices which occur during these months.

In addition, the location of the city makes it more prone to cold and dry air, with relatively gentle winds resulting in less chance for dust and other particulates to be diffused, causing the formation of haze.

Keeping in mind the aforementioned facts, Hefei may be considered a potential site for the observation of pollutants like NO_2 and CHOCHO. However, there have been no studies which have reported on the monitoring of CHOCHO and effects of different meteorological conditions on the retrieval of NO_2 and CHOCHO.

In this study we primarily focused on retrieval of CHOCHO and NO_2 from MAX-DOAS observations. Time series and weekly and diurnal cycles of VCDs of NO_2 and CHOCHO were generated. The dependence of CHOCHO and NO_2 VCDs on different meteorological parameters is discussed. As a quick overview of the arguments discussed above, the current research claims the unique contribution of monitoring the air quality of Hefei, China.

2. Materials and Methods

2.1. Instrument

Ground-based observations of several trace gases were carried out using a remote sensing MAX-DOAS instrument. The instrument contains a spectrometer which has a resolution of 0.6 nm and a spectral range of 300–490 nm. The major source of light employed by the instrument is dispersed sunlight, while spectral observations can be taken at different viewing angles depending on the concerned atmospheric species [21]. An inbuilt processor automatically controls the day-to-day measurements of the instrument. The spectral measurements were recorded at elevation viewing angles of 1°, 2°, 3°, 4°, 5°, 6°, 8°, 10°, 15°, 30°, and 90°. The average time span for single measurement was 60 s.

2.2. Observation Site

Anhui Province is situated in Eastern Central China with Hefei (31.783° N, 117.201° E) as the capital. The climate of the region is subtropical and humid with the existence of four distinct seasons: summer, fall, winter, and spring. The locality is dominated by southeasterly winds during summer and northwesterly during the winter season. Following a general trend, Hefei has experienced a boom in growth and development over the past two decades. At present, the city supports a permanent population of about 7.7 million. For the current study, the MAX-DOAS instrument was fitted on the building of the Hefei Environmental Protection Bureau. The observation site was almost in the center of city. Figure 1 shows the location of the observation site for MAX-DOAS measurements in Hefei, China.

Figure 1. Location of Hefei observation site for multi-axis differential optical absorption spectroscopy (MAX-DOAS) measurements in China.

2.3. DOAS Analysis

The application of DOAS provides differential slant column densities (dSCDs) for a variety of trace gases [22]. The current study analyzes MAX-DOAS spectra by employing QDOAS software developed by the Royal Belgian Institute for Space Aeronomy (BIRA-IASB) [23]. The spectra were employed to correct the measurement spectra prior to further analysis. A variety of absorption cross sections for trace gas [24–28], a low order polynomial, a Fraunhofer reference spectrum, and a Ring spectrum were included in the DOAS fit. Comprehensive details for the DOAS fit settings have been depicted in Table 1.

Table 1. The parameter settings used in MAX-DOAS observations.

Parameter	Data Source	Trace Gases	
		NO$_2$	CHOCHO
Wavelength		425–490 (nm)	438–457 (nm)
NO$_2$	298 K, Vandaele et al. [24]	✓	✓
NO$_2$	220 K, Vandaele et al. [25]	✓	✓
O$_3$	223K, Serdyuchenko et al. [26]	✓	✓
CHOCHO	296K, Volkamer et al. [8]	x	✓
O$_4$	293 K, Thalman and Volkamer [27]	✓	✓
H$_2$O	296 K, HITEMP (High-temperature spectroscopic absorption parameters), Rothman et al. [28]	✓	✓
Ring	Calculated with QDOAS	✓	✓
Polynomial degree		5	5

Figure 2 displays a characteristic DOAS spectral fitting of the spectrum obtained at a viewing angle of 30° on 11 November 2018. The fitting shows obvious absorption structures and low residuals, revealing that the spectral fitting is of good quality.

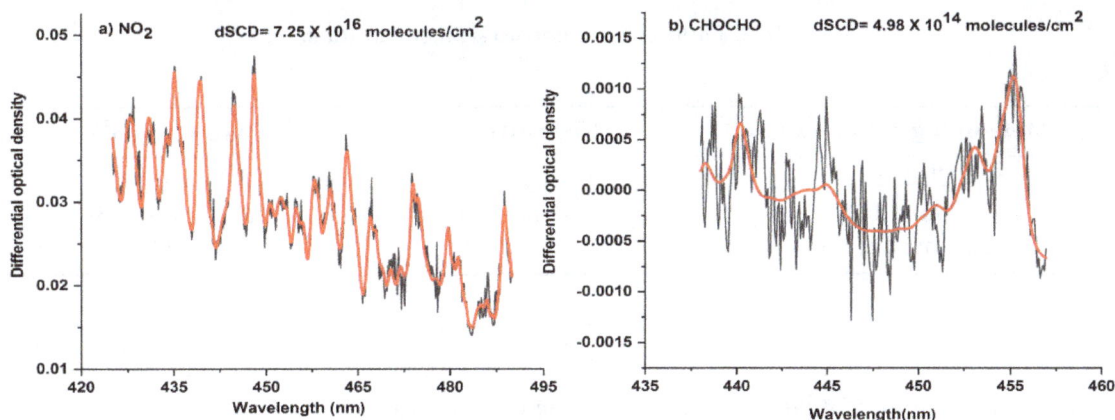

Figure 2. Characteristic DOAS spectral fitting of the spectrum obtained at a viewing angle of 30° on 11 November 2018 for (**a**) NO$_2$ and (**b**) CHOCHO. Legend: dSCD, differential slant column density.

Vertical column densities were generated using an air mass factor (AMF) [29]. For the current study, a differential air mass factor (dAMF) was applied.

$$VCD_{trop} = \frac{dSCD_\alpha}{dAMF_\alpha} \tag{1}$$

The difference in AMF between $\alpha \neq 90°$ and $\alpha = 90°$ is actually referred to as dAMF.

$$dAMF_\alpha = AMF_\alpha - AMF_{90°} \tag{2}$$

$$VCD_{trop} = \frac{dSCD_\alpha}{AMF_\alpha - AMF_{90°}} \tag{3}$$

AMF can be estimated using a geometrical approximation approach [30,31].

$$AMF = \frac{1}{\sin(\alpha)} \tag{4}$$

Equation (4) then becomes:

$$VCD_{trop} = \frac{dSCD_\alpha}{1/sin(\alpha) - 1} \tag{5}$$

3. Results and Discussion

3.1. Meteorological Conditions

The observation period of this study, i.e., July 2018 to December 2018, has been divided into three different categories of meteorological conditions based on meteorological parameters. These three categories are (1) clear days, (2) haze days, and (3) severe haze days. The clear days are days with visibility greater than 10 km and a PM2.5 concentration less than 70 $\mu g/m^3$. The haze days are days with visibility less than 10 km and greater than 5 km and where the PM2.5 concentration is greater than 70 $\mu g/m^3$ and less than 115 $\mu g/m^3$. The severe haze days are days with visibility less than 5 km and a PM2.5 concentration greater than 115 $\mu g/m^3$ [32,33]. Table 2 shows a summary of different meteorological conditions. The data for meteorological parameters like temperature, humidity, and visibility was downloaded from (http://www.wunderground.com/). The data for meteorological parameters was obtained from a weather station installed at Hefei airport. The data for PM2.5 was downloaded from (http://beijingair.sinaapp.com/).

Table 2. Different meteorological conditions.

Meteorological Condition	Visibility (km)	PM2.5 ($\mu g/m^3$)
Clear days	>10	<70
Haze days	>5 and <10	>70 and <115
Severe haze days	<5	>115

3.2. Impact of Meteorological Conditions on DOAS Fit of CHOCHO and NO$_2$

The measurements of MAX-DOAS in the ultraviolet as well as the visible spectral range largely rely on the intensity of sunlight. The excellence of the DOAS fit is characterized by the structure of the residual left after subtracting numerous absorbers using a numerical least square fitting method [22,34] which is usually expressed using the root mean square (RMS). It is a measure of mean "instrument error" which is largely subjected to many specific parameters like limitations of the instruments (such as dark current and spectral resolution, etc.), along with limitations in illustrating the actual state of the atmosphere (for instance aerosols and scattering processes, etc.).

Figure 3 shows the average RMS and dSCD errors during different weather conditions. A higher RMS is observed during severe haze and haze days while on clear days the RMS is low. A similar trend is observed for the dSCD error with higher values during severe haze and haze days and lower values during clear days. These results can be related to the fact that during haze days the intensity of light is low, resulting in a lower signal to noise ratio. The quality of the DOAS fit is affected during severe haze days, which can be observed from the RMS and dSCD errors for both CHOCHO and NO$_2$. This can result in an underestimation of CHOCHO and NO$_2$ levels. By contrast, the RMS and dSCD errors during haze and clear days are in an acceptable range and do not have any significant effect on the levels of CHOCHO and NO$_2$.

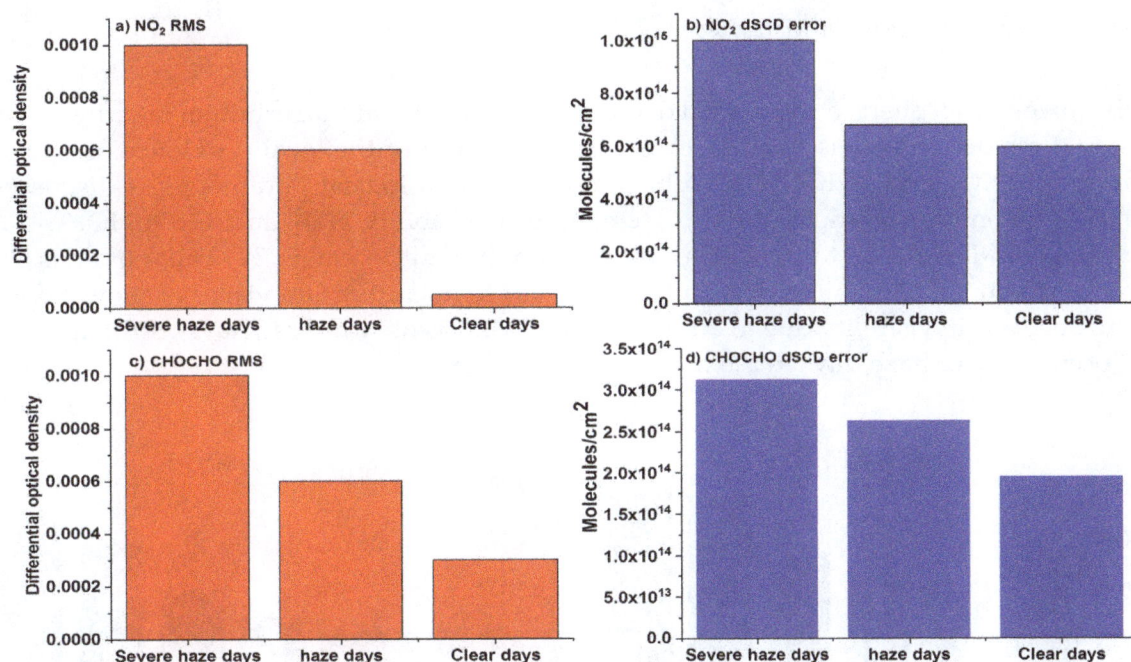

Figure 3. (**a**) NO$_2$ RMS (**b**) NO$_2$ dSCD error, (**c**) CHOCHO RMS and (**d**) CHOCHO dSCD error under different weather conditions.

3.3. Time Series of NO$_2$, CHOCHO, and Meteorological Parameters

Observations of the trace gases NO$_2$ and CHOCHO were performed from July 2018 to 31 December 2018 in Hefei, China. QDOAS software was used for analysis of the spectrum to obtain dSCDs. VCDs were generated from these dSCDs using a geometric approximation approach. Data for meteorological parameters like temperature, humidity, and visibility was downloaded from (http://www.wunderground.com/). Data for PM2.5 was downloaded from (http://beijingair.sinaapp.com/). This website belongs to the national environmental monitoring network. Figure 4 shows time series of NO$_2$ and CHOCHO.

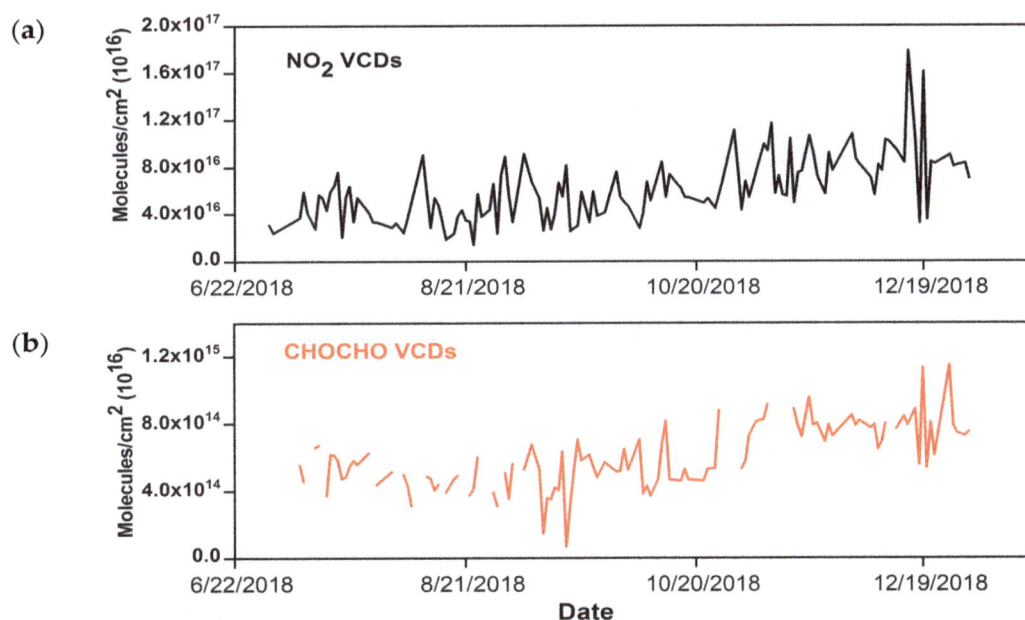

Figure 4. Time series of (**a**) NO$_2$ and (**b**) CHOCHO from 1 July 2018 to 31 December 2018. Legend: VCDs, vertical column densities.

3.4. Dependence of Trace Gases on Meteorological Parameters

In the lower atmosphere, a substantial driving factor of pollutant distribution in terms of chemical behavior and residence time is the meteorological condition of the locality. Hence, the relation of trace gases, i.e., NO_2 and CHOCHO, with different meteorological parameters is discussed here. Figure 5 shows monthly variations of PM2.5, temperature, visibility, NO_2, and CHOCHO. The average concentration of PM2.5 is seen to increase gradually from July to December. A similar trend is observed for the concentration of NO_2 and CHOCHO, whereas air temperature and visibility show a decreasing trend. The decrease in visibility due to an increase in PM2.5 is obviously to have resulted in the more frequent occurrence of haze days in the month of December.

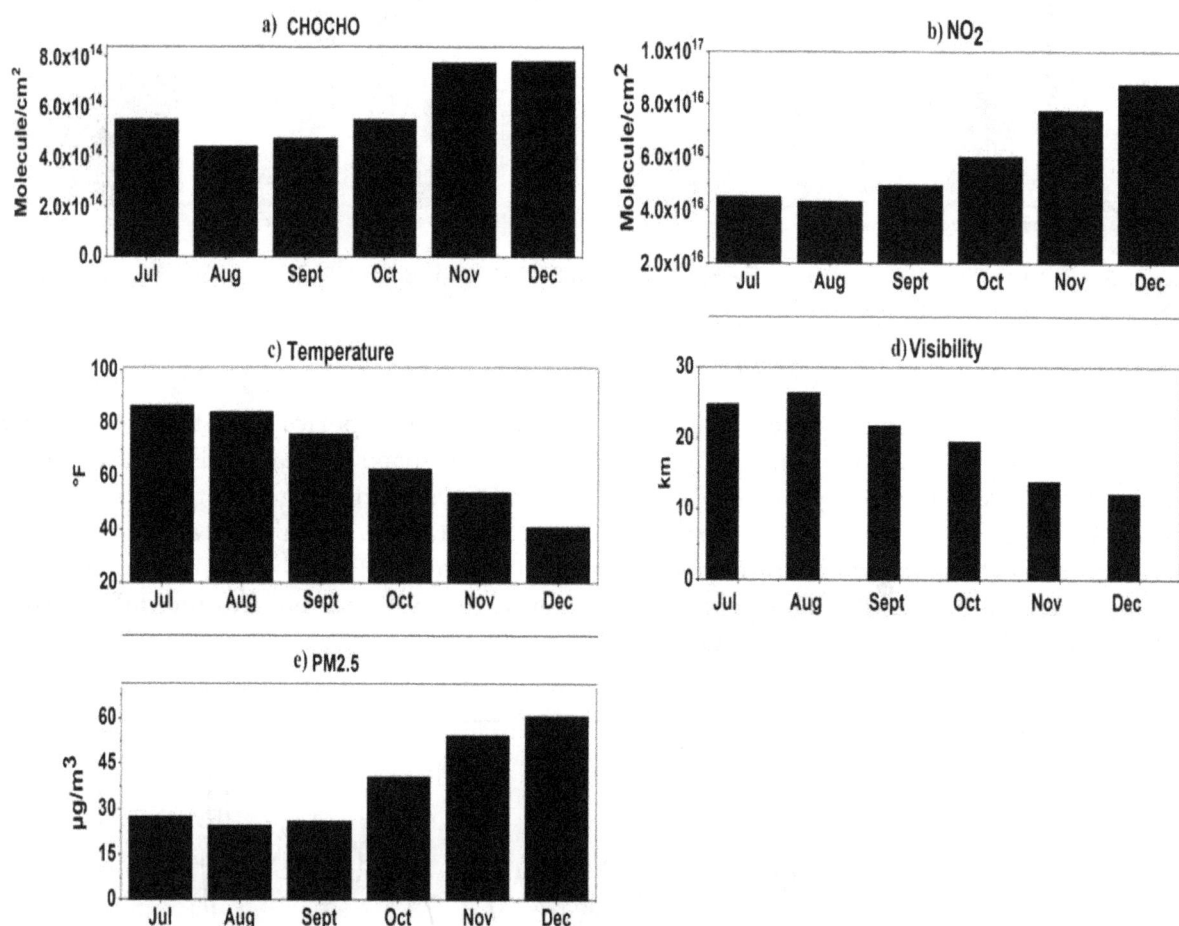

Figure 5. Monthly variations in (**a**) CHOCHO, (**b**) NO_2, (**c**) temperature, (**d**) visibility, and (**e**) PM2.5.

The increasing trend of CHOCHO and NO_2 VCDs can be explained by the fact that photolysis is a major sink of both these trace gases. During the winter months when haze conditions start to occur more frequently due to an increase in PM2.5, visibility decreases. The rate of photolysis decreases, as visibility is low. This results in the accumulation of NO_2 and CHOCHO because their major sink, i.e., the rate of photolysis, is very low. The observed increase in CHOCHO from October may also be related to heating during winter. This finding is consistent with glyoxal retrieved over Beijing and Northern China from OMI (Ozone monitoring instrument) satellite data [35,36]. The increase in the concentration of NO_2 may also be related to an increase in the burning of fossil fuels during the winter months. Figure 6 shows correlation plots of NO_2 and CHOCHO with meteorological parameters.

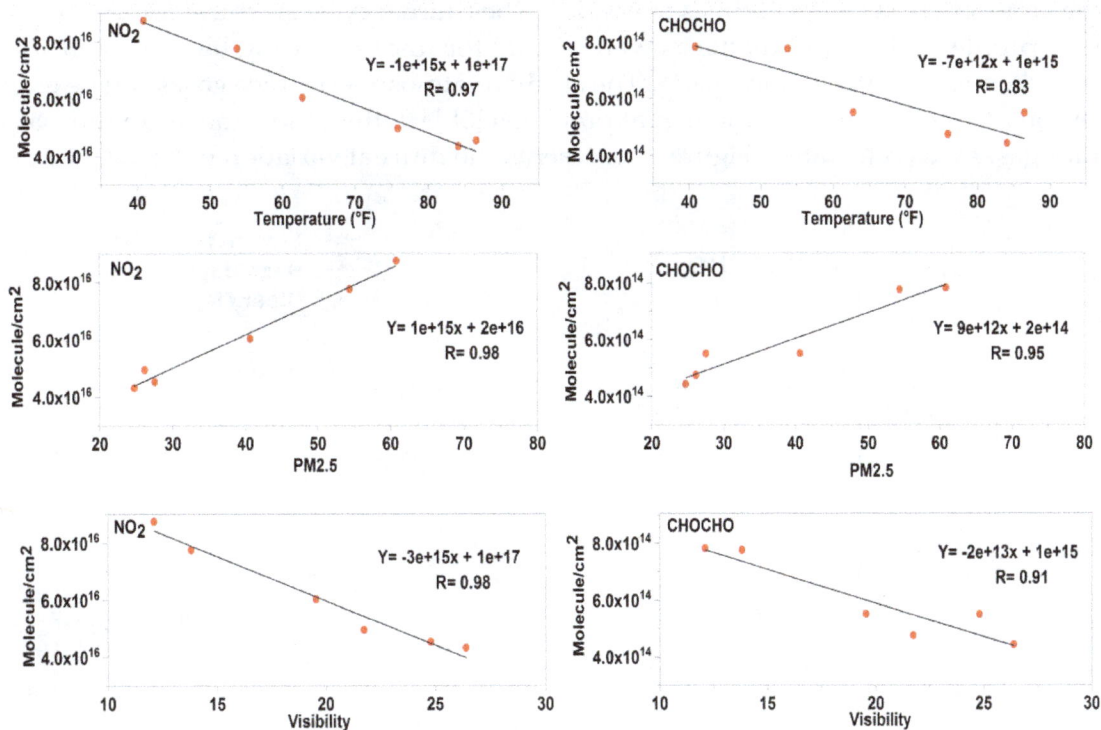

Figure 6. Correlation plots of NO_2 and CHOCHO with PM2.5, temperature, and visibility.

NO_2 shows strong negative correlations with temperature and visibility, producing R values of 0.97 and 0.98, respectively. CHOCHO also exhibits strong negative correlations with temperature and visibility, giving R values of 0.83 and 0.91, respectively. However, both CHOCHO and NO_2 show positive correlations with PM2.5, having R values of 0.95 and 0.98, respectively. Table 3 shows the average concentration of NO_2, CHOCHO, and PM2.5 during clear days, haze days, and severe haze days. It can be observed that the average concentration of NO_2 and CHOCHO is higher during haze days and heavy days compared to clear days, which is due to reduced photolytic activity, with photolytic activity acting as a major sink for NO_2 and CHOCHO. During heavy haze days the values are lower when compared with those of haze days.

The most likely reason for this trend is the fact that the quality of the DOAS fit is affected during severe haze days due to higher RMS and dSCD errors which result in the underestimation of NO_2 and CHOCHO during severe haze days.

Table 3. Average concentrations of NO_2, CHOCHO, PM2.5, and visibility under different meteorological conditions.

Meteorological Condition	NO_2 (molecules/cm^2)	CHOCHO (molecules/cm^2)	Visibility (km)	PM2.5 (μg/m^3)
Clear days	1.92×10^{16}	4.70×10^{14}	22	36
Haze days	4.38×10^{16}	7.62×10^{14}	8.8	103
Severe haze days	3.30×10^{16}	5.25×10^{14}	3.9	151

3.5. Diurnal Variation

Atmospheric trace gases may come from a variety of natural sources of chemical and biological nature as well as from agricultural and industrial practices. The concentration of such gases in the atmosphere and their variation depends upon emission sources, transport routes, and removal mechanisms [37]. In urban centers, the concentration of trace gases is directly or indirectly linked to anthropogenic activities. Specific trends can be seen when looking at the diurnal cycle of these trace gases in the atmosphere at different times of the day. The diurnal cycle is critical for understanding the atmospheric profile of these pollutants as well as vital for understanding the emission sources and atmospheric chemistry of these trace gases. The study of atmospheric trace gases can be employed to address various contemporary environmental problems [38]. In this study, the mean diurnal variation of both trace gases was calculated. Figure 7 represents the diurnal variation of CHOCHO and NO_2.

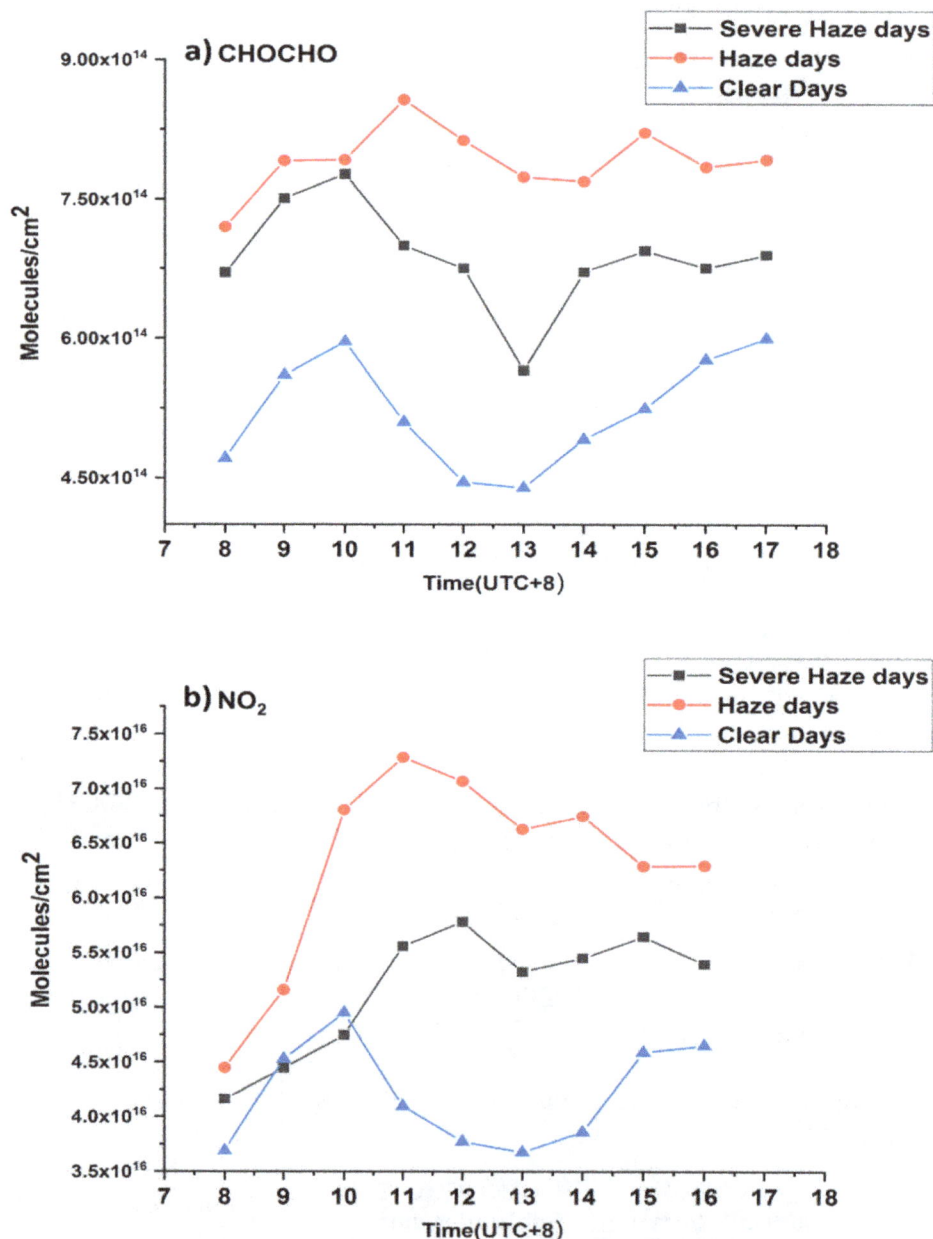

Figure 7. Diurnal cycles for (**a**) CHOCHO and (**b**) NO_2 VCDs monitored in Hefei, China.

Diurnal variation of NO_2 VCDs is observed to follow a typical pattern during clear days. Higher values occur during the morning and evening whereas lower values occur during noontime. In the morning NO_2 levels increase with a peak at around 10 a.m. local time, which can be attributed to increased traffic-related NOx emissions. The lower values during noontime can be attributed to the enhanced rate of photolysis, along with the oxidation of NO_2 by OH radicals. However, this reduction in NO_2 levels is very small because our observation period took place during the winter season, when the rate of photolysis is very low. The level of NO_2 VCDs again starts to rise during the late afternoon because rate of photolysis decreases and NO_2 begins to accumulate as the load of traffic again increases during the evening. It is worth noting that our results are in agreement with other MAX-DOAS measurements for megacities [17–19,39].

For haze and severe haze days there is no reduction in the level of NO_2 during the afternoon. This is obviously because the major sink for NO_2, i.e. photolysis, is absent during these days, which results in a slight increase in residence time of the trace gas. During severe haze days, a lower concentration of NO_2 is observed as compared to haze days. These results may be ascribed to an underestimation of trace gases during severe haze days due to a lower signal to noise ratio.

During clear days, due to a faster photolysis rate, a certain decrease in CHOCHO emissions is observed during the afternoon; haze and heavy haze days do not depict such a trend. CHOCHO concentrations during severe haze days are lower than during haze days, which can be explained by taking into account the fact that DOAS retrieval for CHOCHO is very sensitive and is underestimated during severe haze days.

3.6. Weekly Cycles

Human activities can be categorically divided according to weekly cycles. During weekends, because of reduced industrial activities and transport, the emissions of atmospheric pollutants normally decline [40] in comparison to week days, with peak anthropogenic activities at industrial as well as personal levels (including more use of public and private transport). This factor may have a significant effect on weekly cycles of various atmospheric species. However, in China, because of continuous industrial activity throughout the week and no formal weekly breaks, the weekend effect for various atmospheric species has been observed to be non-significant [15,18,19,41]. Figure 8 shows the results of our weekly cycle. There is no meaningful conclusion which can be drawn from these results. These results are consistent with the findings of previous studies.

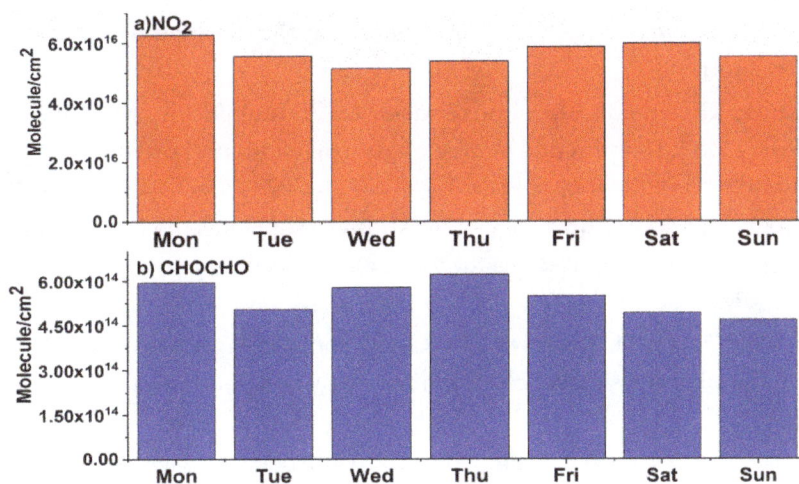

Figure 8. Weekly cycles for (**a**) NO_2 and (**b**) CHOCHO VCDs monitored in Hefei, China.

4. Conclusions

In this work, MAX-DOAS observations for NO_2 and CHOCHO were performed from 1 July 2018 to 31 December 2018 in Hefei, China. Hefei is the capital of Anhui Province and has a population of around 7.80 million. Hefei is becoming a potential site for air quality monitoring. Because of the development and growth of its economy its air quality is deteriorating and haze days are occurring more frequently. There have been fairly sparse studies reporting the monitoring of CHOCHO throughout China. This is first study to observe CHOCHO in Hefei. NO_2 is well known for its significance in tropospheric chemistry. The observation time of our study was divided into three different categories based on meteorological parameters and PM2.5 levels. These categories were named clear days, haze days, and severe haze days. The excellence of the DOAS fit was not good during the severe haze days for both CHOCHO and NO_2. RMS and dSCD errors were higher during severe haze days due to a low signal to noise ratio. This therefore resulted in the underestimation of NO_2 and CHOCHO during severe haze days. NO_2 and CHOCHO showed positive correlations with PM2.5, giving R values of 0.95 and 0.98, respectively. NO_2 showed strong negative correlations with visibility and air temperature, displaying R values of 0.97 and 0.98, respectively. CHOCHO also exhibited strong negative correlations with temperature and visibility, producing R values of 0.83 and 0.91, respectively. These outcomes can be accredited to the fact that photolysis is the main sink for both NO_2 and CHOCHO. An increase in PM2.5 concentration caused a reduction in visibility and hence resulted in the reduction of the rate of photolysis. The average concentrations of NO_2, CHOCHO, and PM2.5 during haze days were higher when compared to those of clear days. Diurnal variations of both CHOCHO and NO_2 showed significant decreasing trends during clear days due to photolysis, while during haze days the trace gases started to accumulate as the residence time of these gases increased. There was no prominent weekly cycle for both trace gases because of continuous industrial activity throughout the week and no formal weekly breaks; the weekend effect for various atmospheric species has been observed to be non-significant.

Author Contributions: Conceptualization, Z.J. and C.L.; formal analysis, Z.J.; methodology, C.L., W.T., and H.L.; resources, C.L.; validation, W.T. and C.X.; writing, review, and editing, Z.J., K.U., and C.L.

Acknowledgments: We are indebted to Aimon Tanvir for their useful communications.

References

1.   Liu, J.; Diamond, J. China's environment in a globalizing world. *Nature.* **2005**, *435*, 1179. [CrossRef] [PubMed]

2.   Hong, Q.; Liu, C.; Chan, K.L.; Hu, Q.; Xie, Z.; Liu, H.; Si, F.; Liu, J. Ship-based MAX-DOAS measurements of tropospheric NO_2, SO_2, and HCHO distribution along the Yangtze River. *Atmos. Chem. Phys.* **2018**, *18*, 5931–5951. [CrossRef]

3.   Stavrakou, T.; Müller, J.F.; Smedt, I.D.; Roozendael, M.V.; Kanakidou, M.; Vrekoussis, M.; Wittrock, F.; Richter, A.; Burrows, J.P. The continental source of glyoxal estimated by the synergistic use of spaceborne measurements and inverse modelling. *Atmos. Chem. Phys.* **2009**, *9*, 8431–8446. [CrossRef]

4.   Fu, T.M.; Jacob, D.J.; Wittrock, F.; Burrows, J.P.; Vrekoussis, M.; Henze, D.K. Global budgets of atmospheric glyoxal and methylglyoxal, and implications for formation of secondary organic aerosols. *Geophys.Res. Atmos.* **2008**, *113*, D15. [CrossRef]

5.   Wittrock, F.; Richter, A.; Oetjen, H.; Burrows, J.P.; Kanakidou, M.; Myriokefalitakis, S.; Volkamer, R.; Beirle, S.; Platt, U.; Wagner, T. Simultaneous global observations of glyoxal and formaldehyde from space. *Geophys.Res. Lett.* **2006**, *33*, 16. [CrossRef]

6.   Myriokefalitakis, S.; Vrekoussis, M.; Tsigaridis, K.; Wittrock, F.; Richter, A.; Brühl, C.; Volkamer, R.; Burrows, J.P.; Kanakidou, M. The influence of natural and anthropogenic secondary sources on the glyoxal global distribution. *Atmos. Chem. Phys.* **2008**, *8*, 4965–4981. [CrossRef]

7.   Sinreich, R.; Volkamer, R.; Filsinger, F.; Frieß, U.; Kern, C.; Platt, U.; Sebastián, O.; Wagner, T. MAX-DOAS detection of glyoxal during ICARTT 2004. *Atmos. Chem. Phys.* **2007**, *7*, 1293–1303. [CrossRef]

8.   Volkamer, R.; Molina, L.T.; Molina, M.J.; Shirley, T.; Brune, W.H. DOAS measurement of glyoxal as an indicator for fast VOC chemistry in urban air. *Geophys.Res. Lett.* **2005**, *32*, 8 . [CrossRef]

9.   Vrekoussis, M.; Wittrock, F.; Richter, A.; Burrows, J.P. Temporal and spatial variability of glyoxal as observed from space. *Atmos. Chem. Phys.* **2009**, *9*, 4485–4504.

10.   Crutzen, P.J. The influence of nitrogen oxides on the atmospheric ozone content. *Q. J. R. Meteorol. Soc.* **1970**, *96*, 320–325. [CrossRef]

11.   Seinfeld, J.H.; Pandis, S.N. *Atmospheric Chemistry and Physics: From Air Pollution to Climate Change,* 2nd ed.; John Willey & Sons. Inc.: New York, NY, USA, 2006.

12. Chaloulakou, A.; Mavroidis, I.; Gavriil, I. Compliance with the annual NO_2 air quality standard in Athens. Required NOx levels and expected health implications. *Atmos. Environ.* **2008**, *42*, 454–465. [CrossRef]

13. Noxon, J.F. Stratospheric NO_2 in the Antarctic winter. *Geophys.Res. Lett.* **1978**, *5*, 1021–1022. [CrossRef]

14. Sahu, L.K.; Tripathi, N.; Yadav, R. Contribution of biogenic and photochemical sources to ambient VOCs during winter to summer transition at a semi-arid urban site in India. *Environ. Pollut.* **2017**, *229*, 595–606. [CrossRef] [PubMed]

15. Ma, J.Z.; Beirle, S.; Jin, J.L.; Shaiganfar, R.; Yan, P.; Wagner, T. Tropospheric NO_2 vertical column densities over Beijing: results of the first three years of ground-based MAX-DOAS measurements (2008–2011) and satellite validation. *Atmos. Chem. Phys.* **2013**, *13*, 1547–1567. [CrossRef]

16. Wang, Y.; Ying, Q.; Hu, J.; Zhang, H. Spatial and temporal variations of six criteria air pollutants in 31 provincial capital cities in China during 2013–2014. *Environ. Int.* **2014**, *73*, 413–422. [CrossRef] [PubMed]

17. Hendrick, F.; Müller, J.F.; Clémer, K.; Wang, P.; Mazière, M.D.; Fayt, C.; Gielen, C.; Hermans, C.; Ma, J.Z.; Pinardi, G.; et al. Four years of ground-based MAX-DOAS observations of HONO and NO_2 in the Beijing area. *Atmos. Chem. Phys.* **2014**, *14*, 765–781. [CrossRef]

18. Chan, K.L.; Hartl, A.; Lam, Y.F.; Xie, P.H.; Liu, W.Q.; Cheung, H.M.; Lampel, J.; Pöhler, D.; Li, A.; Xu, J.; et al. Observations of tropospheric NO_2 using ground based MAX-DOAS and OMI measurements during the Shanghai World Expo 2010. *Atmos. Environ.* **2015**, *119*, 45–58. [CrossRef]

19. Tian, X.; Xie, P.; Xu, J.; Li, A.; Wang, Y.; Qin, M.; Hu, Z. Long-term observations of tropospheric NO_2, SO_2 and HCHO by MAX-DOAS in Yangtze River Delta area, China. *J. Environ. Sci.* **2018**, *71*, 207–221. [CrossRef]

20. Javed, Z.; Liu, C.; Khokhar, M.F.; Xing, C.; Tan, W.; Subhani, M.A.; Rehman, A.; Tanvir, A. Investigating the impact of Glyoxal retrieval from MAX-DOAS observations during haze and non-haze conditions in Beijing. *J. Environ. Sci.* **2019**. [CrossRef]

21. Plane, J.M.; Saiz-Lopez, A. UV-visible differential optical absorption spectroscopy (DOAS). In *Analytical Techniques for Atmospheric Measurement*; Oxford publisher: Oxford, UK, 2006; pp. 147–188.

22. Platt, U.; Stutz, J. Differential absorption spectroscopy. In *Differential Optical Absorption Spectroscopy*; Springer: Berlin/Heidelberg, Germany, 2008; pp. 135–174.

23. Danckaert, T.; Fayt, C.; Van Roozendael, M.; De Smedt, I.; Letocart, V.; Merlaud, A.; Pinardi, G. *QDOAS Software User Manual*; Belgian Institute for Space Aeronomy: Brussels, Belgium, 2013.

24. Vandaele, A.C.; Hermans, C.; Simon, P.C.; Carleer, M.; Colin, R.; Fally, S.; Merienne, M.F.; Jenouvrier, A.; Coquart, B. Measurements of the NO_2 absorption cross-section from 42,000 cm^{-1} to 10,000 cm^{-1} (238–1000 nm) at 220 K and 294 K. *J. Quant. Spectrosc. Radiat. Transf.* **1998**, *59*, 171–184. [CrossRef]

25. Vandaele, A.C.; Hermans, C.; Fally, S. Fourier transform measurements of SO_2 absorption cross sections: II.: Temperature dependence in the 29,000–44,000 cm^{-1} (227–345 nm) region. *J. Quant. Spectrosc. Radiat. Transf.* **2009**, *110*, 2115–2126. [CrossRef]

26. Serdyuchenko, A.; Gorshelev, V.; Weber, M.; Chehade, W.; Burrows, J.P. High spectral resolution ozone absorption cross-sections–Part 2: Temperature dependence. *Atmos. Meas. Tech.* **2014**, *7*, 625–636. [CrossRef]

27. Thalman, R.; Volkamer, R. Temperature dependent absorption cross-sections of O_2–O_2 collision pairs between 340 and 630 nm and at atmospherically relevant pressure. *Phys. Chem. Chem. Phys.* **2013**, *15*, 15371–15381. [CrossRef] [PubMed]

28. Rothman, L.S.; Gordon, I.E.; Barber, R.J.; Dothe, H.; Gamache, R.R.; Goldman, A.; Perevalov, V.I.; Tashkun, S.A.; Tennyson, J. HITEMP, the high-temperature molecular spectroscopic database. *J. Quant. Spectrosc. Radiat. Transf.* **2010**, *111*, 2139–2150. [CrossRef]

29. Solomon, S.; Schmeltekopf, A.L.; Sanders, R.W. On the interpretation of zenith sky absorption measurements. *Geophys. Res. Atmos.* **1987**, *92*, 8311–8319. [CrossRef]

30. Hönninger, G.; Friedeburg, C.V.; Platt, U. Multi axis differential optical absorption spectroscopy (MAX-DOAS). *Atmos. Chem. Phys.* **2004**, *4*, 231–254. [CrossRef]

31. Celarier, E.A.; Brinksma, E.J.; Gleason, J.F.; Veefkind, J.P.; Cede, A.; Herman, J.R. Validation of ozone monitoring instrument nitrogen dioxide columns. *J. Geophys. Res. Atmos.* **2008**, *113*, D15. [CrossRef]

32. Zheng, G.J.; Duan, F.K.; Su, H.; Ma, Y.L.; Cheng, Y.; Zheng, B.; Zhang, Q.; Huang, T.; Kimoto, T.; Chang, D.; et al. Exploring the severe winter haze in Beijing: the impact of synoptic weather, regional transport and heterogeneous reactions. *Atmos. Chem. Phys.* **2015**, *15*, 2969–2983. [CrossRef]

33. Duan, L.; Xiu, G.; Feng, L.; Cheng, N.; Wang, C. The mercury species and their association with carbonaceous compositions, bromine and iodine in PM2.5 in Shanghai. *Chemosphere* **2016**, *146*, 263–271. [CrossRef]

34. Wagner, T.; Burrows, J.P.; Deutschmann, T.; Dix, B.; Friedeburg, C.V.; Frieß, U.; Hendrick, F.; Heue, K.P.; Irie, H.; Iwabuchi, H.; et al. Comparison of box-air-mass-factors and radiances for Multiple-Axis Differential Optical Absorption Spectroscopy (MAX-DOAS) geometries calculated from different UV/visible radiative transfer models. *Atmos. Chem. Phys.* **2007**, *7*, 1809–1833. [CrossRef]

35. Alvarado, L.M.A.; Richter, A.; Vrekoussis, M.; Wittrock, F.; Hilboll, A.; Schreier, S.F.; Burrows, J.P. An improved glyoxal retrieval from OMI measurements. *Atmos. Meas. Tech.* **2014**, *7*, 4133. [CrossRef]

36. Wang, Y.; Tao, J.; Cheng, L.; Yu, C.; Wang, Z.; Chen, L. A Retrieval of Glyoxal from OMI over China: Investigation of the Effects of Tropospheric NO_2. *Remote Sens.* **2019**, *11*, 137. [CrossRef]

37. Rasmussen, R.A.; Khalil, M.A.K. Atmospheric trace gases: trends and distributions over the last decade. *Science* **1986**, *232*, 1623–1624. [CrossRef] [PubMed]

38. Wilkniss, P.E.; Lamontagne, R.A.; Larson, R.E.; Swinnerton, J.W.; Dickson, C.R.; Thompson, T. Atmospheric trace gases in the southern hemisphere. *Nat. Phys. Sci.* **1973**, *245*, 45–47. [CrossRef]

39. Gratsea, M.; Vrekoussis, M.; Richter, A.; Wittrock, F.; Schönhardt, A.; Burrows, J.; Kazadzis, S.; Mihalopoulos, N.; Gerasopoulos, E. Slant column MAX-DOAS measurements of nitrogen dioxide, formaldehyde, glyoxal and oxygen dimer in the urban environment of Athens. *Atmos. Environ.* **2016**, *135*, 118–131. [CrossRef]

40. Cleveland, W.S.; Graedel, T.E.; Kleiner, B.; Warner, J.L. Sunday and workday variations in photochemical air pollutants in New Jersey and New York. *Science* **1974**, *186*, 1037–1038. [CrossRef] [PubMed]

41. Beirle, S.; Platt, U.; Wenig, M.; Wagner, T. Weekly cycle of NO_2 by GOME measurements: a signature of anthropogenic sources. *Atmos. Chem. Phys.* **2003**, *3*, 2225–2232. [CrossRef]

8

Anthropogenic CH_4 Emissions in the Yangtze River Delta Based on a "Top-Down" Method

Wenjing Huang [1,2], Wei Xiao [1,2], Mi Zhang [1], Wei Wang [1], Jingzheng Xu [3], Yongbo Hu [1], Cheng Hu [1,4], Shoudong Liu [1] and Xuhui Lee [1,2,5,*]

[1]   Yale-NUIST Center on Atmospheric Environment, Nanjing University of Information, Science and Technology, Nanjing 210044, China; 20181108059@nuist.edu.cn (W.H.); wei.xiao@nuist.edu.cn (W.X.); zhangm.80@nuist.edu.cn (M.Z.); wangw@nuist.edu.cn (W.W.); 20171103089@nuist.edu.cn (Y.H.); huxxx991@umn.edu (C.H.); lsd@nuist.edu.cn (S.L.)
[2]   NUIST-Wuxi Research Institute, Wuxi 214073, China
[3]   Radio Science Research Institute Inc., Wuxi 214073, China; xu.jingzheng@js1959.com
[4]   Department of Soil, Water, and Climate, University of Minnesota-Twin Cities, St. Paul, MN 55108, USA
[5]   School of Forestry and Environmental Studies, Yale University, New Haven, CT 06511, USA
*   Correspondence: Xuhui.lee@yale.edu (X.L.)

Abstract: There remains significant uncertainty in the estimation of anthropogenic CH_4 emissions at local and regional scales. We used atmospheric CH_4 and CO_2 concentration data to constrain the anthropogenic CH_4 emission in the Yangtze River Delta one of the most populated and economically important regions in China. The observation of atmospheric CH_4 and CO_2 concentration was carried out from May 2012 to April 2017 at a rural site. A tracer correlation method was used to estimate the anthropogenic CH_4 emission in this region, and compared this "top-down" estimate with that obtained with the IPCC inventory method. The annual growth rates of the atmospheric CO_2 and CH_4 mole fractions are 2.5 ± 0.7 ppm year^{-1} and 9.5 ± 4.7 ppb year^{-1}, respectively, which are 9% and 53% higher than the values obtained at Waliguan (WLG) station. The average annual anthropogenic CH_4 emission is 4.37 (± 0.61) $\times 10^9$ kg in the YRD (excluding rice cultivation). This "top-down" estimate is 20–70% greater than the estimate based on the IPCC method. We suggest that possible sources for the discrepancy include low biases in the IPCC calculation of emission from landfills, ruminants and the transport sector.

Keywords: "top-down" method; the Yangtze River Delta; CO_2; CH_4; annual growth rate; anthropogenic CH_4 emissions

1. Introduction

The source apportionment of CH_4 is important for the study of carbon cycle and climate change. The mole fraction of CH_4 in the atmosphere increased by 157% from 1750 to 2011 [1,2]. As the second largest greenhouse gas next to CO_2, CH_4 has a warming potential of 28 times that of CO_2 with a 100-year time horizon [2]. In addition to the greenhouse effect, CH_4 also affects the chemical and photochemical reactions in the atmosphere [3]. The annual growth rate of atmospheric CH_4 was 6.9 ± 2.4 ppb year^{-1} from 2007 to 2017 [4]. However, the source contributions of CH_4 have so far not been accurately quantified, especially at the regional and the city scale [5].

Anthropogenic CH_4 emissions account for 50–65% of the global CH_4 emissions of 5.82 (± 0.5) $\times 10^{11}$ kg year^{-1} [6,7]. Large uncertainties still exist in regional anthropogenic emission estimates. These estimates are usually based on the Intergovernmental Panel on Climate Change (IPCC) inventory method. The IPCC method aggregates the CH_4 emissions generated by different anthropogenic activities and sums up the individual components to the domain of interest. One problem is that activity data, such as landfill and livestock, and emission factors cannot be accurately determined at the regional and the city scale [8–10]. A study in Beijing found that the uncertainty caused by landfill accounts for nearly half of the total uncertainty in the CH_4 emission estimate [11]. Accurate and timely calculations of anthropogenic CH_4 emissions at the regional scale are necessary for assessing the effectiveness of emission reduction policies.

Anthropogenic greenhouse gas emissions can also be estimated from observations of the gaseous concentrations in the atmosphere ("top-down" methods). The "atmospheric method" used in this study is one of the "top-down" approaches. One reason for using the atmospheric method is that many sources of anthropogenic CH_4 cannot be quantified with traditional methods, such as the chamber method [12–14]. The atmospheric method requires simultaneous concentration measurements of the target gas (CH_4) and a tracer gas (usually CO_2) when there is no disturbance from sinks or other natural sources [15]. In an observational study using aircraft profile measurement over a broad region of Alaska and Canada, the concentrations of CH_4 and CO_2 increase synchronously with height, showing a strong positive correlation between the two gases [16]. A similar positive relationship also exists in the surface air in a moderately polluted urban atmosphere of Boulder, USA [16]. The explanation for the positive relationship is that the two gases share common source areas and undergo the same long-range transport [17,18]. The concentration ratio between the two gases were used to estimate the CH_4 emissions in the densely populated urban areas in Southern California, showing that inventory CH_4 emission estimates for these urban areas are lower in comparison to the "top-down" atmospheric estimate [19]. In a study of anthropogenic CH_4 emissions in the Los Angeles megacity, a ground-based remote sensing concentration measurement with 29 different surface targets was used to spatially resolve CH_4:CO_2 emission ratio, once again relying on the linear relationship between the two gaseous concentrations [20].

The atmospheric method is based on a strong correlation between the observed concentration values of two relatively inert gases CH_4 and CO_2. Because the lifetime of these two gases in the atmosphere (7–11 years and 50–200 years, respectively) [21] is much longer than hourly time scales at which the observations are made, the linear slope value of the regression is essentially equivalent to the ratio of their anthropogenic emission strengths. In the applications cited above, the CH_4 emission flux is computed as the product of the concentration regression slope and the anthropogenic CO_2 emission flux, the latter of which can be obtained reliably with the IPCC inventory method [22].

The atmospheric method has been used to track emissions of other gas species besides CH_4. This method was used to infer anthropogenic Hg emissions in Northeast USA, with wintertime Hg and CO_2 concentration data and CO_2 as the tracer [23]. Simultaneous observations of the CO_2 and CO concentrations in a suburban site outside Beijing was used to determine the efficiency of fossil fuel combustion [24]. The concentrations of CO_2 and CO observed in the Asian outflow air combined with a three-dimensional global chemical transport model were used to quantify CO_2 emissions in East Asia [25].

In this paper, we report long-term (five years), near-continuous, and simultaneous observations of atmospheric CH_4 and CO_2 at a lake site near Wuxi, Jiangsu Province, China. Our main objective was to quantify the CH_4 emission in the Yangtze River Delta (YRD) and its interannual variability, using the atmospheric tracer method described above. The second objective was to evaluate the validity of the CH_4 emissions calculated with the "bottom-up" IPCC inventory method against the atmospheric or "top-down" emission estimates.

2. Experiments

2.1. Study Area and Observational Site

The YRD in East China occupies only 2.1% of the land area of China (including Jiangsu Province, Zhejiang Province, Anhui Province and Shanghai), but contributes 1/4 of the total economic output [26].

The Wuxi Municipality is located roughly in the middle of the Yangtze River Delta, with a population of about 5 million. Other major cities in this region are Shanghai, Nanjing, Hangzhou, Ningbo and Hefei.

The observational site is located at the Taihu Lake Ecosystem Observatory of the Chinese Academy of Sciences (31.4197° N, 120.2139° E) in Wuxi. Measurements of the CO_2 and CH_4 concentrations were made on a platform about 200 m from the north shore of the lake (Figure 1). The platform was once also part of the Lake Taihu Eddy Flux Network (site id MLW) [27]. The site is surrounded by water, scattered farmlands and residential buildings. The closest traffic road is about 10 km away. The prevailing wind is northwesterly in the winter (Figure S1) and southerly in the summer. In the winter season, the landscape upwind is mostly rural (Figure S1).

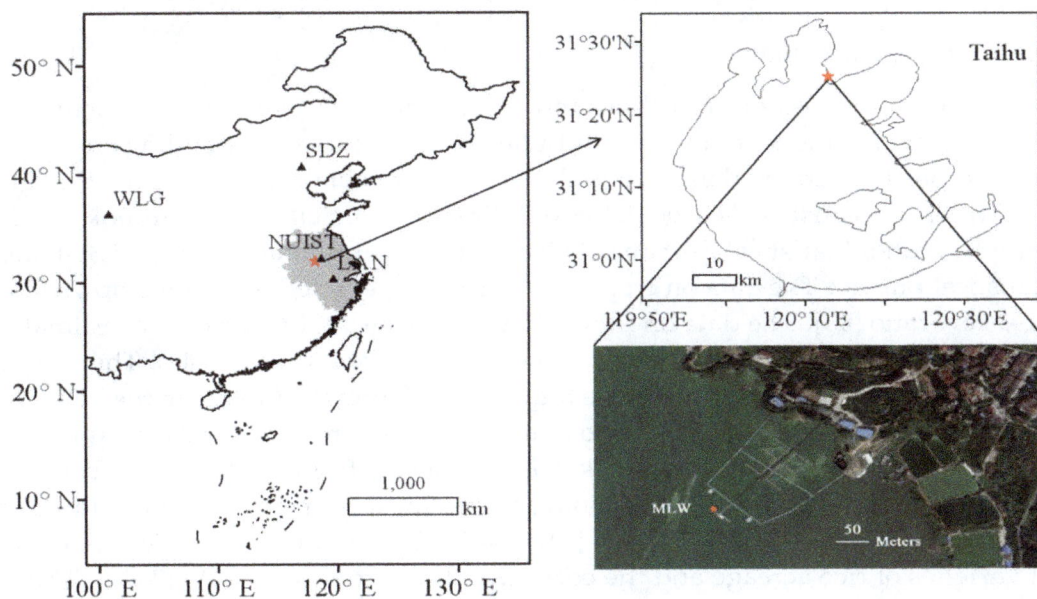

Figure 1. Location of the MLW site (pentagram in red color), Nanjing University of Information Science and Technology (NUIST) and three Chinese WMO/GAW stations, including Lin'an (LAN), Shangdianzi (SDZ) and Waliguang (WLG).

2.2. Trace Gas Analyzer

The analyzer we used was based on wavelength scanning cavity ring-down spectroscopy (model G1301 from 14 May 2012 to 6 July 2013 and model G2301 from 20 October 2013 until now, Picarro Inc., Sunnyvale, CA, USA). The measurement frequency is 1 Hz (sampling rate) and the precision (5-s mean) is 0.15 ppm for CO_2 and 1 ppb for CH_4. The air inlet is placed at a height of 3.5 m above the water surface. The water vapor concentration measured by the instrument was used to remove the water vapor dilution effect so the concentrations of CH_4 and CO_2 are expressed as molar ratio of CH_4 or CO_2 to dry air.

The observation period in this study was from May 2012 to April 2017. A large data gap occurred between July and October 2013 owing to instrument malfunction. In the second half of 2016, some data were lost because of loss of power on the platform.

The data from the analyzer were averaged to half-hour intervals. Standard deviation in half an hour and a five-point moving average method with a threshold of 1.5 times the standard deviation were used to filter the outliers. The daily average was achieved when daily data exceeded 75%.

For model G1301, the CO_2 and CH_4 measurements were calibrated twice (2 May 2012 and 22 June 2012) and the details can be found in the supplementary materials of a previously published paper [28]. Model G2301 was calibrated three times on 11 September 2013, 4 November 2015 and 25 August 2016, respectively. The mole fractions of CO_2 and CH_4 were calibrated against two standard CO_2 gases (concentration: 490 ppm or 491 ppm and 385 ppm, National Primary Standard prepared by the National Institute of Meteorology (NIM), China) and two standard CH_4 gases (2.02 ppm and 3.05 ppm or 3.52 ppm, National Primary Standard) each time. The relative error is 0.62–0.37% for CO_2 and 0.81–0.41% for CH_4. We used a dew-point generator (model 610, LI-COR, Inc., Lincoln, NE, USA) to correct the analyzers humidity measurement. The humidity level of the air coming out of the dew-point generator was regulated at five levels and measurement at each humidity level lasted 15 min. The first several minutes when the measurement was transitional were excluded from the analysis. A linear regression fit generated from saturation vapor mixing ratio and observed vapor mixing ratio resulted in a slope value of 0.97–0.99.

No standards were available for us to trace our calibration gases to the WMO scale. NIM participated in two inter-agency comparison experiments on calibration standards including standards traceable to the WMO scale. The results can be found in references [29,30].

2.3. The IPCC Inventory Calculation

The IPCC emissions inventory is a "bottom-up" approach. It takes emission activity data from different economic sectors into account and multiplies them by the corresponding emission factors to estimate the emission. The activity data used in this study were obtained from China Energy Statistical Yearbook [31–34], China Statistical Yearbook [35–38], China Rural Statistical Yearbook [39–42], Jiangsu Statistical Bureau, Auhui Statistical Bureau, Zhejiang Statistical Bureau, Shanghai Statistical Bureau and Wuxi Statistical Bureau. The data on crop straw burning were derived from crop yields combined with grain-to-straw ratio [43]. The data on firewood usage from 1991 to 2006 were estimated with the ratio of firewood usage in the YRD combined with the total firewood usage [44]. Then a time-varying line of the firewood usage was fitted to estimate the 2012–2015 firewood usage in the YRD. We used the default emission factors provided by IPCC if no domestic values are available, such as industry energy consumption, industry processes and livestock. CH_4 emissions from landfills were based on the first order decay model provided by IPCC, taking into account local climatic conditions, the landfill waste volume, organic carbon content and waste age [11,45,46]. CH_4 emissions from rice paddies accounted for different varieties of rice acreage and the corresponding growth period [47–49]. The proportion of open-pit mining to the total mining volume has increased year by year in the YRD region [50]. The CH_4 fugitive emissions in the YRD mainly come from coal mining in Xuzhou in Jiangsu Province and Huainan, Huaibei, Fuyang, etc. in Anhui Province. The remaining emission factors for other sectors not listed here were obtained from the relevant literature [51,52].

The Monte Carlo method [52,53] was used to obtain uncertainty ranges of the inventory calculations. The uncertainties of the IPCC method arise from the choice of emission factors and from uncertainties in the activity data. These uncertainties were assumed to follow uniform distributions. The range of variations of the emission factors were given by IPCC or domestic values. For the activity data, an uncertainty range of 10% was assumed. A total of 400,000 ensemble members were calculated to determine a probability distribution function and estimate the emission uncertainties. A 95% confidence interval was used to quantify the random errors.

The CO_2 inventory is well quantified and with less uncertainty. For example, in Austria, Norway, the Netherlands, the UK and the USA, the uncertainty of CO_2 emission factors and activity data for main sources is between $\pm 0.5\%$ and $\pm 7\%$ of the means [54]. In comparison, the uncertainty of the CH_4 factors sector is $\pm 20\%$ to $\pm 50\%$ of the means [54]. The overall uncertainty of the CO_2 emission estimates is 2–4%, much smaller than that of CH_4 (17–48%) or N_2O (34–230%) [54–56].

2.4. Application of the Atmospheric Method

We used a geometric mean regression to determine the slope of the CH_4 molar mixing ratio against the CO_2 molar mixing ratio. Because uncertainties exist in both CO_2 and CH_4 concentration measurements, geometric mean regression gives more robust parameter estimates than the ordinary least squares regression [57]. Moreover, there is slight variations between the four seasons according to the national level fuel consumption [58], so we focused on wintertime (December to February inclusive) measurements because plant photosynthesis is minimal and atmospheric CO_2 variations are driven primarily by anthropogenic sources. The annual anthropogenic CH_4 emission flux is the product of the regression slope in winter and the anthropogenic CO_2 flux derived from the IPCC inventory method. We refer to this flux as the "top-down" estimate.

The uncertainty of the atmospheric method comes from two aspects. The first is a result of the regression procedure, and is characterized by the standard deviation of the regression estimate of the geometric mean slope. The second source of uncertainty is caused by the anthropogenic CO_2 emission calculation using the IPCC method. The overall uncertainty of the atmospheric CH_4 emission estimate was calculated with the Monte Carlo method that combines these two sources of uncertainty.

The daytime CO_2 and CH_4 concentrations are indicative of source contributions at the regional scale if there are no direct nearby emissions. This is because the daytime atmospheric boundary layer (ABL) is well mixed and there is no nearby direct emission disturbance. At a suburban site in Xiamen in southeastern China, the vertical profile of CO_2 concentration shows little variation with altitude (below 350 m) between 8:40 and 15:45 [59]. At the Zotino Tall Tower Observatory, CO_2 and CH_4 in the atmosphere becomes well-mixed and their concentrations become nearly indistinguishable at six height levels (4, 52, 92, 157, 227 and 301 m) during day in summer [60]. The ABL height at the MLW site in the winters of 2014–2016, simulated with the Meteoinfo open-source software [61], varies between 570 m and 970 m in the midday period (10:00–17:00; Figure 2). The Meteoinfo program used the Data Assimilation System (GDAS1) as input data, and the predicted ABL height was interpolated spatially to the MLW site. The analysis of Potential Source Contribution Function (PSCF) calculated with the HYSPLIT trajectory model with the 500-m height, the approximate mean height of the mid-point of the ABL, as the end point is shown in Figure 3. The PSCF value can be interpreted as the conditional probability for the specific grid cell, or the ratio of the number of endpoints falling in the grid cell with high concentration (>85th percentile concentration at the receptor site) to the total number of endpoints falling in the grid cell. Grid points having high PSCF values are likely to be potential source regions contributing to the observed concentrations [62]. A weighting function was introduced to deal with the bias brought by PSCF value when the number of endpoints falling in the grid cell was small. The choice of the function was set according to that reported by Sigler and Lee [63]. The calculation was performed for December 2014 with 48-h backward trajectories at 10:00, 13:00 and 16:00 LST each day. In Figure 3, the area with the PSCF value greater than the typical threshold of 0.1 [51,64] is 3.8×10^5 km^2, of which 76% fall in the political boundary of the YRD. As a result, the daytime observational data at the MLW site can represent the source signatures of the YRD, consistent with a similar study conducted in Nanjing [51].

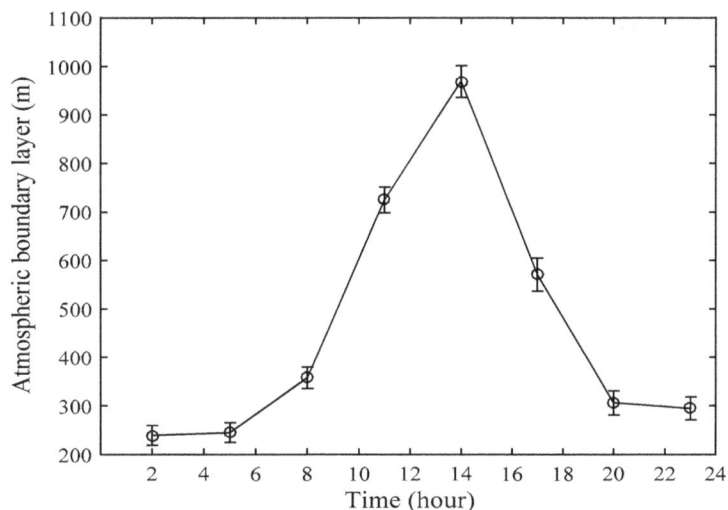

Figure 2. Diurnal variation of the boundary layer height at the MLW site. Error bars are ± 1 standard deviation of the mean.

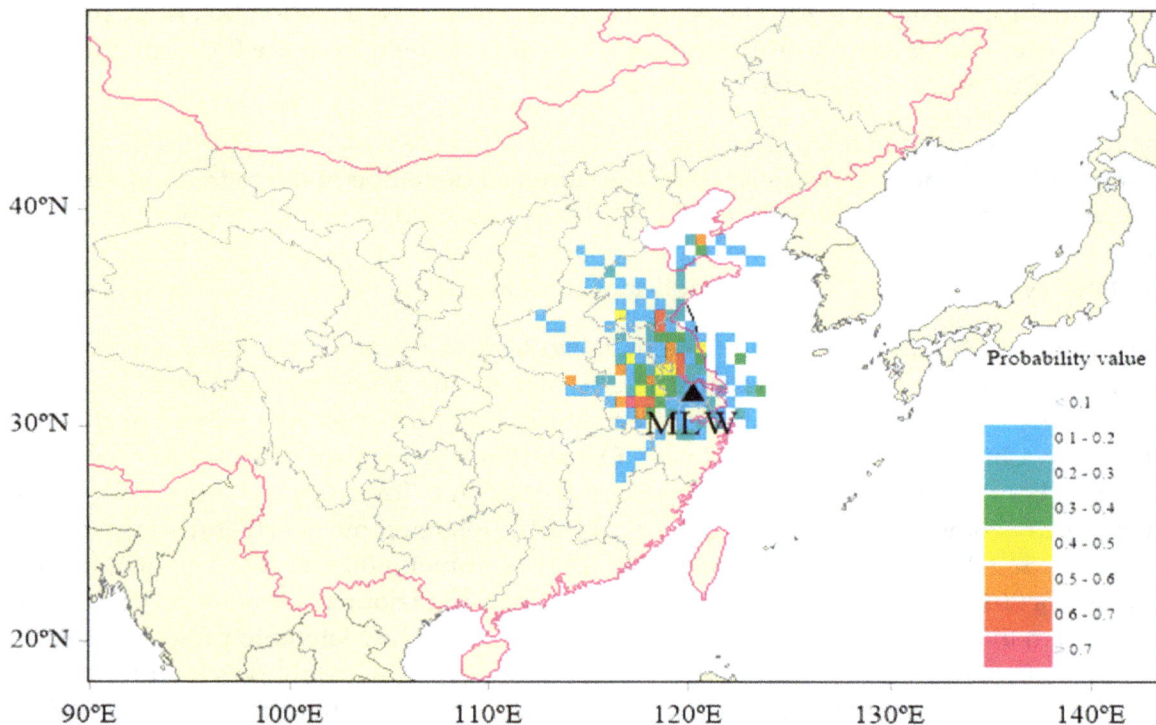

Figure 3. Spatial distribution of the potential source region contributions simulated for December 2014.

The footprint analysis revealed the source region mostly likely to have impacted the daytime measurement at MLW. The actual probability value, or weighting factor, was not used later when we aggregated the inventory emission data to the whole YRD region.

At night (23:00–05:00), an inversion layer typically prevails near the ground surface, with high atmospheric stability. The mean height of the boundary layer is 260 m (Figure 2). Because of the strong stability, the CO_2 and CH_4 emitted by anthropogenic sources are trapped near the surface. The lack of mixing implies that the source areas of the observed concentration may span only several kilometers [65]. In other words, the regression slope represents more the emission ratio of local sources than the emission ratio of regional sources, although the exact spatial representation of nighttime observations needs to be further studied.

3. Results

3.1. Temporal Variations of CO_2 and CH_4 Concentrations

Figure 4 shows the temporal variations of half-hourly atmospheric CH_4 and CO_2 mole fractions during the observation period. The data show significant periodic fluctuations through the 24-h cycle, especially for CO_2. Data for 2014 and 2015 are nearly gap-free and are representative of seasonal variations.

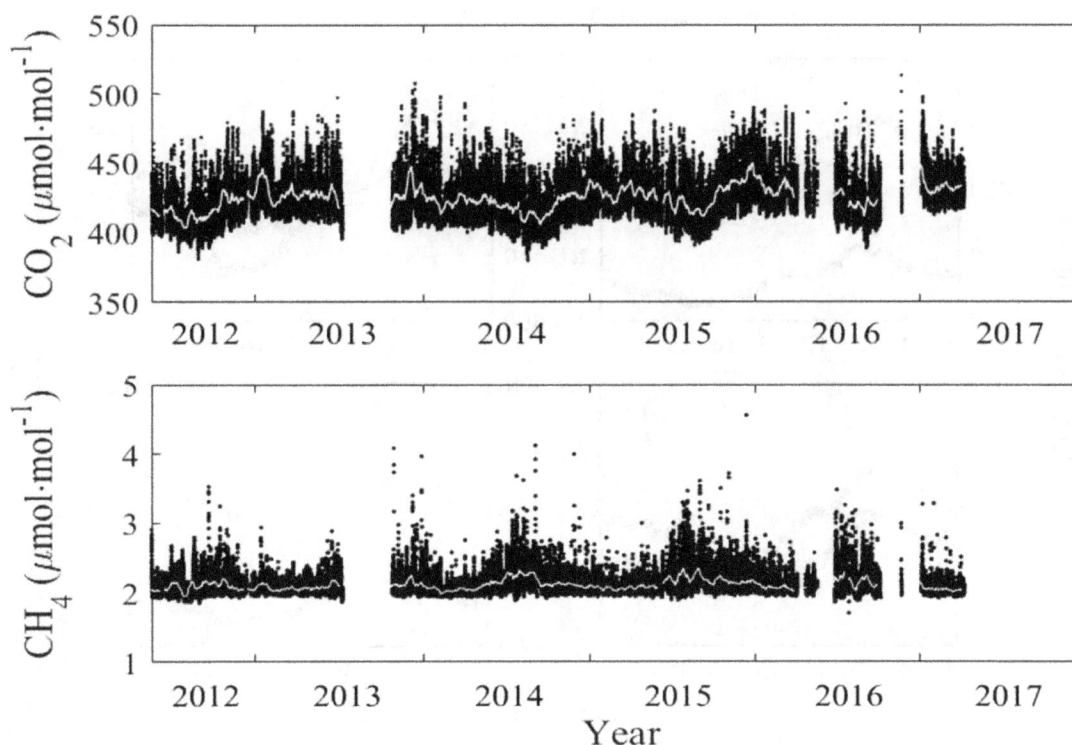

Figure 4. The half-hourly CO_2 and CH_4 mole fractions during the observation period (black dots). The white points are 14-day moving averages.

Figure 5 shows the diurnal composite concentrations for each of the four seasons in 2014. The diurnal variations of CO_2 and CH_4 show similar patterns in all seasons. In the spring (March, April and May) and autumn (June, July and August), the peaks appear at about 07:00 and the troughs appear at around 17:00. In the summer, the peaks occur at 04:00 and the minimum still appear around 17:00. In the winter (December, and January and February in the next year), the diurnal variations are gentler than in the other three seasons.

Figure 6 shows a comparison of our monthly mean CO_2 and CH_4 concentrations with those observed at Nanjing University of Information Science and Technology (NUIST; 32.20° N, 118.72° E), about 170 km to the northwest of the MLW site. The concentrations observed during the same time period at Mt. Waliguan (WLG, 36.28° N, 100.90° E, 3810 m above the mean sea level), a WMO baseline station representing the background atmosphere for Asia, are also shown [66,67]. NUIST is located at the outskirt of Nanjing, surrounded by residential areas and traffic roads and in the vicinity of two industrial complexes [68]. The atmospheric CO_2 molar fraction is highest at NUIST, followed by MLW and lowest at WLG. Among the three sites, the strongest seasonality of the atmospheric CO_2 molar fraction occurs at the MLW site, with low values in the summer and high values in the winter. The atmospheric CH_4 molar fraction at the MLW site shows an opposite seasonality to CO_2, with high values in the summer and low values in the winter.

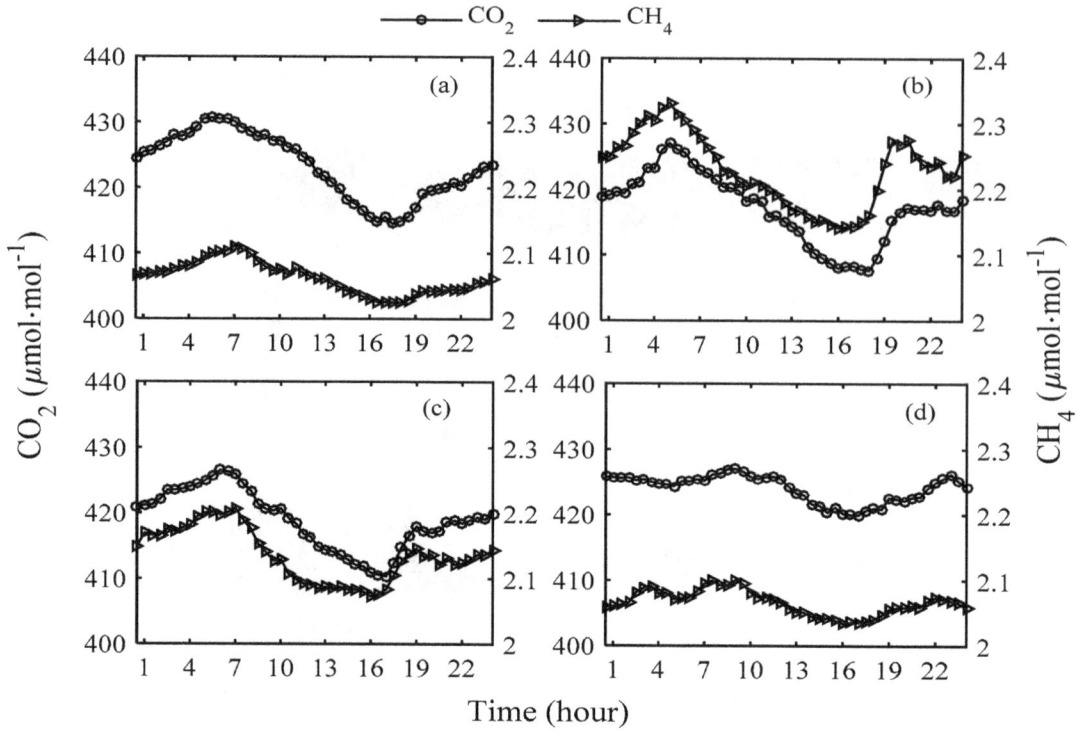

Figure 5. Diurnal variations of the molar fraction of CH_4 and CO_2 in the atmosphere in the four seasons in 2014: (**a**) spring; (**b**) summer; (**c**) autumn; and (**d**) winter.

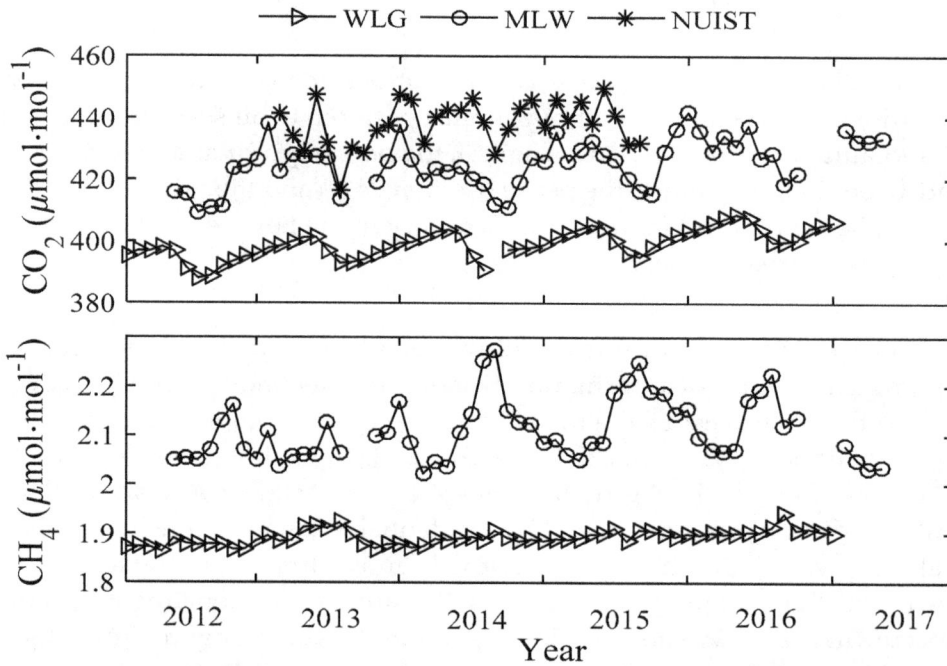

Figure 6. Variations of monthly molar fraction of CO_2 in the atmosphere at Nanjing University of Information Science and Technology (NUIST), WLG and MLW and monthly molar fraction of CH_4 at WLG and MLW from 2012 to 2017.

The monthly time series were deseasonalized to avoid the interference of seasonality of CH_4 and CO_2. A relative anomaly of CH_4 and CO_2 in a particular month of a given year was computed as the difference between the given concentration and the average of all years in that month, divided by the standard deviation of all the concentrations during the research period for that specific month.

Then, the least square method was used to obtain the growth rates of CO_2 and CH_4. At the MLW site, the growth rates are 2.5 ± 0.7 ppm year^{-1} for CO_2 and 9.5 ± 4.7 ppb year^{-1} for CH_4. These rates are higher than those at WLG (2.3 ± 0.2 ppm year^{-1} for CO_2 and 6.2 ± 1.7 ppb year^{-1} for CH_4).

3.2. Diurnal and Inter-Annual Variations of the CH_4 Versus CO_2 Regression Slope

Figure 7 shows the diurnal variations of the CH_4 versus CO_2 regression slope and their linear correlation coefficients (R) for the summer and the winter in 2015. The slope value was determined for each half-hour period of the day, using all the data collected in the same half-hourly period. In the summer, the regression slope is marked by a sudden rise at around 18:00, coinciding with the onset of the surface inversion layer. The summertime R is high at night but has very low values in the afternoon. In the winter, the slope value is generally smaller than the summer value, and its diurnal variation is weak. The wintertime R fluctuates around 0.8. The fact that the regression slope is stable and that the R value is high in the winter is not a surprise. In the winter months, biological sources of CO_2 and CH_4 are much weaker than the anthropogenic sources. For this reason, the data in winter were selected for further study.

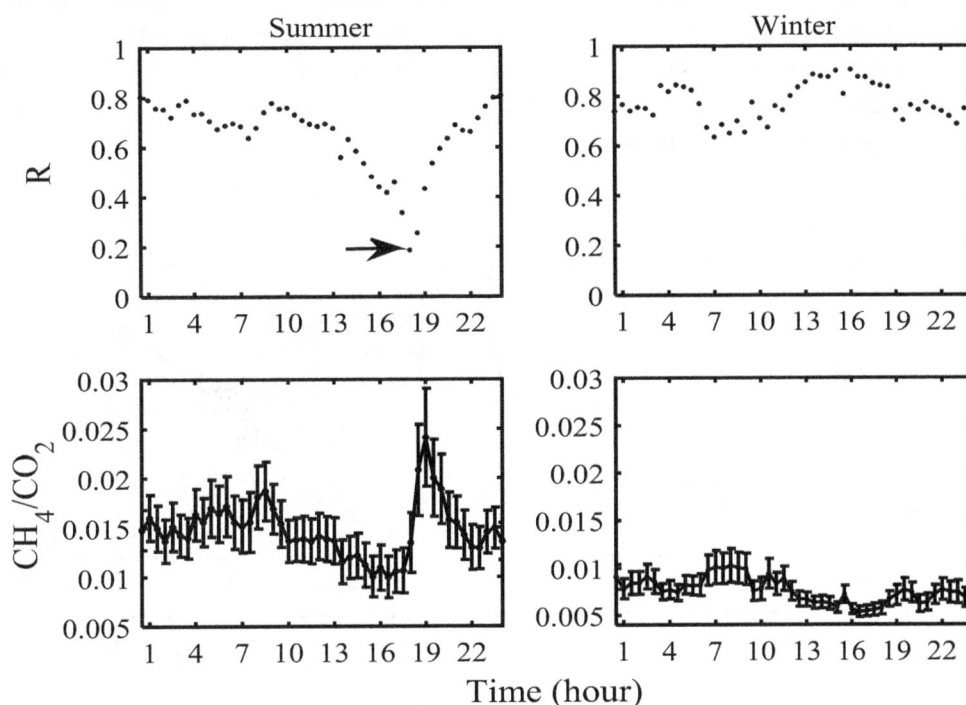

Figure 7. Diurnal variation of CH_4:CO_2 emission ratio in the summer (June, July, and August) in 2014 and in the winter (December, January and February) during 2014–2015. The arrow indicates the linear correlation at the p value of 0.05.

An important biological CH_4 source is wetland, including natural wetlands (offshore, coastal, swamp, lakes and rivers) and constructed wetlands. The total wetland area in the YRD is 5.4×10^4 km^2 (2013 value) [36], or 15% of the total area of the YRD. However, wintertime wetland CH_4 flux is generally weak. A study found that the methane emission rate in a freshwater wetland in Australia is less than 0.01 mmol m^{-2} h^{-1} in the winter, which is much lower than that in the summer (1.3–3.3 mmol m^{-2} h^{-1}) [69]. Similarly, CH_4 emissions from natural wetlands in the YRD are 1.98×10^8 kg [51], far less than anthropogenic emissions in winter (3.10×10^9 kg, Section 3.3). The Xixi wetland, one of the four major wetlands in the YRD, even becomes a weak sink of CH_4 in the winter (0.0019 mg m^{-2} h^{-1}) [70]. Thus, CH_4 emissions from wetlands could be omitted when we compared the "top-down" and "bottom-up" estimates using wintertime observations.

Rice paddies are another important biological CH_4 source, contributing about 3.3–7.0% (18.3×10^9 kg year^{-1} to 8.8×10^9 kg year^{-1}, 1901–2010) [71] of the global CH_4 emissions (5.58×10^{11} kg year^{-1}, 2003–2012) [72]. China accounts for 20% of the rice production area in the world and the rice planting area in the YRD accounts for 18% of the China's total [35–38]. The important conditions for CH_4 production are organic matter applied (such as rice straw) and anoxic soils established in flooded paddies. As a typical monsoon climate zone in southeastern China, the growth period for rice is from May to October [73]. During non-rice growing season in the winter, CH_4 emissions from non-permanently flooded rice paddies are about 4–6% of the emissions in the growth season [74], and were ignored in the comparison with our "top-down" estimation.

Figures 8 and 9 are regression results using winter time observations made during 2012–2016. The regression was done separately for daytime (10:00–17:00 Beijing time) and nighttime (23:00–05:00 Beijing time) periods for consideration of different mixing conditions and source regions during the day and at night. In this regression, each data sample is a daytime or nighttime mean value. Taking the maximum drift over 24 h in CO_2 and CH_4 values (<120 ppb for CO_2 and <1 ppb for CH_4) for each point caused by the observational instrument into consideration, the slope was affected by <0.0001 ppm:ppm, which can be ignored comparing with the standard error caused by the fitting method (0.0005 in 2013). The daytime regression slope fluctuates between 0.0055 ± 0.0006 and 0.0068 ± 0.0005 ppm CH_4 per ppm CO_2 without an obvious interannual trend. The correlation coefficient R is greater than 0.8, and all four winters passed the 0.01 significance test.

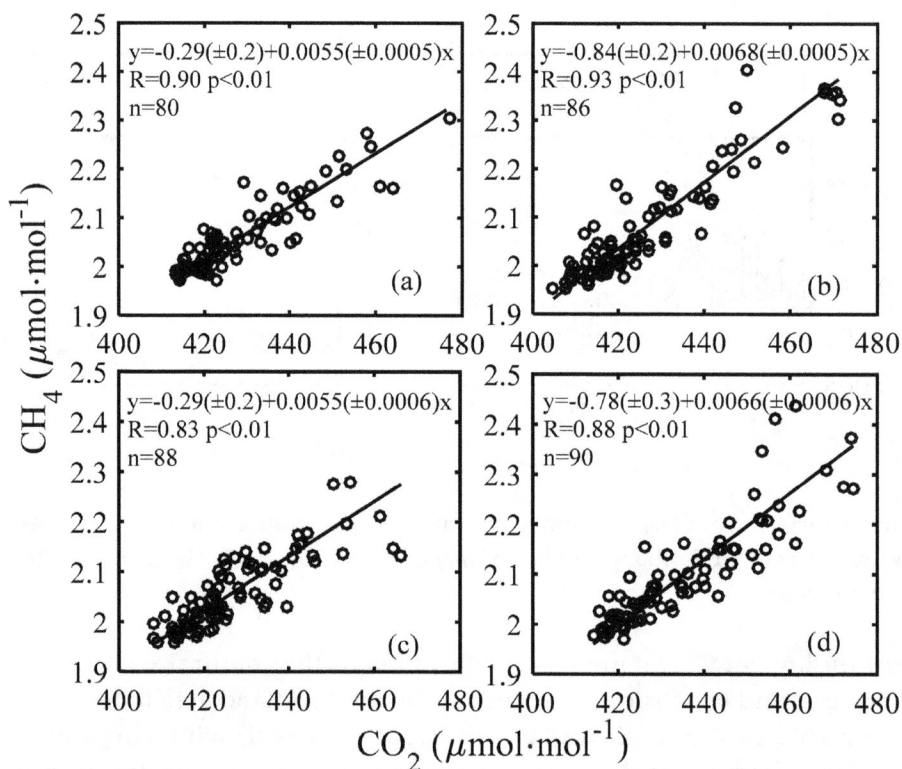

Figure 8. Scatter plots of winter (December–February) daytime CH_4 and CO_2 concentrations at MLW from 2012 to 2016. Each data point represents one daytime mean value. Regression statistics (regression equation, linear correlation R and number of samples) are also shown. Parameter ranges in the parentheses are 95% confidence bounds: (a) December 2012–February 2013; (b) December 2013–February 2014; (c) December 2014–February 2015; and (d) December 2015–February 2016.

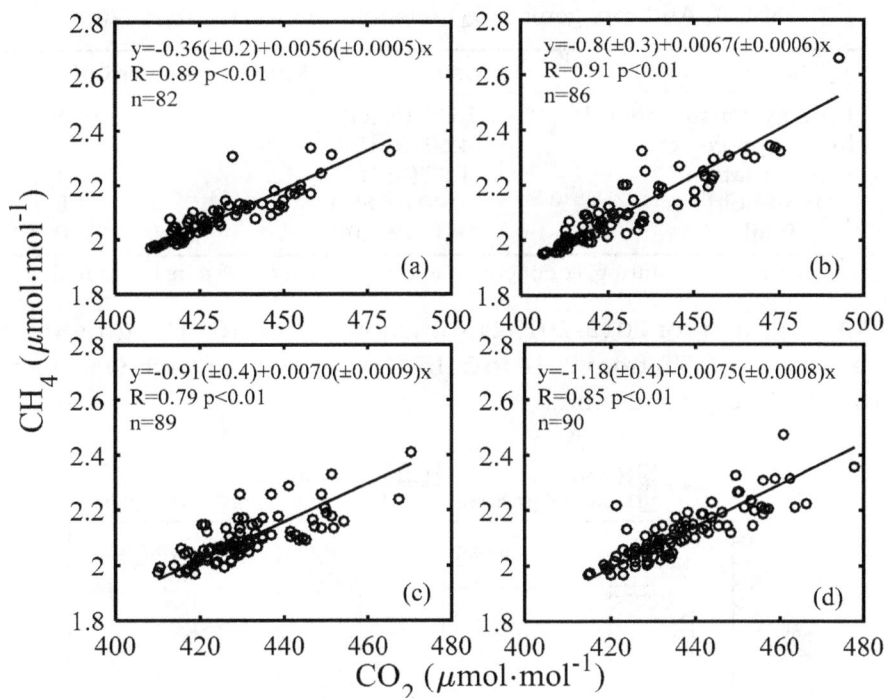

Figure 9. Same as Figure 8 except for winter (December–February) nighttime: **(a)** December 2012–February 2013; **(b)** December 2013–February 2014; **(c)**December 2014–February 2015; and **(d)** December 2015–February 2016.

The nighttime slope value increases from 0.0056 ± 0.0005 ppm/ppm for the winter of 2012–2013 to 0.0075 ± 0.0008 ppm/ppm for the winter of 2015–2016. With the exception of winter 2013–2014, the nighttime slope values are greater than the daytime values, implying more CH_4 emission per mole of CO_2 release by local anthropogenic sources than by regional sources. The correlation also passed the 0.01 significance test but the R values are slightly lower than the daytime R values.

An alternative approach to obtain the emissions ratio is to divide the CH_4 concentration enhancement over a background value by the CO_2 concentration enhancement. Here, we defined the clean background as the concentration at the 5th percentile. The emissions ratio, calculated as $(CH_{4, mean} - CH_{4, 5\%})/(CO_{2, mean} - CO_{2, 5\%})$ for the winter, is in good agreement with the slope of the CH_4 concentration versus CO_2 concentration (Figure S2).

In this study, we assumed that the daytime observations were influenced by sources located in the YRD. We used the EDGARv4.3.2 inventory data (2012) to understand the sensitivity of the emissions ratio to the spatial footprint. We found that by expanding the source region by 100 km from all sides of the YRD boundary, the $CH_4:CO_2$ emissions ratio changed by less than 1%.

3.3. Inventory Results

Table 1 shows the results of the CO_2 emission in the YRD in 2012. Industrial energy consumption is the largest emitter, followed by industrial processes, and the residential sector is the smallest emitter. The emission estimate for the transportation sector has the largest relative uncertainty, mainly due to the lack of accurate data on the annual driving range and the emission factor for different vehicle types and driving conditions [75]. The total emission amount was 19.18×10^{11} kg in 2012. The Monte Carlo method gives an overall uncertainty of 10%. For comparison the total emission amount in 2009 was 15.35×10^{11} kg [51].

Table 1. Anthropogenic CO_2 emissions in the YRD in 2012.

Sector	Emission ($\times 10^{11}$ kg)	Percent of Total (%)
Industrial energy consumption [1]	13.03 ($\pm$ 11%)	67.9
Industrial processes	4.40 ($\pm$ 10%)	23.0
Transportation	1.35 ($\pm$ 18%)	7.0
Household	0.40 ($\pm$ 8%)	2.1
Total	19.18 ($\pm$ 10%)	100

[1] CO_2 emissions in manufacturing, commerce, and construction are also included in this sector.

The annual CO_2 emission for 2012–2015 is shown in Figure 10. The growth rate was 3% from 2012 to 2013 and decreased to 0.5% from 2014 to 2015. The reduced growth was mainly caused by the decrease in the industrial energy emission.

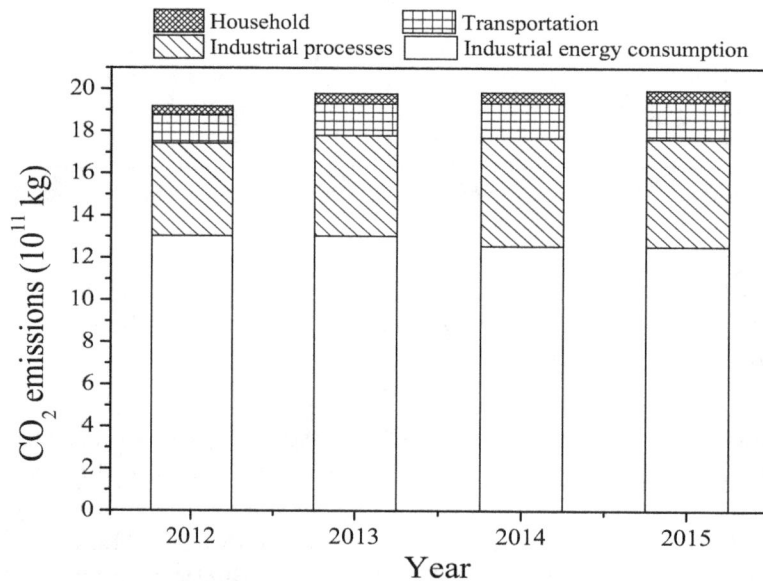

Figure 10. Annual anthropogenic CO_2 emissions from different sectors in the YRD from 2012 to 2015, based on the IPCC inventory method.

Similarly, Table 2 shows the anthropogenic CH_4 emissions in the YRD in 2012. Rice cultivation and coal mining are the major sources of anthropogenic CH_4 emissions. Emission estimates for landfills, wastewater treatment and fuel burning (traffic) emissions have large uncertainties. The total regional emission was 5.78×10^9 kg in 2012, increasing by 9.7% compared to that reported for 2009 [51]. The uncertainty of the total estimate (21%) is much greater than the uncertainty (10%) of the CO_2 emission estimate (Table 1), supporting the use of CO_2 as a tracer gas to calculate the CH_4 emission with the atmospheric method.

Table 2. Anthropogenic CH_4 emissions in the Yangtze River Delta in 2012, based on the IPCC inventory method.

Sector	Emission ($\times 10^9$ kg)	Percent of Total (%)
Rice cultivation	2.68 ($\pm$ 12%)	46.3
Landfill	0.50 ($\pm$ 35%)	8.7
Wastewater treatment	0.28 ($\pm$ 40%)	4.8
Livestock	0.31 ($\pm$ 14%)	5.4
Fuel and Biomass burning	0.32 ($\pm$ 17%)	5.6
Coal mining	1.69 ($\pm$ 30%)	29.2
Total	5.78 ($\pm$ 21%)	100

Figure 11 shows the annual anthropogenic CH_4 emission in the YRD from 2012 to 2015. There was a slight downward trend during the study period, but there was a slight increase from 2014 to 2015.

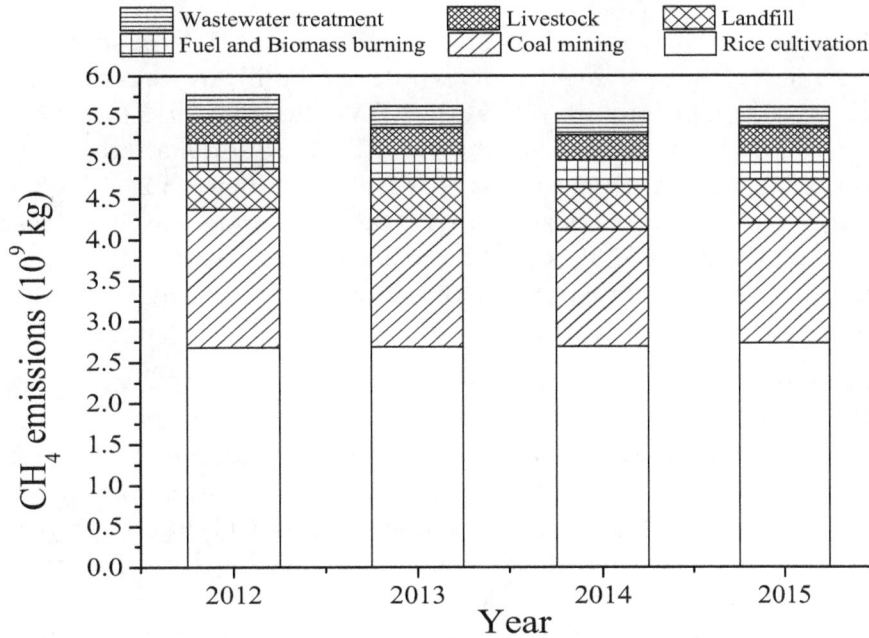

Figure 11. Annual anthropogenic CH_4 emissions from different sectors in the YRD from 2012 to 2015, based on the IPCC inventory method.

3.4. Comparison of CH_4 Emission Estimates between the Methods

Figure 12 shows the comparison of the annual anthropogenic CH_4 emissions obtained with the two methods. In this comparison, emissions from rice cultivation were excluded from the IPCC estimate because no rice is grown in the winter months. The "top-down" atmospheric estimate fluctuates year by year, in the range from 3.84×10^9 kg to 4.89×10^9 kg, which is 1.2–1.7 times the IPCC result.

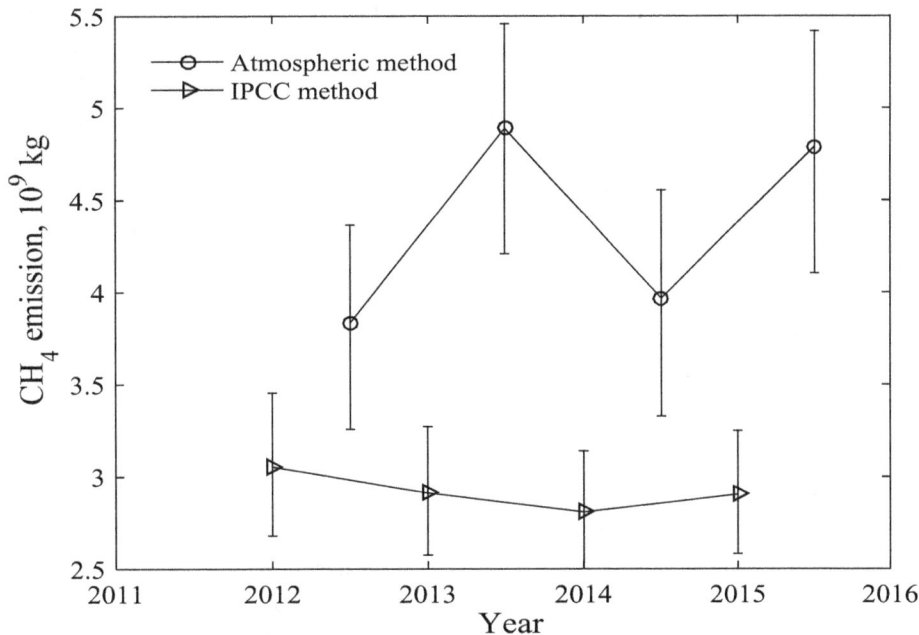

Figure 12. Annual anthropogenic CH_4 emissions (excluding rice cultivation) in the YRD from 2012 to 2015. Error bars indicate the uncertainty at a confidence level of 97.5%.

4. Discussion

4.1. Annual Growth Rates of CH_4 and CO_2 Concentrations

The observation lasted nearly five years. Due to the lack of data in some months, the annual average is not representative, thus the linear increase rate of monthly averages was used to characterize the annual growth rate. The CO_2 growth rate at the MLW site from 2012 to 2017 was 2.5 ± 0.7 ppm year^{-1}, which is 9% higher than observed at the WLG background station (2.3 ± 0.2 ppm year^{-1}). A similar result was obtained from observations near Beijing, showing a 24% higher growth rate than at WLG [76]. The large growth rate is consistent with the anthropogenic CO_2 emissions trend in the YRD with an increase of 30% from 15.35×10^{11} kg in 2009 [51] to 19.95×10^{11} kg in 2015 (Figure 10). In this period, emissions from the transport sector increased by 88%, in line with the increasing car ownership, while emissions from industrial processes were nearly unchanged [51].

The annual growth rate of CH_4 at MLW (9.5 ± 4.7 ppb year^{-1}) is nearly twice that at WLG (6.2 ± 1.7 ppb year^{-1}). The increasing trend at MLW is close to that in the Lin'an station in Zhejiang Province, which is also located in the center of YRD, with a growth rate of 8.0 ± 1.2 ppb year^{-1} from 2009 to 2011 [77]. The growth rate at WLG from 2012 to 2016 was quite similar to global mean growth rate (about 6 ppb year^{-1}, 2010–2014) [78,79]. The annual trend of CH_4 is 4.8 ppb year^{-1} (2008–2013) at the Zhongshan station in Antarctica, which is the least influenced by human activities [80,81]. At the Shangdianzi regional background station in Beijing, atmospheric CH_4 concentration influenced by airmasses passing through the highly developed Beijing Municipality, Tianjin Municipality and Hebei Province increased at a rate of 10 ± 0.1 ppb year^{-1} from 2009 to 2013, whereas atmospheric CH_4 concentrations influenced by airmasses originating from Russia, Mongolia, and the Inner Mongolia Autonomous Region of China increased at a rate of 6 ± 0.1 ppb year^{-1} over the same period [82].

One reason for the high CH_4 growth rate at the MLW site is the steady increase in anthropogenic emissions according to the inventory data, although the "top-down" estimate does not reveal such a trend in anthropogenic emissions. Another reason may be related to the expansion of wetland areas. According to two national surveys of wetland resources, the total wetland area in the YRD obtained by the second census (2009–2013) increased by 54% in comparison with the first (1995–2003), from 3.5×10^4 km^2 to 5.4×10^4 km^2 [36]. About 67% of this growth is contributed by constructed wetlands, which increased by nearly 700%. In natural wetlands around the world, the CH_4 emission flux, ranging from 7.0×10^3 to 2.8×10^4 kg km^{-2} year^{-1} with an average of 2.1×10^4 kg km^{-2} year^{-1} [83,84] is lower than in constructed wetlands (ranging from 1.3×10^3 to 1.5×10^5 kg km^{-2} year^{-1}, with an average of 4.7×10^4 kg km^{-2} year^{-1}) [85]. The annual total emissions by wetlands for the two time periods (1995–2003 and 2009–2013) aere 1.89×10^8 kg year^{-1} (range 1.6×10^8 to 14.8×10^8 kg year^{-1}) and 2.98×10^8 kg year^{-1} (range 2.5×10^8 to 23.3×10^8 kg year^{-1}), respectively, using the Tier 1 approach provided by 2006 IPCC guidelines for moist and warm climates [8]. This method only considers diffusive emissions for natural wetlands during ice-free periods. Emissions from wetlands in two national surveys were 6.1% and 9.6% anthropogenic emission (excluding rice cultivation) in 2012 ($3.1 (\pm 0.5) \times 10^9$ kg, Table 2).

4.2. Comparison of the CH_4/CO_2 Emissions Ratio

Some sources, such as transportation and landfills, release both CH_4 and CO_2, but their CH_4/CO_2 emissions ratios are very different. The average CH_4/CO_2 emissions ratio of vehicle traffic is $4.6 (\pm 0.2) \times 10^{-5}$ ppm/ppm according to observations in traffic tunnels in Switzerland [86], indicating that CH_4 emissions from traffic emissions account for only a small fraction of the anthropogenic emission. In a study of eight cities in China, the traffic emissions ratio is $7.0 (\pm 3.6) \times 10^{-3}$ ppm/ppm [87]. The ratio is much larger than that in Switzerland. In comparison, CH_4 comprises up to 61% (median: 34%) of the total volume of landfill gas, with the remaining 3% to 69% (median: 33%) being CO_2 [88,89]. In the case of wetlands, the emissions ratio varies with the type of wetlands and sometimes is negative because they act as sources of CH_4 but sinks of CO_2 [90–92]. The emissions ratio estimated from the regression slope is a composite signal of all the sources at the regional scale.

Table S1 summarizes the CH_4/CO_2 emissions ratio observed in different parts of the world. All these values were obtained as the regression slope of atmospheric CH_4 concentration against the CO_2 concentration. Some of these slope values were given by the cited literature, while others were estimated from the original concentration data. Our emissions ratio is close to those observed in Los Angeles and Pasadena, USA, but is higher than that observed in the daytime in Nanjing, China. The highest values occur in northern latitude remote sites far away from major industrial activities, such as Barrow, USA and Alert, Canada, due to abundant wetlands at these latitudes (>50% land area) [93]. About 60% of the total CH_4 emission from natural wetlands come from those between latitude 50° N to 70° N [83].

4.3. Comparison between the IPCC Method and the Atmospheric Method

Shen et al. measured atmospheric CO_2 and CH_4 molar fraction at a suburban site in Nanjing from June 2010 to April 2011 [51]. They argued that the regression slope from daytime data represents emissions ratio of sources in the YRD, and the regression slope from nighttime data represents emissions ratio of the local sources in the Nanjing Municipality. Their atmospheric estimate for CH_4 emission for sources in Nanjing is 200% higher than the IPCC estimate [51]. If we assume that the nighttime data in this study are also indicative of local sources in the Wuxi Municipality, our atmospheric estimate of the CH_4 emission in the winter would be 1.8×10^8–2.3×10^8 kg, which is 4.4–5.7 times as large as the value calculated with the IPCC method (after exclusion of rice paddy emissions; Table 3). The IPCC method is far more uncertain at the urban scale than at the regional scale, for several reasons. First, cities do not have clear boundaries and direct emissions calculated based on urban statistics ignore some emissions of cities, such as aviation and waterways, so the total emission will be underestimated [94]. As the spatial scale increases, the dependence of the accounting area on cross-boundary transport of energy and material is smaller, and errors due to indirect emissions are reduced [95]. Second, emission factors at the cities are not measured accurately. Third, wastewater treatment and landfill are the two largest source categories in Wuxi, accounting for 45.6% of the city's total CH_4 emissions; these two sources have the largest uncertainty.

Table 3. Anthropogenic CH_4 emissions in Wuxi in 2012 based on the IPCC inventory method.

Sector	Emission ($\times 10^7$ kg)	Percent of Total (%)
Rice cultivation	3.05 ($\pm$ 13%)	42.8
Landfill	1.81 ($\pm$ 38%)	25.4
Wastewater treatment	1.45 ($\pm$ 40%)	20.3
Livestock	0.48 ($\pm$ 22%)	6.7
Fuel and Biomass burning	0.34 ($\pm$ 21%)	4.8
Coal mining	—	—
Total	7.13 ($\pm$ 26%)	100

In comparison with the IPCC estimate for the YRD in the winter, the "top-down" atmospheric estimate in this study is 1.2–1.7 times the IPCC result. Shen et al.'s estimate of CH_4 emission from the atmospheric method is 20% lower [51] than the IPCC estimate, even though their measurement and our measurement both took place in the same region (YRD) and not too far from each other (170 km apart). One possible reason for the difference is that Shen et al.'s observation site is not far away from industrial complexes and traffic roads. Traffic CH_4/CO_2 emissions ratio is generally much lower than regional emissions ratios, as noted above. According to the "bottom-up" results provided by the Emissions Database for Global Atmospheric Research (EDGAR), the global CH_4/CO_2 emissions ratios in the chemical production sector, the metal production sector, and the public electricity and heat production sector were 1.2×10^{-3}, 6.4×10^{-4} and 7.6×10^{-5} ppm/ppm, respectively in 2012 (http://edgar.jrc.ec.europa.eu/overview.php?v=432&SECURE=123) [96]. The situation is a little different in China. The CH_4/CO_2 emissions ratio is bigger for the chemical production sector (1.8×10^{-3} ppm/ppm), and is smaller for the metal production sector (4.7×10^{-4} ppm/ppm) and the public electricity and heat production sector (3.4×10^{-5} ppb/ppm). Thus, the CH_4/CO_2 emissions ratio for industrial complexes should be lower than regional emissions ratios. Because our measurement was made at a rural system not directly impacted by local traffic and industrial emissions (Figure S1), our emissions ratio should be more representative than that reported in [51].

Another reason for the difference between our study and the study by Shen et al. [51] is different source areas between Nanjing and MLW. Hu et al. [97] simulated the atmospheric CO_2 concentration in Nanjing City using the WRF-STILT model. They found that the NUIST site is mainly affected by the central and eastern regions of Anhui Province and the central and western regions of Jiangsu Province, with greater concentration contribution weights than other regions in the YRD.

Previous studies have also reported that estimates of CH_4 emission based on the "top-down" atmospheric method are higher than the "bottom-up" inventory estimates. In the Los Angeles metropolitan area in California, USA, anthropogenic CH_4 emissions calculated from two "top-down" approaches are 1.3–1.8 and 1.2–1.6 times, respectively, of the "bottom-up" estimates [20,98]. The US EPA inventory and EDGAR, two "bottom-up" methods, are shown to underestimate CH_4 emissions in the United States by a factor of about 1.5 and 1.7, with the underestimation coming from two sectors: livestock and fossil fuel extraction and processing [99]. That "bottom-up" results are generally smaller indicate that there are unknown emission sources or underestimates of the emission capacity of known sources, such as landfills, coal mining and wastewater treatment. Some scholars have found that based on carbon isotope observations, CH_4 emissions from landfills and ruminants are underestimated by the IPCC method [100]. In China, the methane emission rate is reported to be eight times the IPCC emission estimate for natural gas vehicles [78]. If this result is taken into consideration, the anthropogenic CH_4 emissions in the YRD would increase by 2.4% in 2015. In the urban Boston area, USA, natural gas loss rate from transmission, distribution and end use was $2.7 \pm 0.6\%$ of the total delivered gas, which is more than twice the result of the emission inventory [101]. In 2016, the length of natural gas pipeline reached 5.51×10^5 km in China, although the CH_4 concentration measurements along urban street transects in eight Chinese cities show no evidence of pipeline leakage [87].

5. Conclusions

Continuous observation of atmospheric CO_2 and CH_4 mole fraction was made at the MLW station near Wuxi at Lake Taihu from May 2012 to April 2017. These measurements were combined with anthropogenic CO_2 emission data in a "top-down" method to obtain an estimate of the anthropogenic CH_4 emission in the YRD. For comparison, the CH_4 emission was also calculated with the IPCC inventory method. The key results are as follows:

(1) The growth rates of the CO_2 and CH_4 molar fractions at the MLW site were 2.5 ± 0.7 ppm year^{-1} and 9.5 ± 4.7 ppb year^{-1}, respectively, which are 9% and 53% higher than that observed at WLG over the same period.

(2) To avoid the interference of biological sources, we used the wintertime CO_2 and CH_4 concentration data to obtain the CH_4/CO_2 emissions ratio. Results indicate that the emissions ratio fluctuates between 0.0055 ± 0.0006 ppm/ppm (winters of 2012–2013 and 2014–2015) and 0.0068 ± 0.0005 ppm/ppm (winter of 2013–2014). These ratios are similar to those observed in Los Angeles and Pasadena, USA.

(3) According to the "top-down" method, the annual average anthropogenic emission of CH_4 in the YRD from 2012 to 2015 is $4.37 (\pm 0.61) \times 10^9$ kg year^{-1} (excluding rice cultivation), which is 1.2–1.7 times the result from the IPCC inventory.

(4) The "top-down" method also suggests that at the local scale, the IPCC inventory estimate for anthropogenic CH_4 emission in the Wuxi municipality may be biased low by 4.4—5.7 times.

Author Contributions: Data curation, W.H., J.X. and Y.H.; Formal analysis, W.H.; Investigation, W.H.; Methodology, W.H.; Resources, W.X., J.X., Y.H. and S.L.; Supervision, W.X., M.Z., W.W., C.H. and X.L.; Validation, C.H. and X.L.; Writing—original draft, W.H.; and Writing—review and editing, X.L.

Acknowledgments: We are grateful to all of the staff who work at the WMO/GAW stations in China for collecting the data, and to the Greenhouse Gases Research Laboratory of the China Meteorological Administration (CMA) for data analysis. We appreciate Ed Dlugokencky, Andy Crotwell, and Kirk Thoning of NOAA for their help and support of the research measurements.

References

1. World Meteorological Organization (WMO). *Greenhouse Gas Bulletin: The State of Greenhouse Gases in the Atmosphere Based on Global Observations through 2017*; WMO: Geneva, Switzerland, 2018.

2. IPCC. *The IPCC Fifth Assessment Report—Climate Change 2013: The Physical Science Basis*; Working Group I, IPCC Secretariat: Geneva, Switzerland, 2013.

3. Solomon, S.; Intergovernmental Panel on Climate Change. *Working Group I, Climate Change 2007: The Physical Science Basis: Contribution of Working Group I to the Fourth Assessment Report of the Intergovernmental Panel on Climate Change*; Cambridge University Press: Cambridge, UK; New York, NY, USA, 2007.

4. Dlugokencky, E.J. NOAA/ESRL. Available online: www.esrl.noaa.gov/gmd/ccgg/trends_ch4/ (accessed on 24 March 2019).

5. Dlugokencky, E.J.; Bruhwiler, L.; White, J.W.C.; Emmons, L.K.; Novelli, P.C.; Montzka, S.A.; Masarie, K.A.; Lang, P.; Crotwell, A.M.; Miller, J.B.; et al. Observational constraints on recent increases in the atmospheric CH_4 burden. *Geophys. Res. Lett.* **2009**, *36*, 252–260. [CrossRef]

6. Forster, P.; Ramaswamy, V.; Artaxo, P.; Bernsten, T.; Betts, R.; Fahey, D.W.; Haywood, J.; Lean, J.; Lowe, D.C.; Myhre, G.; et al. *Changes in Atmospheric Constituents and in Radiative Forcing. Chapter 2. Climate Change 2007: The Physical Science Basis*; Cambridge University Press: Cambridge, UK, 2007; pp. 129–234.

7. Aydin, M.; Verhulst, K.R.; Saltzman, E.S.; Battle, M.O.; Montzka, S.A.; Blake, D.R.; Tang, Q.; Prather, M.J. Recent decreases in fossil-fuel emissions of ethane and methane derived from firn air. *Nature* **2011**, *476*, 198–201. [CrossRef] [PubMed]

8. IPCC. *2006 IPCC Guidelines for National Greenhouse Gas Inventories, Prepared by the National Greenhouse Gas Inventories Programme*; Eggleston, HS., Buendia, L., Miwa, K., Ngara, T., Tanabe, K., Eds.; IGES: Kanagawa, Japan, 2006.

9. Yue, Q.; Zhang, G. Preliminary estimation of methane emission and its distribution in China. *Geogr. Res.* **2012**, *31*, 1561–1570. (In Chinese) [CrossRef]

10. Boon, A.; Broquet, G.; Clifford, D.J.; Chevallier, F.; Butterfield, D.M.; Pison, I.; Ramonet, M.; Paris, J.D.; Ciais, P. Analysis of the potential of near ground measurements of CO_2 and CH_4 in London, UK for the monitoring of city-scale emissions using an atmospheric transport model. *Atmos. Chem. Phys.* **2016**, *16*, 6735–6756. [CrossRef]

11. Chen, C.; Liu, C.; Li, Z.; Wang, H.; Zhang, Y.; Wang, L. Uncertainty analysis for evaluating methane emissions from municipal solid waste landfill in Beijing. *Environ. Sci.* **2012**, *33*, 208–215. (In Chinese)

12. Pei, Z.; Ou Yang, H.; Zhou, C. A study on carbon fluxes from alpine grassland ecosystem on Tibetan Plateau. *Acta Ecol. Sin.* **2003**, *23*, 231–236. (In Chinese) [CrossRef]

13. Minamikawa, K.; Yagi, K.; Tokida, T.; Sander, B.O.; Wassmann, R. Appropriate frequency and time of day to measure methane emissions from an irrigated rice paddy in Japan using the manual closed chamber method. *Greenh. Gas Meas. Manag.* **2012**, *2*, 118–128. [CrossRef]

14. Winton, R.S.; Richardson, C.J. A cost-effective method for reducing soil disturbance-induced errors in static chamber measurement of wetland methane emissions. *Wetl. Ecol. Manag.* **2016**, *24*, 419–425. [CrossRef]

15. Lee, X. *Fundamentals of Boundary-Layer Meteorology*; Springer International Publishing: Cham, Switzerland, 2018.

16. Conway, T.J.; Steele, L.P. Carbon dioxide and methane in the Arctic atmosphere. *J. Atmos. Chem.* **1989**, *9*, 81–99. [CrossRef]

17. Conway, T.J.; Steele, L.P.; Novelli, P.C. Correlations among atmospheric CO_2, CH_4, and CO in the Arctic, March 1989. *Atmos. Environ. Part A Gen. Top.* **1993**, *27*, 2881–2894. [CrossRef]

18. Hansen, A.D.A.; Conway, T.J.; Strele, L.P.; Bodhaine, B.A.; Thoning, K.W.; Tans, P.; Novakov, T. Correlations among combustion effluent species at Barrow, Alaska: Aerosol black carbon, carbon dioxide, and methane. *J. Atmos. Chem.* **1989**, *9*, 283–299. [CrossRef]

19. Wunch, D.; Wennberg, P.O.; Toon, G.C.; Keppel-Aleks, G.; Yavin, Y.G. Emissions of greenhouse gases from a North American megacity. *Geophys. Res. Lett.* **2009**, *36*, 139–156. [CrossRef]

20. Wong, K.W.; Fu, D.; Pongetti, T.J.; Newman, S.; Kort, E.A.; Duren, R.; Hsu, Y.K.; Miller, C.E.; Yung, Y.L.; Sander, S.P. Mapping CH_4: CO_2 ratios in Los Angeles with CLARS-FTS from Mount Wilson, California. *Atmos. Chem. Phys.* **2015**, *15*, 241–252. [CrossRef]

21. Lelieveld, J.; Crutzen, P.J.; Dentener, F.J. Changing concentration, lifetime and climate forcing of atmospheric methane. *Tellus Ser. B Chem. Phys. Meteorol.* **1998**, *50*, 128–150. [CrossRef]

22. Zhao, Y.; Nielsen, C.P.; Mcelroy, M.B. China's CO_2 emissions estimated from the bottom up: Recent trends, spatial distributions, and quantification of uncertainties. *Atmos. Environ.* **2012**, *59*, 214–223. [CrossRef]

23. Lee, X.; Bullock, O.R.; Andres, R.J. Anthropogenic emission of mercury to the atmosphere in the northeast United States. *Geophys. Res. Lett.* **2001**, *28*, 1231–1234. [CrossRef]

24. Wang, Y.; Munger, J.W.; Xu, S.; McElroy, M.B.; Hao, J.; Nielsen, C.P.; Ma, H. CO_2 and its correlation with CO at a rural site near Beijing: Implications for combustion efficiency in China. *Atmos. Chem. Phys.* **2010**, *10*, 8881–8897. [CrossRef]

25. Suntharalingam, P.; Jacob, D.J.; Palmer, P.I.; Logan, J.A.; Yantosca, R.M.; Xiao, Y.; Evans, M.J. Improved quantification of Chinese carbon fluxes using CO_2/CO correlations in Asian outflow. *J. Geophys. Res.* **2004**, *109*, 159–172. [CrossRef]

26. Zhu, T.; Yang, G.; Su, W.; Wan, R. Coordination evaluation between urban land intensive use and economic society development in the Yangtze River Delta. *Resour. Sci.* **2009**, *31*, 1109–1116. (in Chinese).

27. Lee, X.; Liu, S.; Xiao, W.; Wang, W.; Gao, Z.; Cao, C.; Hu, C.; Hu, Z.; Shen, S.; Wang, Y.; et al. The Taihu eddy flux network: an observational program on energy, water, and greenhouse gas fluxes of a large freshwater lake. *Bull. Am. Meteorol. Soc.* **2014**, *95*, 1583–1594. [CrossRef]

28. Xiao, W.; Liu, S.; Li, H.; Xiao, Q.; Wang, W.; Hu, Z.; Hu, C.; Gao, Y.; Shen, J.; Zhao, X.; et al. A flux-gradient system for simultaneous measurement of the CH_4, CO_2, and H_2O fluxes at a lake-air interface. *Environ. Sci. Technol.* **2014**, *48*, 14490–14498. [CrossRef] [PubMed]

29. Flores, E.; Viallon, J.; Choteau, T.; Moussay, P.; Wielgosz, R.I.; Kang, N.; Kim, B.M.; Zalewska, E.; van der Veen, A.M.H.; Konopelko, L.; et al. International comparison CCQM-K82: Methane in air at ambient level (1800 to 2200) nmol/mol. *Metrologia* **2015**, *52*, 1–129. [CrossRef]

30. Flores, E.; Viallon, J.; Choteau, T.; Moussay, P.; Idrees, F.; Wielgosz, R.I.; Lee, J.; Zalewska, E.; Nieuwenkamp, G.; van der Veen, A.; et al. CCQM-K120 (Carbon dioxide at background and urban level). *Metrologia* **2019**, *56*, 1–178. [CrossRef]

31. State Statistical Bureau. *China Energy Statistical Yearbook 2012*; China Statistical Press: Beijing, China, 2013. (In Chinese)

32. State Statistical Bureau. *China Energy Statistical Yearbook 2013*; China Statistical Press: Beijing, China, 2014. (In Chinese)

33. State Statistical Bureau. *China Energy Statistical Yearbook 2014*; China Statistical Press: Beijing, China, 2015. (In Chinese)

34. State Statistical Bureau. *China Energy Statistical Yearbook 2015*; China Statistical Press: Beijing, China, 2016. (In Chinese)

35. State Statistical Bureau. *China Statistical Yearbook 2012*; China Statistical Press: Beijing, China, 2013. (In Chinese)

36. State Statistical Bureau. *China Statistical Yearbook 2013*; China Statistical Press: Beijing, China, 2014. (In Chinese)

37. State Statistical Bureau. *China Statistical Yearbook 2014*; China Statistical Press: Beijing, China, 2015. (In Chinese)

38. State Statistical Bureau. *China Statistical Yearbook 2015*; China Statistical Press: Beijing, China, 2016. (In Chinese)

39.   State Statistical Bureau. *China Rural Statistical Yearbook 2012*; China Statistical Press: Beijing, China, 2013. (In Chinese)

40.   State Statistical Bureau. *China Rural Statistical Yearbook 2013*; China Statistical Press: Beijing, China, 2014. (In Chinese)

41.   State Statistical Bureau. *China Rural Statistical Yearbook 2014*; China Statistical Press: Beijing, China, 2015. (In Chinese)

42.   State Statistical Bureau. *China Rural Statistical Yearbook 2015*; China Statistical Press: Beijing, China, 2016. (In Chinese)

43.   Cao, G.; Zhang, X.; Zhen, F.; Wang, Y. Estimating the quantity of crop residues burnt in open field in China. *Resour. Sci.* **2006**, *28*, 9–13. (in Chinese). [CrossRef]

44.   Qiu, L.; Yang, G.; Bi, Y. Discussion on the conditions and countermeasures of developing marsh gas in rural areas of west China. *Agric. Res. Arid Areas* **2005**, *23*, 200–204. (In Chinese) [CrossRef]

45.   Scheutz, C.; Kjeldsen, P.; Bogner, J.E.; Visscher, A.D.; Gebert, J.; Hilger, H.A.; Huber-Humer, M.; Spokas, K. Microbial methane oxidation processes and technologies for mitigation of landfill gas emissions. *Waste Manag. Res.* **2009**, *27*, 409–455. [CrossRef] [PubMed]

46.   Cai, B. Analysis of the features of methane emissions from landfills of China in 2012. *Environ. Eng.* **2016**, *34*, 1–4. (In Chinese) [CrossRef]

47.   Chen, D.; Wang, M.; Shang Guan, X.; Huang, J.; Rasmussen, R.A.; Khalil, M.A.K. Methane emission from rice fields in the south-east China. *Adv. Earth Sci.* **1993**, *8*, 47–54. (in Chinese).

48.   Wang, M. Methane emission and mechanisms of methane production, oxidation, transportation in the rice fields. *Chin. J. Atmos. Sci.* **1998**, *22*, 600–612. (In Chinese) [CrossRef]

49.   Min, J.; Hu, H. Calculation of greenhouse gases emission from agricultural production in China. *China Popul. Resour. Environ.* **2012**, *22*, 21–27. (In Chinese) [CrossRef]

50.   Zhao, B. *The Research on the Surface Mine Group Development & Design Theory and Engineering Optimization*; China University of Mining & Technology: Beijing, China, 2015. (In Chinese)

51.   Shen, S.; Dong, Y.; Wei, X.; Liu, S.; Lee, X. Constraining anthropogenic CH_4 emissions in Nanjing and the Yangtze River Delta, China, using atmospheric CO_2 and CH_4 mixing ratios. *Adv. Atmos. Sci.* **2014**, *31*, 1343–1352. [CrossRef]

52.   Yang, D.; Shen, S.; Zhang, M.; Lee, X.; Xiao, W. Uncertainty analysis on the estimation of CO_2 and CH_4 emission inventory over Nanjing and Yangtze River Delta. *J. Meteorol. Sci.* **2014**, *34*, 325–334. (In Chinese) [CrossRef]

53.   Ramírez, A.; Keizer, C.D.; Sluijs, J.P.V.D.; Olivier, J.; Brandes, L. Monte Carlo analysis of uncertaintes in the Netherlands greenhouse gas emission inventory for 1990–2004. *Atmos. Environ.* **2008**, *42*, 8263–8272. [CrossRef]

54.   Rypdal, K.; Winiwarter, W. Uncertainties in greenhouse gas emission inventories evaluation, comparability and implications. *Environ. Sci. Policy* **2001**, *4*, 107–116. [CrossRef]

55.   Amstel, A.R.V.; Olivier, J.G.J.; Ruyssenaars, P.G. Monitoring of greenhouse gases in the Netherlands: Uncertainty and priorities for improvement. In Proceedings of the National Workshop, Bilthoven, The Netherlands, 1 September 1999.

56.   Winiwarter, W.; Rypdal, K. Assessing the uncertainty associated with national greenhouse gas emission inventories: A case study for Austria. *Atmos. Environ.* **2001**, *35*, 5425–5440. [CrossRef]

57.   Wehr, R.; Saleska, S.R. The long-solved problem of the best-fit straight line: Application to isotopic mixing lines. *Biogeosciences* **2016**, *14*, 17–29. [CrossRef]

58.   Rotty, R.M. Estimates of seasonal variation in fossil fuel CO_2 emissions. *Tellus* **1987**, *39*, 184–202. [CrossRef]

59.   Li, Y.; Deng, J.; Mu, C.; Xing, Z.; Du, K. Vertical distribution of CO_2, in the atmospheric boundary layer: Characteristics and impact of meteorological variables. *Atmos. Environ.* **2014**, *91*, 110–117. [CrossRef]

60.   Winderlich, J.; Gerbig, C.; Kolle, O.; Heimann, M. Inferences from CO_2 and CH_4 concentration profiles at the Zotino Tall Tower Observatory (ZOTTO) on regional summertime ecosystem fluxes. *Biogeosciences* **2014**, *10*, 15337–15372. [CrossRef]

61.   Wang, Y. MeteoInfo: GIS software for meteorological data visualization and analysis. *Meteorol. Appl.* **2014**, *21*, 360–368. [CrossRef]

62. Wang, Y.; Zhang, X.; Draxler, R.R. TrajStat: GIS-based software that uses various trajectory statistical analysis methods to identify potential sources from long-term air pollution measurement data. *Environ. Model. Softw.* **2009**, *24*, 938–939. [CrossRef]

63. Sigler, J.M.; Lee, X. Recent trends in anthropogenic mercury emission in the northeast United States. *J. Geophys. Res. Atmos.* **2006**, *111*, 3131–3148. [CrossRef]

64. Zhang, F.; Fukuyama, Y.; Wang, Y.; Fang, S.; Li, P.; Fan, T.; Zhou, L.; Liu, X.; Meinhardt, F.; Emiliani, P. Detection and attribution of regional CO_2 concentration anomalies using surface observations. *Atmos. Environ.* **2015**, *123*, 88–101. [CrossRef]

65. Obrist, D.; Conen, F.; Vogt, R.; Siegwolf, R.; Alewell, C. Estimation of Hg^0 exchange between ecosystems and the atmosphere using ^{222}Rn and Hg^0 concentration changes in the stable nocturnal boundary layer. *Atmos. Environ.* **2006**, *40*, 856–866. [CrossRef]

66. Zhou, L.; Tang, J.; Wen, Y.; Li, J.; Yan, P.; Zhang, X. The impact of local winds and long-range transport on the continuous carbon dioxide record at Mount Waliguan, China. *Tellus B* **2003**, *55*, 145–158. [CrossRef]

67. Zhang, F.; Zhou, L.; Xu, L. Temporal variation of atmospheric CH_4 and the potential source regions at Waliguan, China. *Sci. China Earth Sci.* **2013**, *56*, 727–736. (In Chinese) [CrossRef]

68. Xu, J.; Lee, X.; Xiao, W.; Cao, C.; Liu, S.; Wen, X.; Xu, J.; Zhang, Z.; Zhao, J. Interpreting the $^{13}C/^{12}C$ ratio of carbon dioxide in an urban airshed in the Yangtze River Delta, China. *Atmos. Chem. Phys.* **2017**, *17*, 3385–3399. [CrossRef]

69. Boon, P.I.; Mitchell, A. Methanogenesis in the sediments of an Australian freshwater wetland, Comparison with aerobic decay, and factors controlling methanogenesis. *Fems. Microbiol. Ecol.* **1995**, *18*, 175–190. [CrossRef]

70. Qin, S.; Tang, J.; Pu, J.; Xu, Y.; Dong, P.; Jiao, L.; Guo, J. Fluxes and influencing factors of CO_2 and CH_4 in Hangzhou Xixi wetland, China. *Earth Environ.* **2016**, *44*, 513–519. (In Chinese) [CrossRef]

71. Zhang, B.; Tian, H.; Ren, W.; Tao, B.; Lu, C. Methane emissions from global rice fields: Magnitude, spatiotemporal patterns, and environmental controls. *Glob. Biogeochem. Cycle* **2016**, *30*, 1246–1263. [CrossRef]

72. Saunois, M.; Bousquet, P.; Poulter, B.; Peregon, A.; Ciais, P.; Canadell, J.G.; Dlugokencky, E.J.; Etiope, G.; Bastviken, D.; Houweling, S.; et al. The global methane budget 2000–2012. *Earth Syst. Sci. Data* **2016**, *8*, 697–751. [CrossRef]

73. Wang, X.; Ciais, P.; Li, L.; Ruget, F.; Vuichard, N.; Viovy, N.; Zhou, F.; Chang, J.; Wu, X.; Zhao, H.; et al. Management outweighs climate change on affecting length of rice growing period for early rice and single rice in China during 1991–2012. *Agric. For. Meteorol.* **2017**, *233*, 1–11. [CrossRef]

74. Jiang, C.; Wang, Y.; Zheng, X.; Zhu, B.; Huang, Y.; Hao, Q. Methane and nitrous oxide emissions from three paddy rice based cultivation systems in southwest China. *Adv. Atmos. Sci.* **2006**, *23*, 415–424. [CrossRef]

75. Song, X.; Xie, S. Development of vehicular emission inventory in China. *Environ. Sci.* **2006**, *27*, 1041–1045. (In Chinese) [CrossRef]

76. Liu, Q.; Wang, Y.; Wang, M.; Li, J.; Li, G. Trends of greenhouse gases in recent 10 years in Beijing. *China J. Atmos. Sci.* **2005**, *29*, 267–271. (in Chinese).

77. Fang, S.; Zhou, L.; Masarie, K.A.; Xu, L.; Rella, C.W. Study of atmospheric CH_4 mole fractions at three WMO/GAW stations in China. *J. Geophys. Res. Atmos.* **2013**, *118*, 4874–4886. [CrossRef]

78. WMO Data Summary. WMO World Data Centre for Greenhouse Gases (WDCGG) Data Summary: Greenhouse Gases and Other Atmospheric Gases, No.38. Japan Meteorological Agency. Available online: http://ds.data.jma.go.jp/gmd/wdcgg/pub/products/summary/sum38/sum38.pdf (accessed on 13 May 2018).

79. World Meteorological Organization (WMO). *Greenhouse Gas Bulletin: The State of Greenhouse Gases in the Atmosphere Based on Global Observations through 2016*; WMO: Geneva, Switzerland, 2016.

80. Bian, L.; Gao, Z.; Sun, Y.; Ding, M.; Tang, J.; Schnell, R. CH_4 Monitoring and Background Concentration at Zhongshan Station, Antarctica. *Atmos. Clim. Sci.* **2015**, *6*, 135–144. [CrossRef]

81. Wang, Y.; Bian, L.; Ma, Y.; Tang, J.; Zhang, D.; Zheng, X. Surface Ozone Monitoring and Background Characteristics at Zhongshan Station over Antarctica. *Chin. Sci. Bull.* **2011**, *56*, 1011–1019. [CrossRef]

82. Fang, S.; Tans, P.P.; Dong, F.; Zhou, H.; Luan, T. Characteristics of atmospheric CO_2 and CH_4 at the Shangdianzi regional background station in China. *Atmos. Environ.* **2016**, *131*, 1–8. [CrossRef]

83. Matthews, E.; Fung, I. Methane emission from natural wetlands: Global distribution, area, and environmental characteristics of sources. *Glob. Biogeochem. Cycle* **1987**, *1*, 61–86. [CrossRef]

84. Aselmann, I.; Crutzen, P.J. Global distribution of natural freshwater wetlands and rice paddies, their net primary productivity, seasonality and possible methane emissions. *J. Atmos. Chem.* **1989**, *8*, 307–358. [CrossRef]

85. Mander, Ü.; Dotro, G.; Ebie, Y.; Towprayoon, S.; Chiemchaisri, C.; Nogueira, S.F.; Jamsranjav, B.; Kasak, K.; Tournebize, J.; Mitsch, W.J. Greenhouse gas emission in constructed wetlands for wastewater treatment: A review. *Ecol. Eng.* **2014**, *66*, 19–35. [CrossRef]

86. Popa, M.E.; Vollmer, M.K.; Jordan, A.; Brand, W.A.; Pathirana, S.L.; Rothe, M.; Röckmann, T. Vehicle emissions of greenhouse gases and related tracers from a tunnel study: CO: CO_2, N_2O: CO_2, CH_4: CO_2, O_2: CO_2 ratios, and the stable isotopes ^{13}C and ^{18}O in CO_2 and CO. *Atmos. Chem. Phys.* **2014**, *14*, 2105–2123. [CrossRef]

87. Hu, N.; Liu, S.; Gao, Y.; Xu, J.; Zhang, X.; Zhang, Z.; Lee, X. Large methane emissions from natural gas vehicles in Chinese cities. *Atmos. Environ.* **2018**, *187*, 374–380. [CrossRef]

88. Nagamori, M.; Isobe, Y.; Watanabe, Y.; Wijewardane, N.K.; Mowjood, M.I.M.; Koide, T.; Kawamotok, K. Characterization of Major and Trace Components in Gases Generated from Municipal Solid Waste Landfills in Sri Lanka. In Proceedings of the 14th International Waste Management and Landfill Symposium, Cagliari, Italy, 30 September–4 October 2013.

89. Ma, Z.; LI, H.; Yue, B.; Gao, Q.; Dong, L. Study on emission characteristics and correlation of GHGs CH_4 and CO_2 in MSW landfill cover layer. *J. Environ. Eng. Technol.* **2014**, *V4*, 399–405. (In Chinese) [CrossRef]

90. Brix, H.; Sorrell, B.K.; Lorenzen, B. Are Phragmites-dominated wetlands a net source or net sink of greenhouse gases? *Aquat. Bot.* **2001**, *69*, 313–324. [CrossRef]

91. Hu, H.; Wang, D.; Li, Y.; Chen, Z.; Wu, J.; Yin, Q.; Guan, Y. Greenhouse gases fluxes at Chongming Dongtan phragmites australis wetland and the influencing factors. *Res. Environ. Sci.* **2014**, *27*, 43–50. (In Chinese) [CrossRef]

92. Van, d.B.R. Restoration of former wetlands in the Netherlands; effect on the balance between CO_2 sink and CH_4 source. *Neth. J. Geosci.* **2016**, *82*, 325–331. [CrossRef]

93. Wania, R. Modelling northern peatland surface processes, vegetation dynamics and methane emissions. Ph.D. Thesis, University of Bristol, Bristol, UK, 2007.

94. Chen, C.; Liu, C.; Tian, G.; Wang, H.; Li, Z. Progress in research of urban greenhouse gas emission inventory. *Environ. Sci.* **2010**, *31*, 2780–2787. (in Chinese). [CrossRef]

95. Cai, B. Advance and review of city carbon dioxide emission inventory research. *China Popul. Resour. Environ.* **2013**, *23*, 72–80. (in Chinese). [CrossRef]

96. Janssens-Maenhout, G.; Crippa, M.; Guizzardi, D.; Muntean, M.; Schaaf, E.; Dentener, F.; Bergamaschi, P.; Pagliari, V.; Olivier, J.G.J.; Peters, J.A.H.W.; et al. EDGAR v4.3.2 Global atlas of the three major greenhouse gas emissions for the period 1970–2012. *Earth Syst. Sci. Data Discuss* **2017**. [CrossRef]

97. Hu, C.; Liu, S.; Cao, C.; Xu, J.; Cao, Z.; Li, W.; Xu, J.; Zhang, M.; Xiao, W.; Lee, X. Simulation of atmospheric CO_2 concentration and source apportionment analysis in Nanjing City. *Acta Sci. Circumstantiae* **2017**, *37*, 3862–3875. (In Chinese) [CrossRef]

98. Jeong, S.; Hsu, Y.K.; Andrews, A.E.; Bianco, L.; Vaca, P.; Wilczak, J.M.; Fischer, M.L. A multitower measurement network estimate of California's methane emissions. *J. Geophys. Res.* **2013**, *118*, 11339–11351. [CrossRef]

99. Miller, S.M.; Wofsy, S.C.; Michalak, A.M.; Kort, E.A.; Andrews, A.E.; Biraud, S.C.; Dlugokencky, E.J.; Eluszkiewicz, J.; Fischer, M.L.; Janssens-Maenhout, G.; et al. Anthropogenic emissions of methane in the United States. *Proc. Natl. Acad. Sci. USA* **2013**, *110*, 20018–20022. [CrossRef] [PubMed]

100. Thompson, R.L.; Stohl, A.; Zhou, L.X.; Dlugokencky, E.; Fukuyama, Y.; Tohjima, Y.; Kim, S.Y.; Lee, H.; Nisbet, E.G.; Lowry, D.; et al. Methane emissions in East Asia for 2000-2011 estimated using an atmospheric Bayesian inversion. *J. Geophys. Res.* **2015**, *120*, 4352–4369. [CrossRef]

101. McKain, K.; Down, A.; Raciti, S.M.; Budney, J.; Hutyra, L.R.; Floerchinger, C.; Herndon, S.C.; Nehrkorn, T.; Zahniser, M.S.; Jackson, R.B.; et al. Methane emissions from natural gas infrastructure and use in the urban region of Boston, Massachusetts. *Proc. Natl. Acad. Sci. USA* **2015**, *112*, 1941–1946. [CrossRef] [PubMed]

9

Sub-Mode Aerosol Volume Size Distribution and Complex Refractive Index from the Three-Year Ground-Based Measurements in Chengdu, China

Chi Zhang [1,2], Ying Zhang [1], Zhengqiang Li [1,*], Yongqian Wang [3], Hua Xu [1], Kaitao Li [1], Donghui Li [1], Yisong Xie [1] and Yang Zhang [3]

[1] State Environmental Protection Key Laboratory of Satellite Remote Sensing, Institute of Remote Sensing and Digital Earth, Chinese Academy of Sciences, Beijing 100101, China; zhangchi@radi.ac.cn (C.Z.); zhangying02@radi.ac.cn (Y.Z.); xuhua@radi.ac.cn (H.X.); likt@radi.ac.cn (K.L.); lidh@radi.ac.cn (D.L.); xieys@radi.ac.cn (Y.X.)

[2] University of Chinese Academy of Sciences, Beijing 100049, China

[3] College of Resources and Environment, Chengdu University of Information Technology, Chengdu 610103, China; wyqq@cuit.edu.cn (Y.W.); zhangyang@cuit.edu.cn (Y.Z.)

* Correspondence: lizq@radi.ac.cn.

Abstract: Chengdu is a typical basin city of Southwest China with rare observations of remote sensing measurements. To assess the climate change and establish a region aerosol model, a deeper understanding of the separated volume size distribution (VSD) and complex refractive index (CRI) is required. In this study, we employed the sub-mode VSD and CRI in Chengdu based on the three years observation data to investigate the sub-mode characteristics and climate effects. The annual average fraction of the fine-mode aerosol optical depth (AOD_f) is 92%, which has the same monthly tendency as the total AOD. But the coarse-mode aerosol optical depth (AOD_c) has little variation in different months. There are four distinguishing modes of VSD in Chengdu; the median radii are 0.17 μm $\pm$ 0.05, 0.31 μm $\pm$ 0.12, 1.62 μm $\pm$ 0.45, 3.25 μm $\pm$ 0.99, respectively. The multi-year average and seasonal variations of fine- and coarse-mode VSD and CRI are also analyzed to characterize aerosols over this region. The fine-mode single scattering albedos (SSAs) are higher than the coarse-mode ones, which suggests that the coarse-mode aerosols have a stronger absorbing effect on solar light than the small-size aerosol particles in Chengdu.

Keywords: Chengdu; aerosol; sub-mode volume size distribution; sub-mode complex refractive index; remote sensing

1. Introduction

As one component of the terrestrial atmosphere, aerosol is an important factor in global climate change, with direct effects and indirect effects. Direct effects include changing the radiation balance of the Earth-atmosphere system by absorbing and scattering sunlight [1,2]. On the other hand, aerosol serves as cloud condensation nuclei (CCN) involved in cloud microphysics processes, referred to as indirect effects [3]. Moreover, as one major constituent of haze, aerosol particles endanger public health [4].

Chengdu, located in the central region of the Sichuan basin (Chengdu Plain) with a population of ~16 million (the resident population of Chengdu was counted in 2016), is the economic center and transportation hub in southwest China. Due to the special topography of the basin, the average wind speed is low [5,6]. Coupled with the development economy in Chengdu, a large amount of pollutants associated with regional industrial emissions are easy to accumulate. Abundant water vapor is conducive to the formation of aerosol particulates, which has made Chengdu become one of the regions in China with serious haze pollution [7–10]. The special topography and climate in Sichuan basin leads to a kind of wet, rainy and cloudy weather condition [11]. Therefore, it is difficult to obtain a large number of measurements through optical remote sensing observations in Chengdu. Sampling method of atmospheric particulates is one kind of Chengdu air pollution study [7,9,10,12], which fails to acquire the aerosol's larger-scale spatial characteristics. The spatial characteristics of aerosols can be obtained by remote sensing, one kind of non-contact method [13–15]. For instance, some studies analyzed the spatial-temporal distribution of aerosol optical depth (AOD) in Sichuan basin using satellite remote sensing [16,17]. Besides, aerosol optical parameters obtained by the ground-based remote sensing observation were statistically analyzed in this area [18–21].

Although the above-mentioned research studies focus on the total-column aerosol properties by remote sensing methods, it still remains a big challenge to obtain the characteristics of sub-mode aerosol (fine- and coarse-mode) in Chengdu. The fine- and coarse-mode particles are related to the different sources and compositions. Fine-mode aerosols are mainly dominated by anthropogenic emissions and are related to the fog or low-altitude cloud dissipated events [22,23]. Coarse-mode particles are determined by the sea salt, dust and other natural sources. Therefore, the knowledge of aerosol sub-mode properties plays a role in the research on regional climate and the improvement of the aerosol model used in the satellite retrieval algorithms.

In this study, the sub-mode volume size distribution (VSD) and complex refractive index (CRI) in Chengdu area were investigated based on three years of measurements of the Sun-sky radiometer Observation NETwork (SONET). Section 2 presents the observation data and the method to separate VSD and CRI into sub mode (fine and coarse mode). Section 3 shows the seasonal characteristics of sub-mode VSD and CRI. The sub-mode optical properties (AOD, single scattering albedo (SSA)) and the climate effect of sub-mode aerosol in Chengdu are discussed in Section 4. The results are summarized in Section 5.

2. Data and Method

2.1. Observation Site and Data

SONET Chengdu site (Figure 1 red dot) is located in Chengdu, Sichuan province, a developing urban area. The CE318 (Cimel Electronique, Pairs France) is an automatic instrument for long-term continuous observation of direct solar radiation and diffused sky radiation in the Chengdu site with nine individual spectral channels: 340 nm, 380 nm, 440 nm, 500 nm, 675 nm, 870 nm, 1020 nm, 1640 nm and 940 nm. The parameters include AOD, Angstrom Exponent (AE), VSD, CRI and SSA, and so on [24–26]. To ensure data quality, SONET instruments are calibrated by a routine process of laboratory and field calibration experiments. The direct sun measurements are calibrated every year in a field, located on Ling Mountain (~1600 m, MSL), and compared with a master instrument that is regularly calibrated by Langley plot method with high precision [27,28]. For the calibration of sky radiance measurements, SONET adopts the method of vicarious/transfer calibration, which is different from the AErosol RObotic NETwork (AERONET) calibration method [29].

The accuracy assessment of SONET products is based on the Distributed Regional Aerosol Gridded Observational Network (DRAGON)–Korea–United States Air Quality Study (KORUS-AQ) 2016 campaign [30]. The aerosol optical and microphysical parameters were inverted by SONET and AERONET algorithm at the same time. The differences of two network products can reflect the accuracy and data acquisition of SONET products. It turned out that the Level 1.5 data amount of two kinds of products are all above 85%, and the average AOD difference between SONET and AERONET is 0.002 ± 0.0001, less than AERONET AOD uncertainty [28]. The VSD difference between SONET and AERONET is 1.5% ± 26% (radius range from 0.1 to 7 μm) and 18% ± 85% (radius more than 7 μm) [28]. The real part of CRI has a difference of 0.007 ± 0.04 from AERONET, and the imaginary part has a difference value of 18% ± 46%. The average difference on SSA is slightly higher, but other parameters are close to or less than AERONET normal uncertainties [28].

The products of SONET are graded into three level (Level 1.0, Level 1.5, Level 2.0), defined following the AERONET data level protocols of version 2.0. In detail, Level 1.0 is raw data calculated from measurement and calibration coefficient. The Level 1.5 is based on Level 1.0 with automatic cloud screening procedures [31]. Level 2.0 has the additional application of pre- and post-calibration coefficient and expert checking. This paper is based on Level 2.0 data of SONET Chengdu site from June 2013 to December 2016.

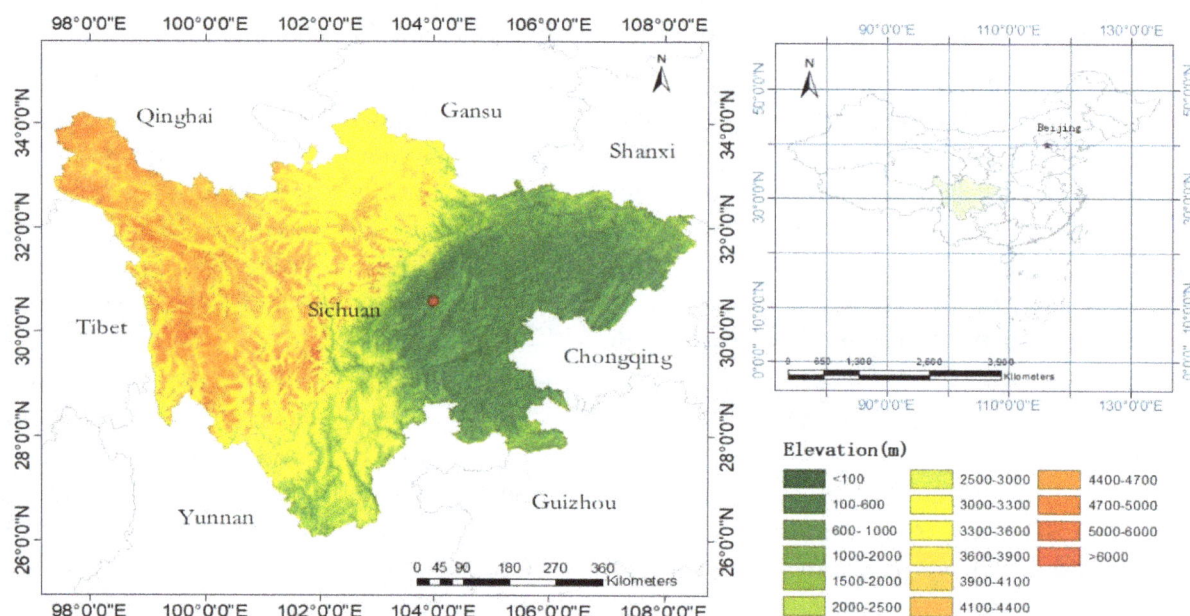

Figure 1. Topographic map of Sichuan province and location of the Sun-sky radiometer Observation NETwork (SONET) Chengdu site (marked with red dot). Data source is ASTER GDEM (Advanced Spaceborne Thermal Emission and Reflection Radiometer Global Digital Elevation Model).

2.2. Method

In this paper, the sub-mode VSDs and CRIs were retrieved by the aerosol products [24,26], following by Zhang et al. (2016) and Zhang et al. (2017). The method developed by Cuesta et al. (2008) is employed to separate VSD into a single Log-Normal Modes (LNM). Each LNM parameters have three parameters: the modal concentration, geometric standard deviation and modal radius [32,33]. The sub-mode VSD can be modeled by the following function:

$$\frac{dV(r)}{d\ln r} = \sum_{i=1}^{m} \frac{C_i}{\sqrt{2\pi}|\ln \sigma_i|} \exp\left[-\frac{1}{2}\left(\frac{\ln r - \ln r_i}{\ln \sigma_i}\right)^2\right] \tag{1}$$

where C_i $(\mu m^3/\mu m^2)$ is the volume modal concentration, $r_i(\mu m)$ is the median radius, σ_i is standard deviation, m is the total number of modes, $dV/d\ln r$ $(\mu m^3/\mu m^2)$ is aerosol volume particle size distribution.

The modes of VSD with radius less than 1 μm can be considered as the fine mode and others belong to the coarse mode.

CRI can describe the scattering and absorption properties of atmospheric particulates. Most research analyzed aerosol total-columnar CRI, but fine- and coarse-mode particles are associated with different composition and source of pollution. So, in this study we recalculated the complex refractive indices for both of fine and coarse mode, following Zhang et al. (2017). For the calculation of the sub-mode CRI, we also choose the same radius limit (1 μm) as the sub-mode VSD. The separated results of CRI have their own fine and coarse modes in different wavelength as follow:

$$n_{f/c}(\lambda) = n_{f/c} \quad \lambda = 440, 675, 870, 1020 \text{ nm} \tag{2}$$

$$k_{f,c}(\lambda) = \begin{cases} k_{f,c\ 440} & \lambda = 440 \text{ nm} \\ k_{f,c} & \lambda = 675, 870, 1020 \text{ nm} \end{cases} \tag{3}$$

where n is real part of CRI, k is imaginary part of CRI, λ denotes the standard wavelengths of AERONET products, the subscripts f and c represent the fine and coarse modes, respectively.

The input parameters are the VSD, spectral AOD, and absorbing AOD. The initial guesses of sub-mode CRIs are from the inversion CRIs of measurements [34]. The effective CRIs are corresponding to each VSD bin, following the volume average rule [35]:

$$n(r) = \frac{n_f V_f(r) + n_c V_c(r)}{V_f(r) + V_c(r)} \tag{4}$$

$$k(\lambda, r) = \frac{k_f V_f(r) + k_c V_c(r)}{V_f(r) + V_c(r)} \tag{5}$$

Then the fine- and coarse-mode CRIs are found by iterative fitting of the input AODs and the calculated AODs by the CRIs (Equations (4) and (5)) and VSDs.

With regard to the test of error estimation, the error of real part of CRI is less than 0.046 and that of imaginary part is less than 0.003 in three typical modal (WS: water soluble, BB: biomass burning, DU: dust). As this algorithm applied to AERONET measurements, the total uncertainties are $\Delta n_{f/c} = 0.11, \Delta k_{f/c} = 78\%$ by considering all possible input of AERONET parameter errors together.

3. Results

3.1. Fine- and Coarse-Mode AOD

O'Neil et al. (2003) developed a spectral deconvolution algorithm (SDA) that utilizes spectral total extinction AOD data with the assumption of bimodal aerosol size distribution to infer the fine and coarse mode contributions to atmospheric AOD. SONET employs the algorithm's ability to separated coarse and fine mode AOD that used in Figure 2 [25,36]. As illustrated in Figure 2, we find that the fine-mode AOD (AOD_f) has the same monthly tendency with the total AOD. The annual AOD_f percent is 92%, which varies from 86% to 96%. Nevertheless, the coarse-mode AOD (AOD_c) has little variation in different month. It is demonstrable that the fine-mode aerosols are the principal pollutant in Chengdu area, which lead to the change of AOD. In summer, rainfall affected by the typical subtropical monsoon basin climate is significantly higher than that in winter [37]. The increasing precipitation can play a role in the removal of atmospheric pollutants. As a result, AOD decreases in summer, which reaches the lowest value of 0.64 in June. However, the average AOD in summer is 0.89, which still acts as a pollutant. There are two main reasons: firstly, the average surface wind speed all over the basin is low under the control of subtropical anticyclone in summer [38], which makes aerosol transport to other regions difficult. In addition, high temperature can cause the increasing formation of secondary organic aerosol particles [39]. In winter, the large variation of temperature from day to night leads to the rapid condensation of water vapor, which is conducive to the concentration of water and particulate matter. And the cold air is not conductive to the diffusion of aerosol because of the enclosed basin [16]. Therefore, AOD is the highest in winter (Winter: AOD = 1.12).

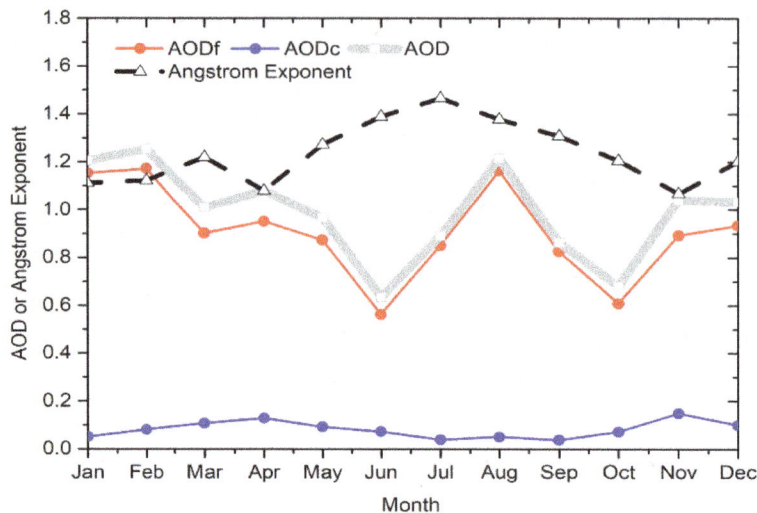

Figure 2. Monthly mean of aerosol optical depth (AOD), AOD_f, AOD_c at 440 nm and Angstrom Exponent (440–870 nm) in SONET Chengdu site from 2013 to 2016.

AE (440–870 nm) changes little with the value more than 1.0, which is similar to the previous research [12,20]. Convective precipitation occurs frequently in summer that the majority of coarse particles have been eliminated by wet settlement, but the fine particles remained in atmosphere with long suspension times [20,40–42]. Also, in Figure 2, it also can be illustrated by that the difference between AOD_f and AOD is lowest in summer. The average AE is less than 1.2 in each month of spring. The main reason for the high AOD and low AE in spring is the long-distance transport of dust pollution from North China [43,44]. In addition, another reason may be the wind speed increasing, by which it can be speculated that large particles are emitted to the atmosphere or transported from other regions by strong wind [20,45].

The AOD can be separated into fine- and coarse-mode and the different mode has the obviously distinguishing extinction for solar light in Chengdu. Therefore, it is important to explore the more sub-mode properties to research on the climate effect of different mode aerosols.

3.2. Sub-Mode VSD

VSD is one of the important aerosol microphysical properties. Although the particle size changes constantly, atmospheric aerosol particles exist in three-mode VSD stably. According to particle radius, the modes can be divided into nuclear mode (less than 0.1 µm), accumulation mode (0.1–1 µm) and coarse mode (larger than 1 µm) [46]. The process of nuclear mode is concentrated in nanometer scale, and the accumulation mode and coarse mode can be shown in VSDs [47,48]. According to Eck et al. (2012), large fine mode-dominated aerosols (submicron radius) were observed after the fog or low-altitude cloud dissipated events. As cloud condensation nuclei or ice nuclei, the smaller coarse mode-dominated aerosols (supermicron radius) are involved in the fog/cloud formation and dissipation [49]. Li, et al. (2014) found that an unusual increase of submicron fine modes is an important mechanism for haze growth in the polluted region. As identified in Figure 3, there are four modes of the aerosol VSD in Chengdu area and the corresponding parameters are listed in Table 1. The different color lines are the average of each mode and the values of N in the legend are the amounts of data involved in averaging. The modes can be clearly distinguished. As the median radius is less than 1µm, there are two peaks: the fine mode and submicron fine (SMF) mode. Furthermore, there are two peaks as the radius more than 1 µm: supermicron coarse (SMC) mode and coarse mode. In the figure, the fine mode has an almost equal amount of data as the coarse mode, but the fine-mode with one addition, which is one record with only one-peak VSD. The SMC mode has the least data amount of all (N = 133). The median radius of each mode is obviously different. As to the fine mode, the average median radius is 0.17 µm ± 0.05 and the volume modal concentration is 0.10 ± 0.07, that is, the largest value of all modes. The median radii of the SMF modes are mainly in the range of 0.2–0.5 µm, which can indicate fog dissipation and haze growth, as previously mentioned. The SMC mode has a median radius of 1.62 µm $\pm$ 0.45 and the volume modal concentration of 0.07 $\pm$ 0.05. The coarse mode has an average median radius of 3.25 µm $\pm$ 0.99.

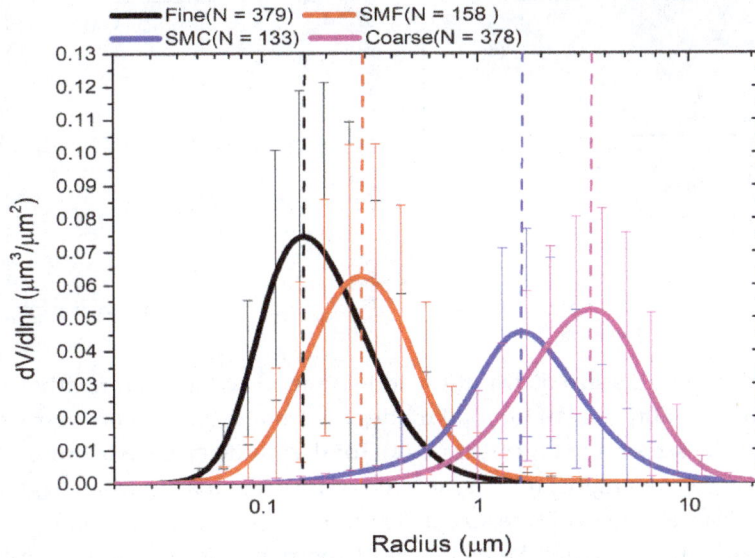

Figure 3. The average sub-mode volume size distribution in Chengdu.

Table 1. The parameters of sub-mode volume size distribution.

	R (µm)	σ	V (µm³/µm²)
Fine	0.17 $\pm$ 0.05	1.55 $\pm$ 0.18	0.10 $\pm$ 0.07
SMF	0.31 $\pm$ 0.12	1.69 $\pm$ 0.28	0.09 $\pm$ 0.05
SMC	1.62 $\pm$ 0.45	1.83 $\pm$ 0.35	0.07 $\pm$ 0.05
Coarse	3.25 $\pm$ 0.99	1.82 $\pm$ 0.29	0.09 $\pm$ 0.05

3.3. Fine- and Coarse-Mode VSD and CRI

The primary focus of Section 3.2 shows that the VSD can be separated into four distinguishing modes. However, in general, the natural aerosols, which are predominately coarse mode particles (r > 1 µm), and combustion-produced and anthropogenic emissions particles, which are predominately fine mode particles (r < 1 µm), of various mixed relative fractions are the mixtures in the aerosols. Furthermore, the fine mode and coarse mode particles are from different components. Also, at the same time, the CRI can only be separated into fine- or coarse-mode due to technical and precision limitations. Therefore, we focus on the fine- and coarse-mode VSD and CRI in Chengdu in this section.

In Figure 4, we present the multi-year average separated CRIs and the breakdown results of VSDs. It can be seen the fine- and coarse-mode VSD are well separated. The pictures (a) and (b) in the first row are the average values retrieved from the ground-based Sun-sky radiometer over multiple years. The second and third rows are the average of fine- and coarse-mode VSD and CRI. The sub-mode real part of CRI has non variation of wavelengths referring to Equation (2). The total CRI and sub-mode CRI are from different algorithms [21,32,34], but the sub-mode CRI is related to the volume modal concentration in the iterative algorithm of estimation of CRI for fine and coarse mode [34].

The fine-mode volume modal concentration is clearly higher than the coarse-mode one. The fine- and coarse-mode real parts of CRI exhibit little difference (n_f = 1.43, n_c = 1.46). However, the coarse mode has a lower imaginary part of CRI at 440 nm than that of fine mode (k_{f440} = 0.0106, k_{c440} = 0.0072). At longer wavelengths, the imaginary part of CRI has little variation between fine- and coarse-mode (k_f = 0.0121, k_c = 0.0112).

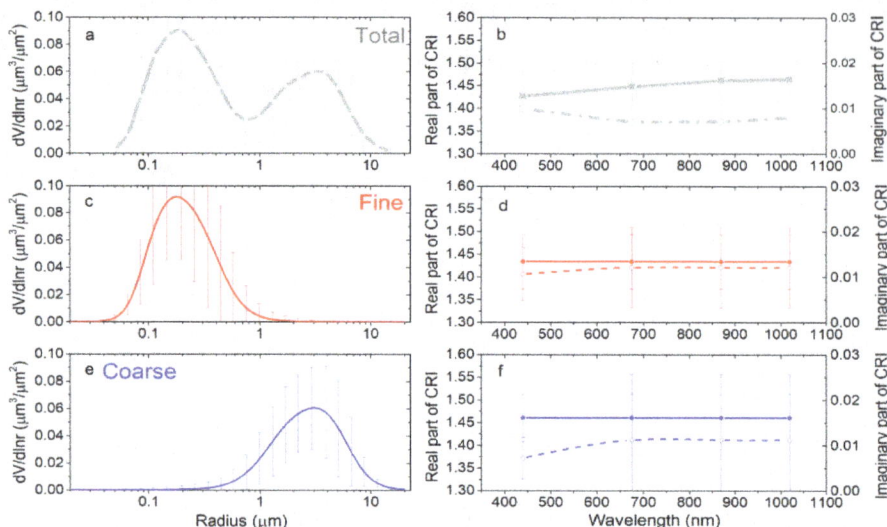

Figure 4. The separated volume size distribution and sub-mode complex refractive index of multi-year average. The data source of (**a**) and (**b**) for total modes are from the inversion algorithm of SONET, and others (**c–f**) are from the sub-mode algorithms. (**a**) Total volume size distribution (**b**) Total complex refractive index in four wavelengths; (**c**) Fine-mode volume size distribution; (**d**) Fine-mode complex refractive index; (**e**) Coarse-mode volume size distribution; (**f**) Coarse-mode complex refractive index; The solid lines in figure (**b**), (**d**), (**f**) are the real parts of complex refractive index, and the dash lines are the imaginary parts. The corresponding parameters are listed in Table 2.

Figure 5 shows the fine-mode VSDs and CRIs in different seasons. The corresponding parameters are listed in Table 2. The typical bimodal or multimodal VSDs in all seasons imply a fine-coarse mixed–size distribution in the Chengdu area, similar to the urban-industrial aerosol type [50]. The fine-mode volume concentration is higher in summer, followed by winter, spring and autumn (Table 2). Furthermore, the fine-mode median radii are higher in summer (0.21) and winter (0.21). This can be explained by the increasing precipitation in summer and the low wind speed in winter that would lead to the high relative humidity and hygroscopic growth [22,23].

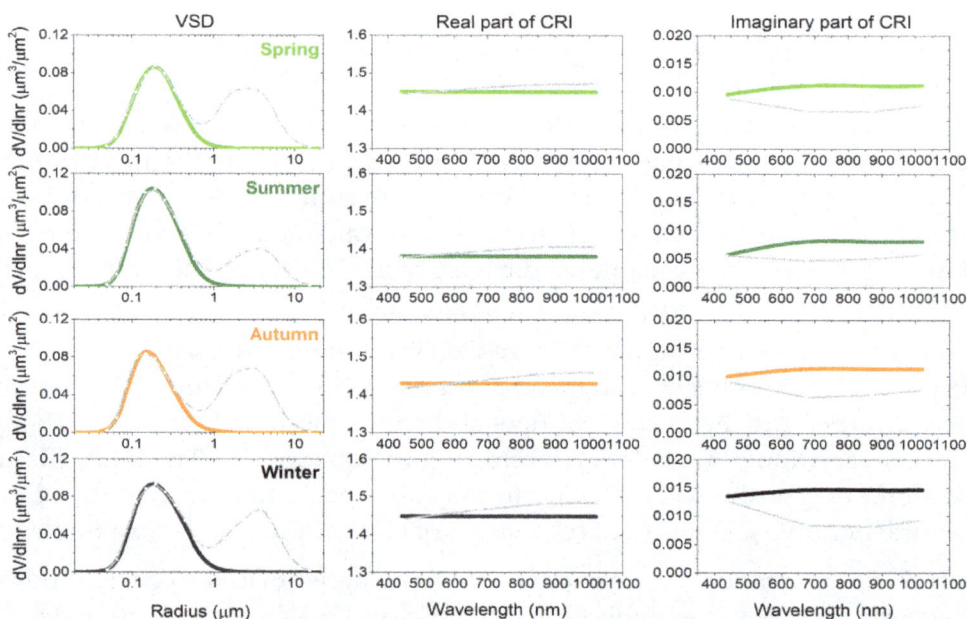

Figure 5. Seasonal variation of fine-mode volume size distribution (VSD) and complex refractive index (CRI) in SONET Chengdu site. The grey line refers to the total parameters. The corresponding parameters are listed in Table 2.

Table 2. The separated volume size distribution and sub-mode complex refractive index in different seasons.

Aerosol Properties	Mode	Parameter	Multi-Year Average	Spring	Summer	Autumn	Winter
Real Part of Refractive Index	Fine	λ(440–1020 nm)	1.43 ± 0.06	1.45 ± 0.05	1.38 ± 0.04	1.43 ± 0.06	1.45 ± 0.06
	Coarse	λ(440–1020 nm)	1.46 ± 0.05	1.47 ± 0.05	1.41 ± 0.03	1.45 ± 0.04	1.48 ± 0.05
	Total	λ(440 nm)	1.43 ± 0.06	1.44 ± 0.05	1.38 ± 0.04	1.42 ± 0.06	1.44 ± 0.06
		λ(675 nm)	1.45 ± 0.05	1.46 ± 0.05	1.39 ± 0.03	1.44 ± 0.05	1.47 ± 0.05
		λ(870 nm)	1.46 ± 0.05	1.47 ± 0.05	1.41 ± 0.03	1.46 ± 0.04	1.48 ± 0.05
		λ(1020 nm)	1.46 ± 0.05	1.47 ± 0.05	1.41 ± 0.03	1.46 ± 0.04	1.49 ± 0.05
Imaginary Part of Refractive Index	Fine	λ(440 nm)	0.011 ± 0.006	0.010 ± 0.006	0.006 ± 0.004	0.010 ± 0.005	0.014 ± 0.005
		λ(675–1020 nm)	0.012 ± 0.009	0.011 ± 0.014	0.008 ± 0.005	0.011 ± 0.006	0.015 ± 0.006
	Coarse	λ(440 nm)	0.007 ± 0.005	0.007 ± 0.007	0.005 ± 0.003	0.007 ± 0.004	0.008 ± 0.003
		λ(675–1020 nm)	0.011 ± 0.014	0.012 ± 0.002	0.008 ± 0.006	0.013 ± 0.022	0.012 ± 0.007
	Total	λ(440 nm)	0.010 ± 0.006	0.009 ± 0.006	0.006 ± 0.003	0.009 ± 0.004	0.013 ± 0.005
		λ(675 nm)	0.007 ± 0.005	0.007 ± 0.007	0.005 ± 0.003	0.006 ± 0.004	0.008 ± 0.003
		λ(870 nm)	0.007 ± 0.005	0.007 ± 0.009	0.005 ± 0.003	0.007 ± 0.004	0.008 ± 0.003
		λ(1020 nm)	0.008 ± 0.006	0.008 ± 0.010	0.006 ± 0.004	0.008 ± 0.004	0.009 ± 0.004
Volume Particle Size Distribution	Fine	R_f	0.20 ± 0.05	0.20 ± 0.04	0.21 ± 0.05	0.18 ± 0.03	0.21 ± 0.05
		Std_f	0.55 ± 0.07	0.53 ± 0.06	0.54 ± 0.07	0.55 ± 0.06	0.56 ± 0.08
		V_f	0.15 ± 0.08	0.13 ± 0.06	0.17 ± 0.09	0.13 ± 0.07	0.16 ± 0.09
	Coarse	R_c	2.86 ± 0.52	2.56 ± 0.43	3.08 ± 0.50	2.57 ± 0.41	3.05 ± 0.47
		Std_c	0.62 ± 0.06	0.64 ± 0.06	0.60 ± 0.06	0.63 ± 0.05	0.61 ± 0.07
		V_c	0.11 ± 0.06	0.12 ± 0.06	0.07 ± 0.05	0.12 ± 0.06	0.11 ± 0.05

The CRI can reflect the aerosol chemical composition. The real part indicates the aerosol refractivity and scattering characteristics. Specifically, the real part of CRI of water is generally considered to be 1.33 [51], much lower than other dry matters. Therefore, the real part can also reflect the water content in aerosol. The total real part of CRI in summer is the lowest, which is associated with the high humidity and water content in summer. Moreover, the fine-mode real part of CRI is also the lowest (n_f = 1.38). As shown in Table 2, the fine-mode real parts of CRIs are all lower than the coarse-mode ones, which suggests that the water content in fine-mode particles play a leading role in Chengdu aerosols.

The imaginary part of CRI is related to the absorption characteristics of aerosol. The fine-mode imaginary part of CRI generally indicates that the fine-mode absorption component, that is, Black carbon (BC) or Brown carbon (BrC) [52].

Figure 6 presents the coarse-mode VSDs and CRIs in different seasons. The coarse-mode volume concentration is obviously less than the fine-mode. In particular, the volume concentration in summer gets the lowest (V_c = 0.069), which is associated with the wet removal of coarse particles. In summer, the coarse-mode real part is the lowest value over the four seasons, but also higher than the fine-mode one in Figure 5 that suggests the coarse particles are weakly hygroscopic (n_f = 1.38, n_c = 1.41). For all seasons, the coarse-mode imaginary part of CRI is quite constant (Table 2). In contrast, the fine-mode imaginary part has great seasonal variation. That demonstrates the coarse particle components are relatively stable. The spectral difference of the imaginary part (between 440 nm and longer wavelength) has a discrepancy between fine- and coarse-mode. It should be mentioned that the imaginary part reflects the aerosol absorbing property, and its spectral pattern can reveal the relative fractions of absorbing aerosols, that is, BC and DU [52–54].

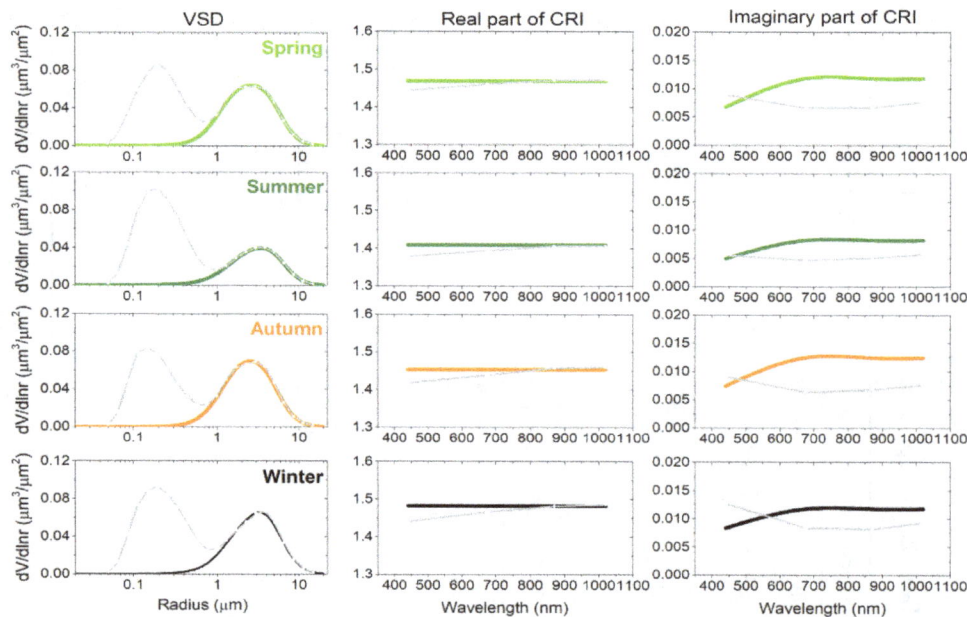

Figure 6. Seasonal variation of coarse-mode VSD and CRI in SONET Chengdu site. The grey line refers to the total parameters. The corresponding parameters are listed in Table 2.

4. Discussion

In order to evaluate the overall analysis scheme, a numerical experiment was used to assess the performance of sub-mode results. In Figure 7, we illustrated the recovery of AOD in four wavelengths by the sub-mode VSD and CRI. It can be seen that the correlation coefficients are all larger than 0.98. The absolute deviations in four wavelengths are 0.04, 0.02, 0.03 and 0.03, corresponding relative standard deviations are 0.03, 0.03, 0.06, 0.09, respectively. These biases are basically close to the claimed uncertainties of SONET products (AOD), which demonstrate that the sub-mode VSD and CRI results are acceptable in understanding of optical closure.

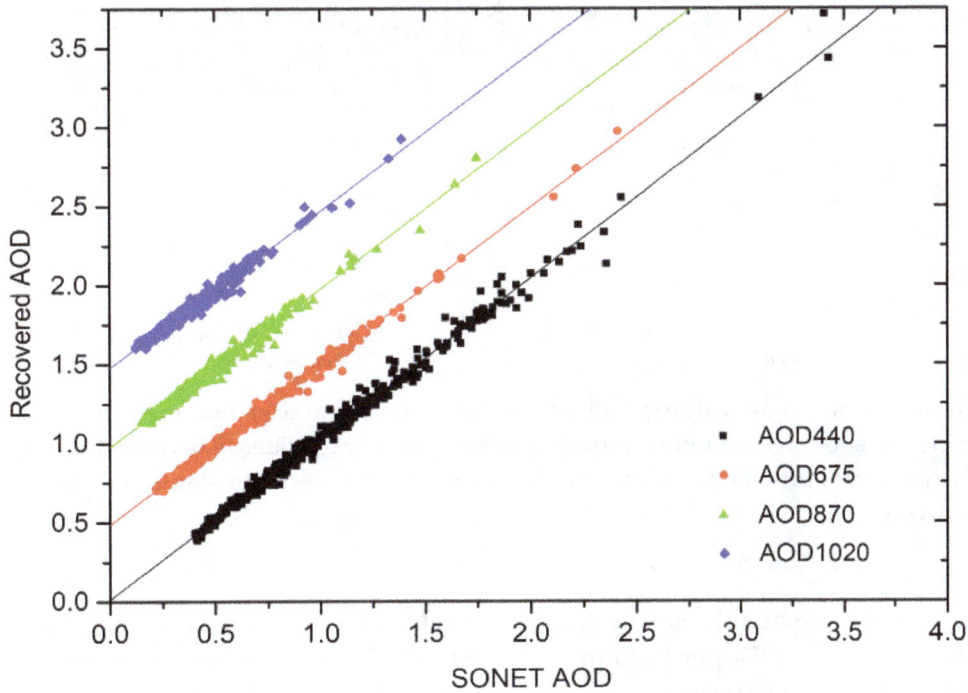

Figure 7. Recovery of AOD at different wavelength based on the separated volume size distribution and sub-mode complex refractive index. $N = 380$ and curves (at 675 nm, 870 nm, 1020 nm) are shifted for a better viewing.

For further details of the climate effect, the sub-mode SSAs were taken into account. SSAs could reflect the aerosol absorption and scattering of solar lights, which is an important parameter in climate modeling [55,56]. Hansen et al. (1997) noted that a change in SSA from 0.9 to 0.8 can change the radiative forcing from negative to positive depending on the reflectance of the underlying surface and the altitude of the aerosols. Moreover, strongly absorbing aerosols may have a large impact on the regional climate and heating the atmosphere [57,58].

In Figure 8, the sub-mode SSA is calculated by the separated VSD and sub-mode CRI under ignoring the influence of nonsphericity on dust aerosols. The total SSAs are all larger than 0.9 in the four seasons and the fine-mode SSAs are closed to the total SSAs that indicates the scattering properties are mainly dominated by fine-mode aerosol. Higher SSA at 675 nm could indicate the main absorption component BrC [52,53], but the high value of SSA indicates the absorbing effect is obvious less than scattering effect of solar lights. The SSA spectral trends from 675 nm to 870 nm show different patterns: the increasing pattern is usually corresponding to the absorbing coarse particles (e.g., Dust); the decreasing pattern is usually corresponding to the absorbing fine particles (e.g., BC, BrC) [52]. The fine-mode SSAs decrease and the coarse-mode SSAs increase with the increase of wavelength, which indicates the different absorbing component in different modes. Furthermore, the coarse-mode SSAs are lower than that of the fine-mode that suggests the absorption of coarse-mode particles is likely stronger than that of the small size aerosol particles. In this regard, the absorbing component in large-size aerosols (such as mineral dust, biomass burning, etc.) is possibly more than that of small-size aerosols.

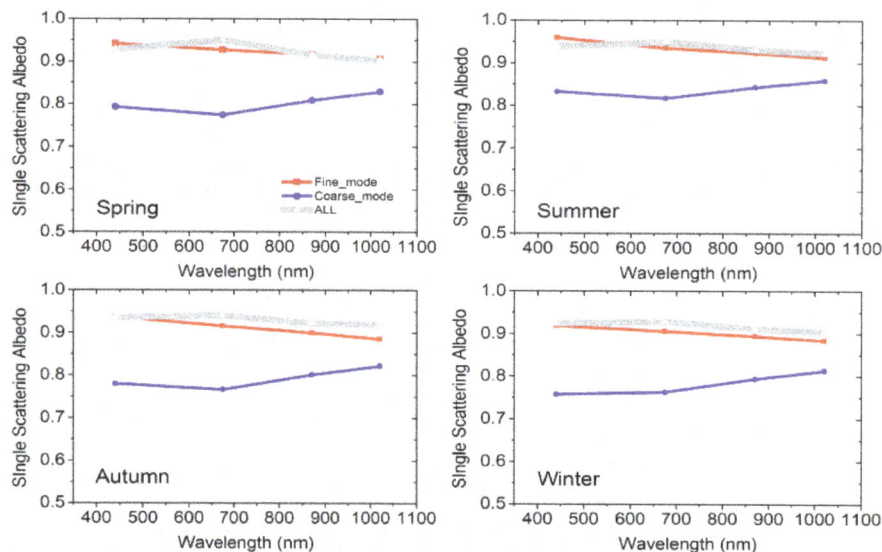

Figure 8. The sub-mode single scattering albedo in different season based on the separated volume size distribution and sub-mode complex refractive index (ignore the influence of nonsphericity on dust aerosols). The grey, red, blue lines are the whole-, fine- and coarse-mode single scattering albedos (SSAs), respectively.

5. Conclusions

In this study, we investigated the sub-mode VSD and CRI for fine- or coarse-mode in the Chengdu area. The annual average of AOD_f percentage is 92%, which has the same monthly tendency with the total AOD, but AOD_c has little variation in different months. The typical bimodal or multimodal VSDs employ a fine-coarse mixed–size distribution in Chengdu area. There are four distinguishing modes of VSD in Chengdu that the median radii are 0.17 μm ± 0.05, 0.31 μm ± 0.12, 1.62 μm ± 0.45, 3.25 μm ± 0.99, respectively.

The fine-mode annual average volume modal concentration is clearly higher than the coarse-mode one. The fine-mode volume concentration and median radius are higher in summer and winter. In particular, the coarse-mode volume concentration gets lowest in summer ($V_c = 0.069$), which is associated with the wet removal of large-size particles. For multi-year average results of CRI, the fine-mode and coarse-mode real parts show little difference. However, the coarse mode has a lower imaginary part at 440 nm than the fine-mode ($k_{f440} = 0.0106$, $k_{c440} = 0.0072$). At longer wavelengths, the imaginary part of CRI has little variation between fine and coarse mode ($k_f = 0.0121$, $k_c = 0.0112$). In summer, both fine- and coarse-mode real parts get the lowest value respectively because of the high humidity ($n_f = 1.38$, $n_c = 1.41$). For all seasons, the coarse-mode imaginary part of CRI is quite constant, but the fine-mode imaginary part has great seasonal variations. It indicates the coarse particle components are relatively stable.

In order to assess the performance of the sub-mode results, we illustrated the recovery of AOD by the sub-mode VSD and CRI. It can be seen that all the correlation coefficients are larger than 0.98. The sub-mode SSAs are calculated by sub-mode VSD and CRI under the condition of neglecting the non-sphericity. The total SSAs are all larger than 0.9 in the four seasons and the fine-mode SSAs are closed to the total SSAs that indicates the scattering properties are mainly dominated by fine-mode aerosols. The coarse-mode SSAs are lower than fine-mode, which suggests the absorbing effect of coarse-mode particles is likely stronger than that of the small size aerosol particles.

Author Contributions: Z.L. conceived and designed this study, and participated in drafting and revising the article. C.Z. substantially contributed to the analysis and interpretation of data, and drafted the articles. H.X. reviewed and edited the manuscript. Y.Z. and Y.W. undertook a part of the instrument maintenance job. D.L., K.L. and Y.X. contributed to the calibration of the sun-sky radiometer and analyses of the observation data. Y.Z. do formal analysis and investigation.

Acknowledgments: This work was supported by the National Key Research and Development Program of China under Grant 2016YFE0201400, the National Natural Science Foundation of China under Grant 41671364,41671367, 41771396, and Science and Technology Service network initiative (STS) Project of Chinese Academy of Sciences (KFJ-STS-QYZD-022).

References

1. Twomey, S. Pollution and the planetary albedo. *Atmos. Environ. (1967)* **1974**, *8*, 1251–1256. [CrossRef]
2. Kazil, J.; Stier, P.; Zhang, K.; Quaas, J.; Kinne, S.; O'donnell, D.; Rast, S.; Esch, M.; Ferrachat, S.; Lohmann, U. Aerosol nucleation and its role for clouds and Earth's radiative forcing in the aerosol-climate model ECHAM5-HAM. *Atmos. Chem. Phys.* **2010**, *10*, 10733–10752. [CrossRef]
3. Albrecht, B.A. Aerosols, cloud microphysics, and fractional cloudiness. *Science* **1989**, *245*, 1227–1231. [CrossRef] [PubMed]
4. Pöschl, U. Atmospheric Aerosols: Composition, Transformation, Climate and Health Effects. *Angew. Chem. Int. Ed.* **2005**, *44*, 7520–7540. [CrossRef] [PubMed]
5. Liao, W.; Wang, X.; Fan, Q.; Zhou, S.; Chang, M.; Wang, Z.; Wang, Y.; Tu, Q. Long-term atmospheric visibility, sunshine duration and precipitation trends in South China. *Atmos. Environ.* **2015**, *107*, 204–216. [CrossRef]
6. Pan, L.; Che, H.; Geng, F.; Xia, X.; Wang, Y.; Zhu, C.; Chen, M.; Gao, W.; Guo, J. Aerosol optical properties based on ground measurements over the Chinese Yangtze Delta Region. *Atmos. Environ.* **2010**, *44*, 2587–2596. [CrossRef]
7. Tao, J.; Cheng, T.; Zhang, R.; Cao, J.; Zhu, L.; Wang, Q.; Luo, L.; Zhang, L. Chemical composition of $PM_{2.5}$ at an urban site of Chengdu in southwestern China. *Adv. Atmos. Sci.* **2013**, *30*, 1070–1084. [CrossRef]
8. Tao, M.; Chen, L.; Wang, Z.; Tao, J.; Che, H.; Wang, X.; Wang, Y. Comparison and evaluation of the MODIS Collection 6 aerosol data in China. *J. Geophys. Res. Atmos.* **2015**, *120*, 6992–7005. [CrossRef]
9. Deng, L.; Qian, J.; Liao, R.; Tong, H. Pollution characteristics of atmospheric particulates in Chengdu from August to September in 2009 and their relationship with meteorological conditions. *China Environ. Sci.* **2012**, *32*, 1433–1438.
10. Li, X.; Yang, Z.; Fu, P.; Yu, J.; Lang, Y.-c.; Liu, D.; Ono, K.; Kawamura, K. High abundances of dicarboxylic acids, oxocarboxylic acids, and α-dicarbonyls in fine aerosols ($PM_{2.5}$) in Chengdu, China during wintertime haze pollution. *Environ. Sci. Pollut. Res.* **2015**, *22*, 12902–12918. [CrossRef]
11. Yong, L.; Allen, P.A.; Densmore, A.L.; Qiang, X. Evolution of the Longmen Shan foreland basin (western Sichuan, China) during the Late Triassic Indosinian orogeny. *Basin Res.* **2003**, *15*, 117–138. [CrossRef]
12. Wang, Q.; Cao, J.; Shen, Z.; Tao, J.; Xiao, S.; Luo, L.; He, Q.; Tang, X. Chemical characteristics of $PM_{2.5}$ during dust storms and air pollution events in Chengdu, China. *Particuology* **2012**, *11*, 70–77. [CrossRef]
13. Hsu, N.C.; Tsay, S.-C.; King, M.D.; Herman, J.R. Deep Blue Retrievals of Asian Aerosol Properties during ACE-Asia. *IEEE Trans. Geosci. Remote Sens.* **2006**, *44*, 3180–3195. [CrossRef]
14. Remer, L.; Kaufman, Y.; Tanré, D.; Mattoo, S.; Chu, D.; Martins, J.; Li, R.; Ichoku, C.; Levy, R.; Kleidman, R. The MODIS Aerosol Algorithm, Products and Validation. *J. Atmos. Sci.* **2005**, *62*, 947–973. [CrossRef]
15. Remer, L.; Mattoo, S.; Levy, R.; Munchak, L. MODIS 3 km aerosol product: Algorithm and global perspective. *Atmos. Meas. Tech.* **2013**, *6*, 1829–1844. [CrossRef]
16. Liu, X.; Chen, Q.; Che, H.; Zhang, R.; Gui, K.; Zhang, H.; Zhao, T. Spatial distribution and temporal variation of aerosol optical depth in the Sichuan basin, China, the recent ten years. *Atmos. Environ.* **2016**, *147*, 434–445. [CrossRef]
17. Shi, G.; Liu, R.; Wang, D.Y.; Yang, F. Evaluation of the MODIS C6 Aerosol Optical Depth Products over Chongqing, China. *Atmosphere* **2017**, *8*, 227. [CrossRef]
18. Lin, M.; Tao, J.; Chan, C.-Y.; Cao, J.-J.; Zhang, Z.-S.; Zhu, L.-H.; Zhang, R.-J. Characterization of Regression Relationship between Recent Air Quality and Visibility Changes in Megacities at Four Haze Regions of China. *Aerosol Air Qual. Res.* **2012**, *12*, 1049–1061. [CrossRef]
19. Tao, J.; Zhang, L.; Cao, J.; Hsu, S.-C.; Xia, X.; Zhang, Z.; Lin, Z.; Cheng, T.; Zhang, R. Characterization and source apportionment of aerosol light extinction in Chengdu, southwest China. *Atmos. Environ.* **2014**, *95*, 552–562. [CrossRef]
20. Tao, R.; Che, H.; Chen, Q.; Tao, J.; Wang, Y.; Sun, J.; Wang, H.; Zhang, X. Study of Aerosol Optical Properties Based on Ground Measurements over Sichuan Basin, China. *Aerosol Air Qual. Res.* **2014**, *14*, 905–915. [CrossRef]
21. Dubovik, O.; King, M.D. A flexible inversion algorithm for retrieval of aerosol optical properties from Sun

and sky radiance measurements. *J. Geophys. Res.* **2000**, *105*, 20673–20696. [CrossRef]

22. Eck, T.F.; Holben, B.N.; Reid, J.; Giles, D.; Rivas, M.; Singh, R.P.; Tripathi, S.; Bruegge, C.; Platnick, S.; Arnold, G. Fog-and cloud-induced aerosol modification observed by the Aerosol Robotic Network (AERONET). *J. Geophys. Res. Atmos.* **2012**, *117*, 107–116. [CrossRef]

23. Li, Z.; Eck, T.; Zhang, Y.; Zhang, Y.; Li, D.; Li, L.; Xu, H.; Hou, W.; Lv, Y.; Goloub, P. Observations of residual submicron fine aerosol particles related to cloud and fog processing during a major pollution event in Beijing. *Atmos. Environ.* **2014**, *86*, 187–192. [CrossRef]

24. Dubovik, O.; Smirnov, A.; Holben, B.; King, M.; Kaufman, Y.; Eck, T.; Slutsker, I. Accuracy assessments of aerosol optical properties retrieved from Aerosol Robotic Network (AERONET) Sun and sky radiance measurements. *J. Geophys. Res. Atmos.* **2000**, *105*, 9791–9806. [CrossRef]

25. O'Neill, N.; Eck, T.; Holben, B.; Smirnov, A.; Dubovik, O.; Royer, A. Bimodal size distribution influences on the variation of Angstrom derivatives in spectral and optical depth space. *J. Geophys. Res.* **2001**, *106*, 9787–9806. [CrossRef]

26. Dubovik, O.; Sinyuk, A.; Lapyonok, T.; Holben, B.N.; Mishchenko, M.; Yang, P.; Eck, T.F.; Volten, H.; Muñoz, O.; Veihelmann, B. Application of spheroid models to account for aerosol particle nonsphericity in remote sensing of desert dust. *J. Geophys. Res. Atmos.* **2006**, *111*. [CrossRef]

27. Holben, B.N.; Eck, T.; Slutsker, I.; Tanre, D.; Buis, J.; Setzer, A.; Vermote, E.; Reagan, J.; Kaufman, Y.; Nakajima, T. AERONET—A federated instrument network and data archive for aerosol characterization. *Remote Sens. Environ.* **1998**, *66*, 1–16. [CrossRef]

28. Li, Z.; Xu, H.; Li, K.; Li, D.; Xie, Y.; Li, L.; Zhang, Y.; Gu, X.; Zhao, W.; Tian, Q. Comprehensive Study of Optical, Physical, Chemical, and Radiative Properties of Total Columnar Atmospheric Aerosols over China: An Overview of Sun–Sky Radiometer Observation Network (SONET) Measurements. *Bull. Am. Meteorol. Soc.* **2018**, *99*, 739–755. [CrossRef]

29. Li, Z.; Blarel, L.; Podvin, T.; Goloub, P.; Buis, J.P.; Morel, J.P. Transferring the calibration of direct solar irradiance to diffuse-sky radiance measurements for CIMEL Sun-sky radiometers. *Appl. Opt.* **2008**, *47*, 1368–1377. [CrossRef]

30. Holben, B.N.; Kim, J.; Sano, I.; Mukai, S.; Eck, T.F.; Giles, D.M.; Schafer, J.S.; Sinyuk, A.; Slutsker, I.; Smirnov, A. An overview of mesoscale aerosol processes, comparisons, and validation studies from DRAGON networks. *Atmos. Chem. Phys.* **2018**, *18*, 1–23. [CrossRef]

31. Smirnov, A.; Holben, B.; Eck, T.; Dubovik, O.; Slutsker, I. Cloud-screening and quality control algorithms for the AERONET database. *Remote Sens. Environ.* **2000**, *73*, 337–349. [CrossRef]

32. Zhang, Y.; Li, Z.; Zhang, Y.; Chen, Y.; Cuesta, J.; Ma, Y. Multi-peak accumulation and coarse modes observed from AERONET retrieved aerosol volume size distribution in Beijing. *Meteorol. Atmos. Phys.* **2016**, *128*, 537–544. [CrossRef]

33. Cuesta, J.; Flamant, P.H.; Flamant, C. Synergetic technique combining elastic backscatter lidar data and sunphotometer AERONET inversion for retrieval by layer of aerosol optical and microphysical properties. *Appl. Opt.* **2008**, *47*, 4598–4611. [CrossRef] [PubMed]

34. Zhang, Y.; Li, Z.; Zhang, Y.; Li, D.; Qie, L.; Che, H.; Xu, H. Estimation of aerosol complex refractive indices for both fine and coarse modes simultaneously based on AERONET remote sensing products. *Atmos. Meas. Tech.* **2017**, *10*, 1–17. [CrossRef]

35. Heller, W. Remarks on Refractive Index Mixture Rules. *J. Phys. Chem.* **1965**, *69*, 1123–1129. [CrossRef]

36. O'Neill, N.; Eck, T.; Smirnov, A.; Holben, B.; Thulasiraman, S. Spectral discrimination of coarse and fine mode optical depth. *J. Geophys. Res. Atmos.* **2003**, *108*. [CrossRef]

37. Chen, Y.; Xie, S.-D. Characteristics and formation mechanism of a heavy air pollution episode caused by biomass burning in Chengdu, Southwest China. *Sci. Total Environ.* **2014**, *473*, 507–517. [CrossRef] [PubMed]

38. Liu, Y.M.; Wu, G.X.; Liu, H.; Liu, P. Condensation heating of the Asian summer monsoon and the subtropical anticyclone in the Eastern Hemisphere. *Clim. Dyn.* **2001**, *17*, 327–338. [CrossRef]

39. Kroll, J.H.; Seinfeld, J.H. Chemistry of secondary organic aerosol: Formation and evolution of low-volatility organics in the atmosphere. *Atmos. Environ.* **2008**, *42*, 3593–3624. [CrossRef]

40. Gobbi, G.P.; Kaufman, Y.J.; Koren, I.; Eck, T.F. Classification of aerosol properties derived from AERONET direct sun data. *Atmos. Chem. Phys.* **2007**, *7*, 8713–8726. [CrossRef]

41. Kaufman, Y.J. Aerosol optical thickness and atmospheric path radiance. *J. Geophys. Res. Atmos.* **1993**, *98*, 2677–2692. [CrossRef]

42. King, M.D.; Byrne, D.M.; Herman, B.M.; Reagan, J.A. Aerosol Size Distributions Obtained by Inversion of Spectral Optical Depth Measurements. *J. Atmos. Sci.* **1978**, *35*, 2153–2167. [CrossRef]

43. Gong, S.L.; Zhang, X.Y.; Zhao, T.L.; Mckendry, I.G.; Jaffe, D.A.; Lu, N.M. Characterization of soil dust aerosol in China and its transport and distribution during 2001 ACE-Asia: 2. Model simulation and validation. *J. Geophys. Res.* **2003**, *108*. [CrossRef]

44. Wang, Y.; Sun, Y.; Xin, J.; Li, Z.; Wang, S.; Wang, P.; Hao, W.M.; Nordgren, B.L.; Chen, H.; Wang, L. Seasonal variations in aerosol optical properties over China. *Atmos. Chem. Phys. Discuss.* **2008**, *8*, 8431–8453. [CrossRef]

45. Che, H.; Xia, X.; Zhu, J.; Li, Z.; Dubovik, O.; Holben, B.; Goloub, P.; Chen, H.; Estelles, V.; Cuevas-Agulló, E. Column aerosol optical properties and aerosol radiative forcing during a serious haze-fog month over North China Plain in 2013 based on ground-based sunphotometer measurements. *Atmos. Chem. Phys.* **2014**, *14*, 2125–2138. [CrossRef]

46. Willeke, K.; Whitby, K. Atmospheric Aerosols: Size Distribution Interpretation. *J. Air Pollut. Control Assoc.* **1975**, *25*, 529–534. [CrossRef]

47. Kulmala, M.; Vehkamäki, H.; Petäjä, T.; Maso, M.D.; Lauri, A.; Kerminen, V.M.; Birmili, W.; Mcmurry, P.H. Formation and growth rates of ultrafine atmospheric particles: A review of observations. *J. Aerosol Sci.* **2004**, *35*, 143–176. [CrossRef]

48. Liu, X.; Li, J.; Qu, Y.; Han, T.; Hou, L.; Gu, J.; Chen, C.; Yang, Y.; Liu, X.; Yang, T. Formation and evolution mechanism of regional haze: A case study in the megacity Beijing, China. *Atmos. Chem. Phys.* **2013**, *13*, 4501–4514. [CrossRef]

49. Hammer, E.; Gysel, M.; Roberts, G.C.; Elias, T.; Hofer, J.; Hoyle, C.R.; Bukowiecki, N.; Dupont, J.C.; Burnet, F.; Baltensperger, U. Size-dependent particle activation properties in fog during the ParisFog 2012/13 field campaign. *Atmos. Chem. Phys.* **2014**, *14*, 9475–9516. [CrossRef]

50. Su, X.; Cao, J.; Li, Z.; Li, K.; Xu, H.; Liu, S.; Fan, X. Multi-Year Analyses of Columnar Aerosol Optical and Microphysical Properties in Xi'an, a Megacity in Northwestern China. *Remote Sens.* **2018**, *10*, 1169. [CrossRef]

51. Dubovik, O.; Holben, B.; Eck, T.F.; Smirnov, A.; Kaufman, Y.J.; King, M.D.; Tanré, D.; Slutsker, I. Variability of Absorption and Optical Properties of Key Aerosol Types Observed in Worldwide Locations. *J. Atmos. Sci.* **2002**, *59*, 590–608. [CrossRef]

52. Li, Z.; Gu, X.; Wang, L.; Li, D. Aerosol physical and chemical properties retrieved from ground-based remote sensing measurements during heavy haze days in Beijing winter. *Atmos. Chem. Phys.* **2013**, *13*, 10171–10183. [CrossRef]

53. Wang, L.; Li, Z.-Q.; Li, D.-H.; Li, K.-T.; Tian, Q.J.; Li, L.; Zhang, Y.; Lv, Y. Retrieval of Dust Fraction of Atmospheric Aerosols Based on Spectra Characteristics of Refractive Indices Obtained from Remote Sensing Measurements. *Spectrosc. Spectral Anal.* **2012**, *32*, 1644.

54. Schuster, G.L.; Dubovik, O.; Holben, B.N.; Clothiaux, E.E. Inferring black carbon content and specific absorption from Aerosol Robotic Network (AERONET) aerosol retrievals. *J. Geophys. Res. Atmos.* **2005**, *110* [CrossRef]

55. Bodhaine, B.A. Aerosol absorption measurements at Barrow, Mauna Loa, and the South Pole. *J. Geophys. Res. Atmos.* **1995**, *100*, 8967–8975. [CrossRef]

56. Lee, K.H.; Li, Z.; Man, S.W.; Xin, J.; Wang, Y.; Hao, W.M.; Zhao, F. Aerosol single scattering albedo estimated across China from a combination of ground and satellite measurements. *J. Geophys. Res. Atmos.* **2007**, *112* [CrossRef]

57. Surabi, M.; James, H.; Larissa, N.; Yunfeng, L. Climate effects of black carbon aerosols in China and India. *Science* **2002**, *297*, 2250–2253.

58. Hansen, J.E.; Sato, M.; Ruedy, R. Radiative forcing and climate response. *J. Geophys. Res. Atmos.* **1997**, *102*, 6831–6864. [CrossRef]

10

Hurricane Boundary Layer Height Relative to Storm Motion from GPS Dropsonde Composites

Yifang Ren [1], Jun A. Zhang [2,3,*], Stephen R. Guimond [4,5] and Xiang Wang [6]

[1] The Center of Jiangsu Meteorological Service, Nanjing 21008, China; renyifang2006@126.com
[2] Cooperative Institute for Marine and Atmospheric Studies, University of Miami, Miami, FL 33149, USA
[3] NOAA/AOML/Hurricane Research Division, Miami, FL 33149, USA
[4] Joint Center for Earth Systems Technology, University of Maryland Baltimore County, Baltimore, MD 21250, USA; stephen.guimond@nasa.gov
[5] NASA Goddard Space Flight Center (GSFC), Greenbelt, MD 20771, USA
[6] Centre of Data Assimilation for Research and Application, Nanjing University of Information Science & Technology, Nanjing 210044, China; wangxiang@nuist.edu.cn
* Correspondence: jun.zhang@noaa.gov

Abstract: This study investigates the asymmetric distribution of hurricane boundary layer height scales in a storm-motion-relative framework using global positioning system (GPS) dropsonde observations. Data from a total of 1916 dropsondes collected within four times the radius of maximum wind speed of 37 named hurricanes over the Atlantic basin from 1998 to 2015 are analyzed in the composite framework. Motion-relative quadrant mean composite analyses show that both the kinematic and thermodynamic boundary layer height scales tend to increase with increasing radius in all four motion-relative quadrants. It is also found that the thermodynamic mixed layer depth and height of maximum tangential wind speed are within the inflow layer in all motion-relative quadrants. The inflow layer depth and height of the maximum tangential wind are both found to be deeper in the two front quadrants, and they are largest in the right-front quadrant. The difference in the thermodynamic mixed layer depth between the front and back quadrants is smaller than that in the kinematic boundary layer height. The thermodynamic mixed layer is shallowest in the right-rear quadrant, which may be due to the cold wake phenomena. The boundary layer height derived using the critical Richardson number (R_{ic}) method shows a similar front-back asymmetry as the kinematic boundary layer height.

Keywords: atmospheric boundary layer; tropical cyclone; storm motion; asymmetry; hurricane; aircraft; dropsonde

1. Introduction

The atmospheric boundary layer (ABL) is the turbulent layer close to the Earth's surface that is influenced by surface friction. Physical processes in the ABL play an important role in regulating the atmosphere at both weather and climate scales, including hurricanes. The top of the ABL (i.e., the boundary layer height) is a key parameter that determines the vertical distribution of turbulent mixing in numerical models that require a boundary layer parameterization. In numerical models where turbulent fluxes are parameterized using a first-order K-profile method, the boundary layer height is usually defined as the height at which the bulk Richardson number (R_{ic}) reaches a threshold (typically 0.25), where R_{ic} is defined as the ratio of buoyancy to shear forcing [1–3]. Note that this Richardson number method was also previously used in observational studies to determine the boundary layer height in non-hurricane conditions [4–8].

Under non-hurricane conditions when vertical profiles of turbulent intensity and/or flux are measured, the boundary layer height is usually taken as the height where the magnitude of the turbulence parameter is much smaller (~95%) than that in the surface layer. Note that the surface layer height is usually taken as 10% of the ABL height [9]. Without turbulence data, the height of the temperature inversion layer is generally used to define the boundary layer height, in particular, under unstable conditions [9–12]. Since passive scalars are accumulated in the ABL, large gradients of temperature and water vapor occur at the inversion layer capping the ABL [13,14]. Above the inversion layer, a rising air parcel typically becomes neutrally buoyant.

The top of the stable or neutral boundary layer is more difficult to determine than the unstable convective boundary layer. The boundary layer height is typically defined using the vertical gradient of virtual potential temperature with a threshold [4,15–17], when flux data are not available. This method has been widely used when radiosonde data are available. The parcel method developed by Holzworth [18] was also used in previous studies to estimate the boundary layer height using radiosonde data [15,19,20]. Remote sensing data (e.g., lidars, radars, sodas, and wind profilers), although sometimes ambiguous, can also be used to determine the boundary layer height when other types of data are not available [21–24].

In hurricane conditions, it is even more difficult to acquire direct turbulence measurements due to safety issues and measurement constraints. Multi-level turbulent intensity and flux data that can be used to estimate the boundary layer height are scarce in hurricane conditions. The only in situ data of this kind was collected in regions either 100 km away from the hurricane eyewall or in weak tropical storms [25–28]. Note that a new airborne Doppler radar dataset from the Imaging Wind and Rain Airborne Profiler (IWRAP) is able to provide three-dimensional winds, deep into the hurricane boundary layer, at high resolution (~200 m horizontal and 30 m vertical), which will be useful for characterizing the boundary layer height using dynamical metrics [29,30].

Dropsonde data have been used to estimate the hurricane boundary layer height scales in previous studies [31]. Zhang et al. [31] pointed out differences between the kinematic boundary layer height denoted by the height of the maximum tangential wind speed (h_{vtmax}) or inflow layer depth (h_{inflow}), and the thermodynamic boundary layer height denoted by the mixed layer depth (z_i). Of note, h_{inflow} is taken as the height of strong inflow layer where the inflow strength is equal to 10% of the maximum inflow [32]. Furthermore, z_i is taken as the height where the vertical gradient of the virtual potential temperature (theta-v) is equal to 3 K km^{-1}, a widely used approach in non-hurricane ABL studies [4]. Ming et al. [33] confirmed the result of Zhang et al. [31] by analyzing a smaller number of dropsonde data in Pacific typhoons.

Previous modeling studies on air-sea coupling processes of individual tropical cyclones (TCs) showed an asymmetric distribution of the boundary layer structure relative to the storm motion, suggesting that the boundary layer is more stable in the right-rear quadrant [34,35]. Although a significant number of observational studies have documented the mean boundary layer structure of individual hurricanes [27,36–41], the variability of the ABL structure as a function of azimuth remains to be explored. The objective of the present study is to investigate the asymmetry of the boundary layer structure in a storm-motion-relative framework with a focus on the boundary layer height scales. This study is a first attempt to analyze the motion-induced asymmetry of the boundary layer heights in a climatological sense by compositing dropsonde data collected in multiple hurricanes. The results of this study will be useful for improving our understanding of the low-level structure of hurricanes and providing composite analyses that can be used for model evaluation and physics improvements [42].

This paper is organized as follows. Section 2 describes the dropsonde data and the composite analysis methodology. The dropsonde composites used to compute the boundary layer heights are shown in Section 3, which is followed by the discussion and conclusion in Section 4.

2. Data and Composite Methodology

Dropsondes are typically deployed from research or reconnaissance aircraft at a height of ~3 km (i.e., 700 hPa), with a descending rate of ~ 10 m s^{-1} and a vertical sampling resolution of ~7 m. During the process of descent, atmospheric profiles of pressure, temperature, relative humidity, and horizontal winds are collected by the dropsonde. The accuracy of pressure, temperature, relative humidity and horizontal wind speed are ±1.0 hPa, ±0.2 °C, $\pm5\%$, and ±0.5 ms^{-1}, respectively [43]. The dropsonde data used in this study are from the Hurricane Research Division (HRD) archives and have been quality controlled using the Editsonde and/or ASPEN programs.

In this study, a total of 7326 GPS dropsonde profiles collected by research aircraft in 37 TCs from 1998 to 2015 are analyzed. The storm intensity is obtained from the hurricane best track database produced by the National Hurricane Center (NHC). The storm tracks processed by HRD, which combine the best track storm locations with aircraft center fixes, are used in the analysis. We linearly interpolate the storm intensity, center and motion from the track data to the dropsonde time. For the composite analysis of the inner-core structure, the dropsonde profiles must meet the following requirements: (i) They have measurements of wind speed, temperature, and humidity from flight-level all the way to the surface with no data gaps, (ii) the maximum sustained wind speed (i.e., storm intensity) from NHC's best track is larger than 64 kt (Cat 1 strength), and (iii) the sondes are deployed within four times the radius of maximum wind speed (RMW). A total of 1916 dropsonde profiles are used in the final composite analysis after this screening. Note that the RMW data used in this study are calculated based on the flight-level data following [44]. Table 1 summarizes the storm information and numbers of dropsondes used in this work.

Table 1. Storm information and number of dropsondes (1916).

ID	Storm	Intensity Range (kt)	No. of Sondes
0298	BONNIE	98.6–100.0	101
0498	DANIELLE	65.0–73.1	43
0598	EARL	79.6–83.0	3
0798	GEORGES	65.0–131.1	106
1398	MITCH	139.1–139.6	6
0399	BRET	101.9–111.4	9
0599	DENNIS	67.1–90.0	19
0899	FLOYD	81.3–122.5	23
1399	IRENE	65.0	4
1002	ISIDORE	66.1–110.0	14
1003	FABIAN	100.0–125.0	154
1303	ISABEL	79.6–140.0	271
0304	CHARLEY	64.1–120.4	49
0604	FRANCES	87.6–124.7	128
0904	IVAN	105.0–145.0	158
1104	JEANNE	85.0–105.0	11
0405	DENNIS	71.6–84.5	21
1205	KATRINA	100.0–150.0	60
1805	RITA	123.4–146.5	13
0906	HELENE	80.0–98.3	42
0708	GUSTAV	75.3–111.7	23
1708	PALOMA	79.1–125.0	20
0309	BILL	105.1–115.0	37
0710	EARL	71.8–121.1	41
1310	KARL	71.6–76.9	9
0911	IRENE	72.7–104.6	91
1811	RINA	81.8–100.0	7
0912	ISAAC	64.0–70.0	42
1812	SANDY	65.0–74.5	90
1013	INGRID	65.0–70.9	35
0114	ARTHUR	67.8–82.4	46
0314	BERTHA	64.6–70.0	23
0414	CRISTOBAL	65.0–67.0	28
0614	EDOUARD	72.5–101.5	73
0814	GONZALO	106.9–125.0	40
0415	DANNY	105.7–106.4	10
1115	JOAQUIN	75.0–120.0	66

The dropsonde distribution relative to the storm center is shown in Figure 1, with observation locations rotated with respect to the storm motion direction to the top of the figure. Figure 1 shows nearly evenly distributed data at motion-relative azimuths at all radii, although more dropsonde data are located close to the RMW (i.e., r/RMW = 1). The dropsonde data are composited in four motion-relative quadrants defined clockwise from the storm motion direction, as right-front, right-rear, left-rear and left-front (Figure 2). The storm characteristics in terms of frequency distribution, including the storm intensity V_{max}, RMW, storm translational direction, and storm speed are presented in Figure 3. Storm intensities range from 65–150 kt, sizes in terms of RMW range from 10–65 km, and translational speed ranges from 2–20 ms^{-1}. The mean storm intensity for the whole sample is 105.5 kt, the mean RMW is 28.5 km, and the mean storm translational speed is 5.27 ms^{-1}.

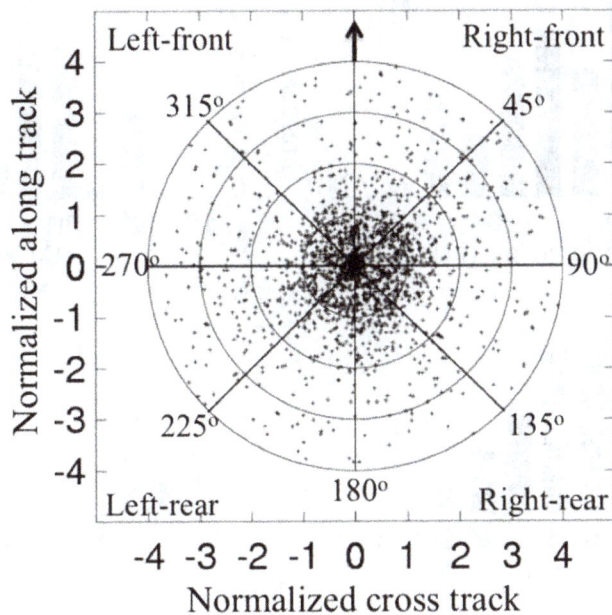

Figure 1. Storm-relative two-dimensional distribution of dropsonde surface observation locations. Cross- and along-track positions are normalized by the radius of maximum wind at the time of observation. The arrow indicates the storm motion direction.

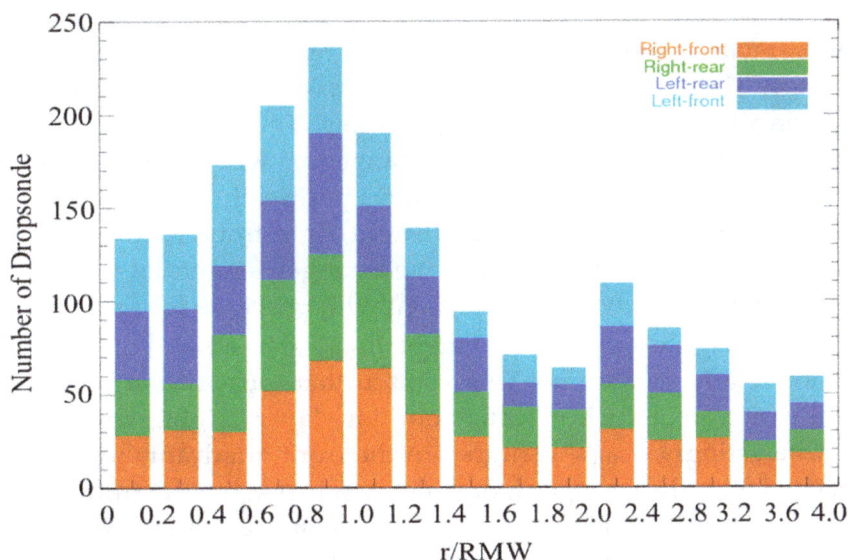

Figure 2. Radial distribution of dropsonde counts per bin as a function of normalized distance by radius of maximum wind speed (RMW).

In this study, we use the same composite methodology as Zhang et al. [31] to construct the radial-vertical profiles of virtual temperature (θ_v), tangential wind, radial wind, and Richardson number in four quadrants relative to the storm motion. When compositing the data, the radial bin width is 0.2 r^* ($r^* = r/RMW$) for the inner core ($r^* < 2$), and it is 0.4 r^* for the outer part. The data are also bin averaged vertically at 10 m resolution. The final averaged data are also smoothed with three passes of a 1-2-1 filter, instead of five passes as in Zhang et al. [31]. The data sampling sizes for different quadrants are displayed in Figure 2 as a function of normalized radius and height. The largest sample size is located in the vicinity of the eyewall as expected. Figure 2 also indicates that the data samples for the four motion-relative quadrants are similar.

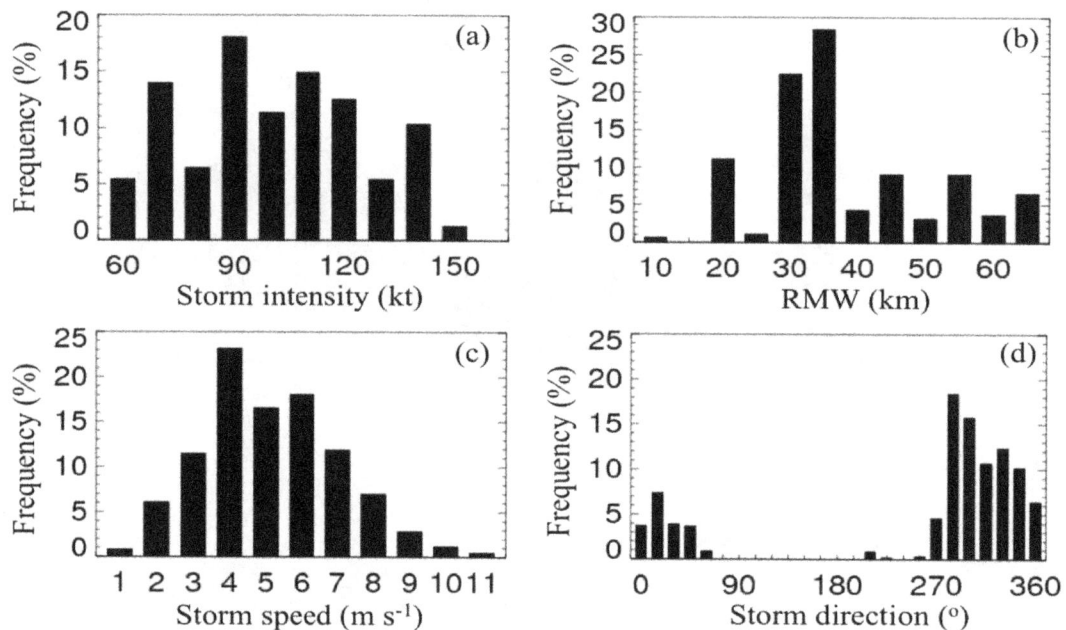

Figure 3. Frequency distribution of dropsondes according to the corresponding (**a**) storm intensity, (**b**) radius of maximum wind speed (RMW), (**c**) storm speed, and (**d**) storm direction rotated clockwise with 0° pointing to the north.

3. Results

The storm-motion-relative normalized radius-height representation of the tangential wind speed is displayed in Figure 4. The tangential wind first increases with height to a maximum value then decreases with height in all four motion-relative quadrants. This tangential wind maximum has been called the boundary layer jet [45,46]. The maximum tangential wind speed is largest in the right-front quadrant and it is smallest at the right-rear quadrant. Specifically, the peak values of the tangential wind speed are 53.5, 55.8, 52.6, and 44.3 ms^{-1} for left-front, right-front, left-rear and right-rear quadrant, respectively. The fact that the front quadrants have stronger tangential wind speeds than the rear quadrants in a storm-relative framework is consistent with previous theoretical studies [45–48].

It is also evident from Figure 4 that the jet height, h_{vtmax}, has a trend of decreasing with smaller r* values in all four motion-relative quadrants. This result further supports the findings of Zhang et al. [31] in terms of the radial variation of the hurricane boundary layer height, which may be applied in the asymmetric sense. Interestingly, h_{vtmax} is higher in the front quadrants than in the rear quadrants, which may be associated with the stronger tangential wind speed in the front quadrants. Specifically, h_{vtmax} increases from ~600 m at a radius of $r^* = 1$ to ~1200 m at a radius of $r^* = 2$, then slightly increases with increasing radius in the front two quadrants. In the left-rear quadrant, h_{vtmax} is ~100 m smaller than in the front two quadrants at r* = 1, and increases with radius to ~800 m at r* = 2, then increases to ~900 m at r* = 4. In the right-rear quadrant, h_{vtmax} is the lowest among all four quadrants, and it gradually increases from ~400 m at radius of $r^* = 1$ to ~600 m at radius of $r^* = 2$, and increases to ~800 m at r* = 4.

The normalized radius-height representation of the radial velocity for the four motion-relative quadrants is displayed in Figure 5. The peak values of the radial wind speed are −18.5, −18.7, −13.5 and −15.4 ms^{-1} for left-front, right-front, left-rear and right-rear quadrant, respectively. These peak inflow values are located at ~100 m altitude between r* = 1 − 2 and closer to r* = 1, which is consistent with the axisymmetric structure documented by Zhang et al. [31]. In addition, a pronounced outflow of 5–10 m s^{-1} can be seen in all the four quadrants above the inflow layer. The inflow is stronger in the front quadrants compared to the rear quadrants, which is consistent with previous theoretical and numerical studies [46–48]. Observational studies of individual hurricanes also showed similar front-back wind asymmetry [37,38,49].

The inflow layer depth, h_{inflow}, which is depicted by the white line in Figure 5, shows a decreasing trend with decreasing radius in all four motion-relative quadrants. It is also evident from Figure 5 that, h_{inflow} is larger in the front two quadrants than in the rear two quadrants. Specifically, h_{inflow} evolves similarly in left-front, right-front and right-rear quadrants from an altitude of ~600 m to an altitude of 900–1200 m with highest heights in the right-front quadrant and lowest heights in the right-rear quadrant. However, h_{inflow} remains nearly constant (~500 m) with a slightly negative trend in the left-rear quadrant. This result suggests that the asymmetric distribution of h_{inflow} follows that of h_{vtmax}. From Figures 4 and 5, the maximum tangential wind speed in the eyewall region is close to the top of the inflow layer in all quadrants.

The thermodynamic mixed layer depth, z_i, is depicted by the white line in Figure 6 that shows the normalized radius-height representation of the vertical gradient of theta-v ($d\theta_v/dz$). In all quadrants, the boundary layer is unstable near the surface as indicated by the negative value of $d\theta_v/dz$. Above this very shallow unstable layer, the boundary layer becomes nearly neutral up to the mixed layer depth and then becomes stable. There is a strong stable layer inside the RMW above the mixed layer at heights of 600–2000 m with $d\theta_v/dz > 5$ K/km. This strong stable layer is shallowest at the right-rear quadrant among the four motion-relative quadrants, which may be tied to the cold wake phenomena observed in the sea surface temperature (SST) field [50–52]. On average when the storm motion of a hurricane is ~6 ms^{-1}, which is the average value of our data, the typical SST close to the cold wake region at the right-rear quadrant is of the order of 1–2 K smaller than the front quadrants based on in-situ data [53]. The smaller SST in the right-rear quadrant would stabilize the low-level boundary layer due to a reduction in surface enthalpy fluxes.

In all four quadrants, z_i decreases with distance toward the storm center, in a similar manner as the kinematic boundary layer heights. This result again supports that of Zhang et al. [31] and Zhang et al. [54]. From Figures 4–6, it is clear that z_i is much smaller than h_{inflow} and h_{vtmax} in all quadrants, the kinematic and thermodynamic boundary layer heights largely depart from each other, as noted by Zhang et al. [31]. This structure is different from that of the ABL in non-hurricane conditions. Close to the eyewall region (r* < 1.5), z_i is nearly symmetric with a value of ~200 m. Outside the eyewall, the left-front quadrant has the largest z_i while the other three quadrants have similar magnitudes of z_i. In the outer radii (r* > 3), z_i is largest in the left-front quadrant and is smallest in the right-rear quadrant. There is a weak front-back asymmetry in z_i, which is similar to the asymmetric distribution of the kinematic boundary layer height, but the mixed layer depth difference is less than ~100 m.

The storm-motion-relative normalized radius-height representation of the bulk Richardson number is displayed in Figure 7. Here the height of $R_{ic} = 0.25$ is taken as the top of the boundary layer (h_{Ric}), which is depicted by the solid white line in Figure 7. The front-back difference in h_{Ric} is clearly shown in Figure 7 with the front two quadrants displaying a deeper boundary layer, consistent with the other height scales investigated earlier. Combining Figures 4–7, it appears that h_{Ric} lies between the thermodynamic mixed layer depth and the kinematic boundary layer height, which agrees with the axisymmetric structure documented by Zhang et al. [31]. The left-front quadrant has the deepest boundary layer while the right-rear quadrant has the shallowest boundary layer, in terms of h_{Ric}.

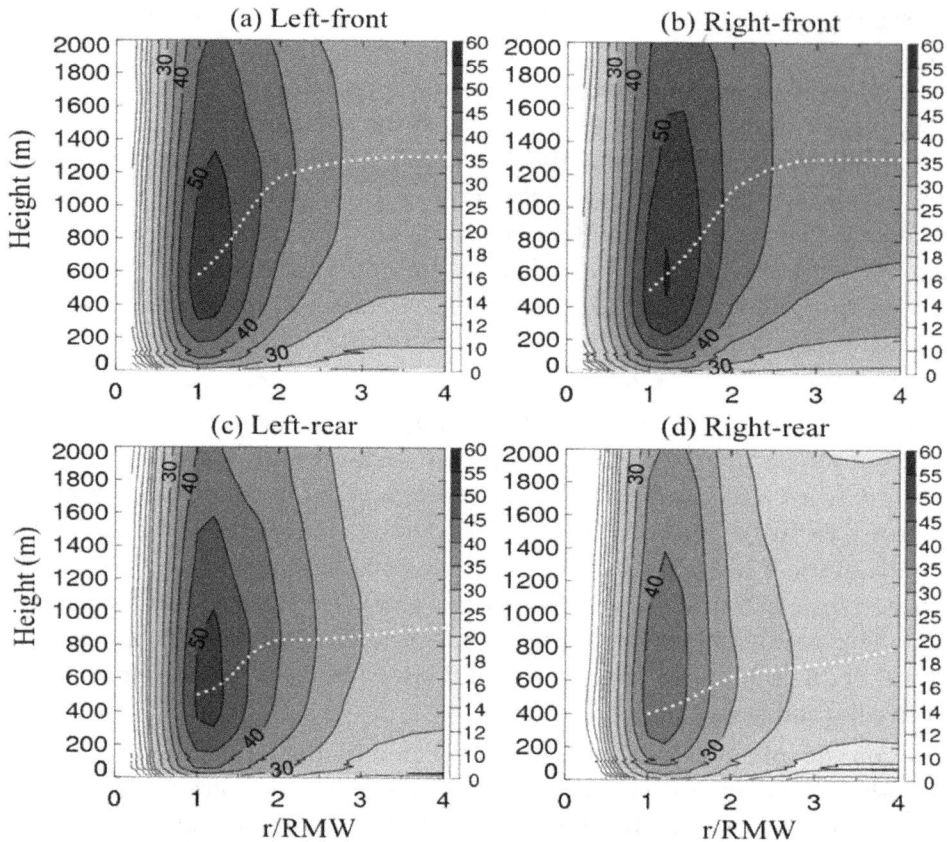

Figure 4. Composite analysis result of the relative tangential wind velocity as a function of altitude and the normalized radius to the storm center for the four quadrants relative to the motion direction. The panels show the left-front (**a**), right-front (**b**), left-rear (**c**) and right-rear (**d**) quadrants. The white dashed line in each panel depicts the height of the maximum tangential wind speed varying with radius.

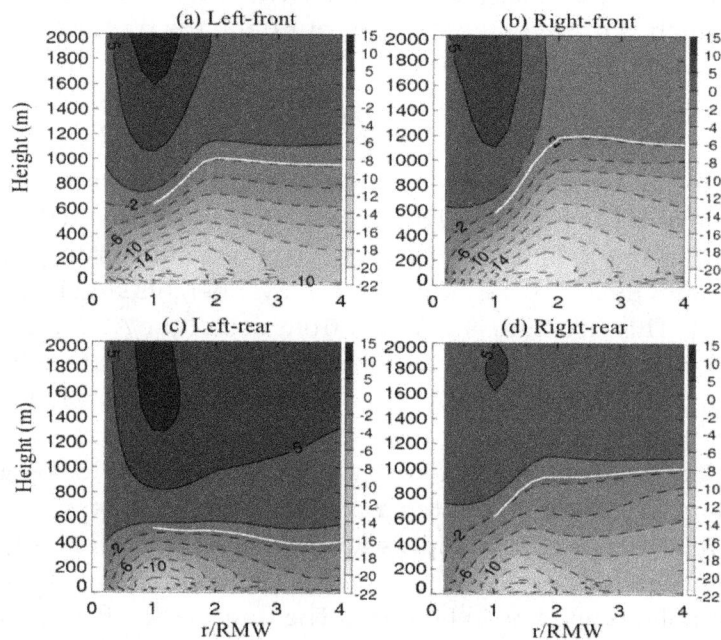

Figure 5. Same as in Figure 4 but the results are for the relative radial wind velocity as a function of altitude and the normalized radius to the storm center for the four quadrants relative to the shear direction. The panels show the left-front (**a**), right-front (**b**), left-rear (**c**) and right-rear (**d**) quadrants. The white line in each panel represents the height of 10% peak inflow.

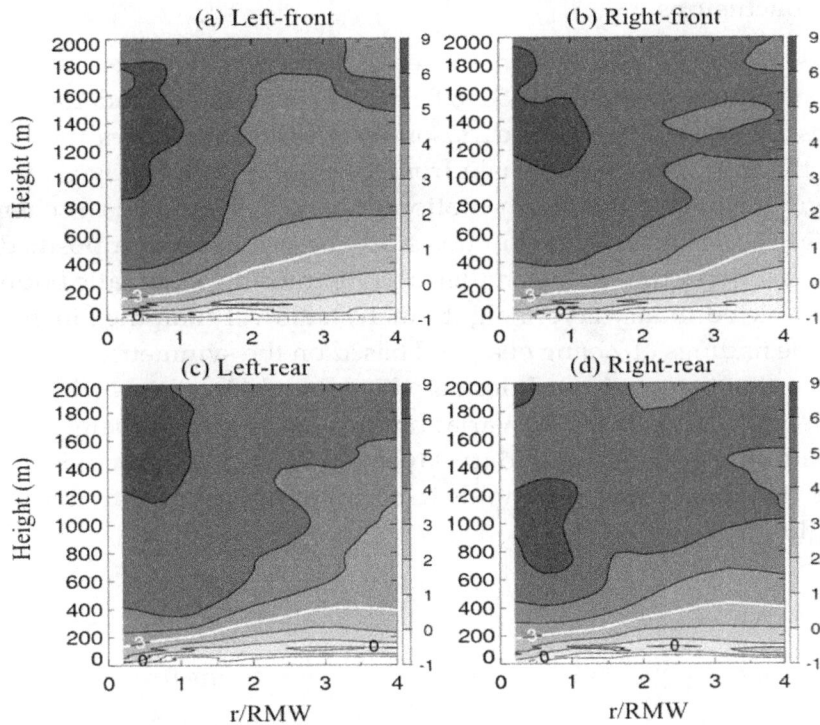

Figure 6. Same as in Figure 4 but the results show the lapse rate of the virtual potential temperature. The panels show the left-front (**a**), right-front (**b**), left-rear (**c**) and right-rear (**d**) quadrants. The thick white line denotes the contour. The contour denotes the constant contour of $d\theta_v/dz = 3\,\mathrm{K\,km^{-1}}$.

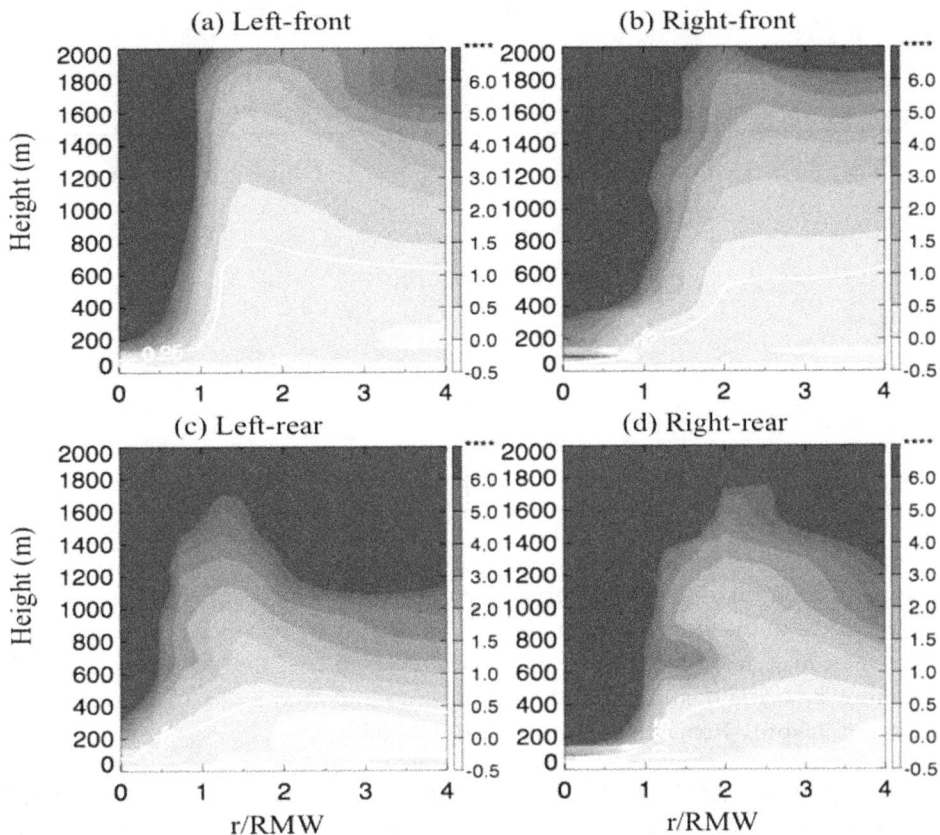

Figure 7. Same as in Figure 4 but the results are for the Richardson numbers as a function of altitude and the normalized radius to the storm center. The panels show the left-front (**a**), right-front (**b**), left-rear (**c**) and right-rear (**d**) quadrants. The white line shows the contour of 0.25.

4. Discussion and Conclusions

The ABL plays an important role in the energy transport processes related to hurricane intensification and maximum intensity [32,55–60] and it is essential to understand the ABL structure. This paper analyzes a total of 1916 GPS dropsondes within four times the RMW distance from 37 hurricanes over the tropical Atlantic basin from 1998 to 2015 to study the characteristic height scales of the ABL with respect to the storm-motion direction. Figure 8 is a schematic diagram that summarizes the height scales investigated in this study based on the composite dropsonde analysis. The results show a clear departure between thermodynamic and kinematic boundary layer heights with the thermodynamic boundary layer height much shallower than the kinematic boundary layer height. Supporting the findings of Zhang et al. [31] based on the symmetric analysis of the dropsonde data, our results show that the hurricane boundary layer height increases with increasing radius in a storm-relative framework. This observed variation of boundary layer height with radius supports the theoretical scaling of a rotating boundary layer [46,59,61–64] in an axisymmetric framework. Our results indicate that this scaling also holds in a motion-relative asymmetric distribution of the boundary layer height.

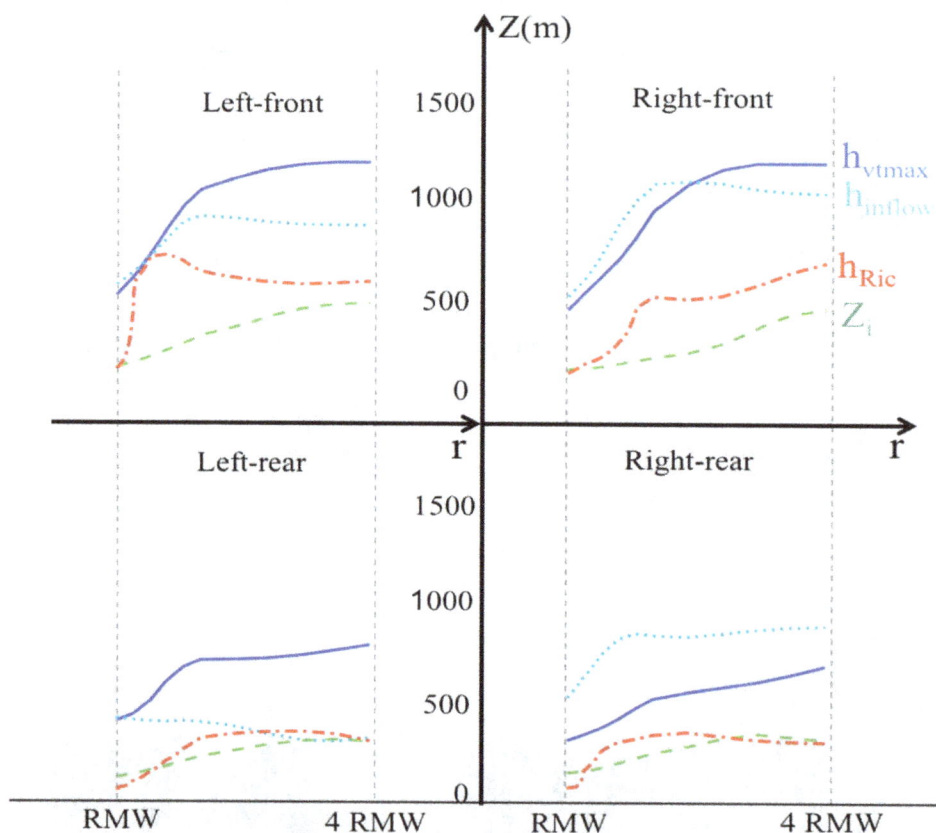

Figure 8. Schematic diagram of the characteristic height scales of the hurricane boundary layer for the four quadrants relative to the storm motion. The height scales are based on the composite analysis of the dropsonde data. h_{inflow} is the inflow layer depth (cyan dotted line); z_i is the mixed layer depth (green dashed line); h_{vtmax} is the height of the maximum tangential wind speed (blue solid line) and; h_{Ric} is the height of the bulk Richardson number value of 0.25.

In the eyewall region, all height scales show relatively small asymmetric distribution. The weak asymmetry is found in the kinematic boundary layer heights (h_{vtmax} and h_{inflow}) that are slightly smaller in the left-rear quadrant than in other quadrants. All height scales demonstrate similar front-back asymmetry outside the eyewall region, in that the two front motion-relative quadrants have a deeper boundary layer. The front-back difference in the thermodynamic boundary layer height is smaller than that in the kinematic boundary layer height. The mixed layer depth being smallest in the right-rear quadrant may be due to the SST cooling effect in this quadrant where the cold wake is located, following the argument of previous modeling studies [35,52].

The boundary layer height derived using the critical Richardson number (R_{ic}) method shows a similar asymmetry as the kinematic boundary layer height, confirming that turbulence in the hurricane boundary layer is mainly shear driven. Consistent with Zhang et al. [31], our results suggest that the hurricane research community should use the kinematic height scales to represent the top of the boundary layer instead of the thermodynamic mixed layer depth typically used in non-hurricane conditions. Kepert [65] and Kepert et al. [66] discussed the limitation of using thermodynamic mixed layer depth to represent the boundary layer height in numerical models, which is in agreement with our observational findings here.

Our results in terms of the asymmetry of maximum tangential and radial wind speeds in the boundary layer generally agree with previous theoretical and numerical studies [46–48] Our result in terms of front-back asymmetry in the velocity fields also agrees with previous observational case studies such as the ABL structure in Hurricanes Mitch (1998), George (1998), Isabel (2003) and Daniel (2008) as shown by Kepert [37,38]. Although previous studies did not focus on the boundary layer height asymmetry, their results indicate a similar structure of the inflow and inflow layer depth as well as the height of the boundary layer jet. The surface wind asymmetry is consistent with Uhlhorn et al. [67] who analyzed stepped frequency microwave radiometer (SFMR) data, as well as Klotz and Jiang [68] who analyzed satellite data. The inflow asymmetry near the surface is consistent with the result of Zhang and Uhlhorn [69], who studied the characteristics of surface inflow angle in both axisymmetric and asymmetric framework.

Note that Zhang et al. [54] investigated the asymmetric structure of the hurricane boundary layer in relation to the environmental vertical wind shear. They focused on investigating how boundary layer thermodynamic structure is tied to the upper-level convection in an energy-cycling paradigm. They also studied the asymmetry of the boundary layer height scales relative to the environmental shear direction and found that the kinematic boundary layer height scales are larger in the downshear direction than in the upshear direction, while the thermodynamic boundary layer height scale is slightly larger left of the shear than right of the shear. The boundary layer being deeper in the downshear quadrants may be tied to the asymmetric distribution of convection that is usually initiated in the downshear-right quadrant and propagates to downshear-left quadrant [70]. The boundary layer is deeper in the quadrants where stronger convection occurs. Following this argument, storm-motion induced asymmetry of the boundary layer height scales may also be linked to asymmetric distribution of convective activity relative to storm motion, although storm motion and environmental wind shear are very different parameters.

The asymmetric structure above the boundary layer (i.e., convection) in tropical cyclones with respect to storm motion have been extensively studied using radar observations in both case studies and composite analysis studies. For instance, Jorgensen et al. [71] utilized flight-level data and found maximum upward mass transport in slow-moving storms occurred to the right of motion, with equal amounts occurring in the front and rear of the inner core. Our composite analysis supports strong convection occurs in the right-front quadrant, as both the maximum tangential wind speed and maximum inflow strength occur in this quadrant. Marks et al. [72] found that the maximum upward vertical velocities were in the left-front quadrant of Hurricane Norbert (1984). Marks [73] found that the maximum precipitation of Hurricane Allen (1980) in the eyewall region was in the left-front quadrant, while the maximum precipitation of Hurricane Elena (1985) was in the right-front quadrant. Through examining the patterns of reflectivity in the eyewall region of Hurricane Olivia (1994), Reasor et al. [74] found that the maximum radar reflectivity was located in the left quadrants relative to the motion direction, which was consistent with the structure in Hurricane Gloria (1985) documented by Franklin et al. [75]. Our composite analysis shows that the strongest inflow in the boundary layer is located in the front quadrants, which support large vertical motion and strong convection due to mass continuity. Overall, it is hypothesized that the asymmetry of boundary layer inflow and jet strength are tied to the asymmetry of convection and precipitation.

Furthermore, the composite analysis result showing the front-back asymmetry of the boundary layer heights supports the theoretical argument that the boundary-layer convergence is larger in the front quadrants relative to the storm motion direction [45,47]. The surface wind asymmetry related to storm-motion effect induces the asymmetry of surface drag forcing, which in turn affects the distribution of boundary layer convergence and convection. Larger wind speed observed in the front quadrants induces larger turbulent mixing in the boundary layer, which supports larger kinematic boundary layer height according to dynamic scaling [42,45].

Of note, the combined effects of storm motion and environmental wind shear on the asymmetry of the boundary layer structure remains to be understood due to the limitation of data sampling size at the current stage. Future studies will combine the dropsonde and Doppler radar data to investigate the linkage between the boundary layer and convective processes and their asymmetry relative to both the storm motion and environmental wind shear. The extent of asymmetry of the boundary layer structure to the storm motion speed will also be evaluated when more observational data are available than those used in the present study.

Author Contributions: Conceptualization, J.A.Z., X.W.; methodology, Y.R., J.A.Z.; software, X.W.; validation, Y.R., J.A.Z., S.R.G. and X.W.; formal analysis, Y.R., X.W.; resources, J.A.Z., X.W.; data curation, Y.R.; writing—original draft preparation, Y.R., J.A.Z., S.R.G., X.W.; writing—review and editing, J.A.Z., S.R.G., X.W.; visualization, X.W.; supervision, J.A.Z.; project administration, Y.R.; funding acquisition, J.A.Z., S.R.G., X.W.

Acknowledgments: The authors would like to thank all the scientists and crew members who have been involved in the hurricane field program and operational reconnaissance hurricane missions to help collect the dropsonde data used in this study. This paper is a follow-up work of the first author's Master thesis and she would like to acknowledge Prof. Xiaolei Zou and Prof. Ming Cai for their helpful discussions. We also wish to thank the reviewers for their comments that led to improvement of the paper.

References

1.   Troen, I.B.; Mahrt, L. A simple model of the atmospheric boundary layer; sensitivity to surface evaporation. *Bound. Layer Meteorol.* **1986**, *37*, 129–148. [CrossRef]

2.   Hong, S.-Y.; Pan, H.-L. Nonlocal Boundary Layer Vertical Diffusion in a Medium-Range Forecast Model. *Mon. Weather Rev.* **1996**, *124*, 2322–2339. [CrossRef]

3.   Hong, S.-Y.; Noh, Y.; Dudhia, J. A New Vertical Diffusion Package with an Explicit Treatment of Entrainment Processes. *Mon. Weather Rev.* **2006**, *134*, 2318–2341. [CrossRef]

4.   Zeng, X.; Brunke, M.A.; Zhou, M.; Fairall, C.; Bond, N.A.; Lenschow, D.H. Marine Atmospheric Boundary Layer Height over the Eastern Pacific: Data Analysis and Model Evaluation. *J. Clim.* **2004**, *17*, 4159–4170. [CrossRef]

5.   Balsley, B.B.; Frehlich, R.G.; Jensen, M.L.; Meillier, Y. High-Resolution In Situ Profiling through the Stable Boundary Layer: Examination of the SBL Top in Terms of Minimum Shear, Maximum Stratification, and Turbulence Decrease. *J. Atmos. Sci.* **2006**, *63*, 1291–1307. [CrossRef]

6.   Hennemuth, B.; Lammert, A. Determination of the Atmospheric Boundary Layer Height from Radiosonde and Lidar Backscatter. *Bound. Layer Meteorol.* **2006**, *120*, 181–200. [CrossRef]

7.   Sicard, M.; Pérez, C.; Rocadenbosch, F.; Baldasano, J.M.; García-Vizcaino, D. Mixed-layer depth determination in the Barcelona coastal area from regular lidar measurements: methods, results and limitations. *Bound. Layer Meteorol.* **2006**, *119*, 135–157. [CrossRef]

8.   Georgoulias, A.K.; Papanastasiou, D.K.; Melas, D.; Amiridis, V.; Alexandri, G.; Georgoulias, A. Statistical analysis of boundary layer heights in a suburban environment. *Theor. Appl. Clim.* **2009**, *104*, 103–111. [CrossRef]

9. Stull, R.B. *An Introduction to Boundary-layer Meteorology*; Kluwer Academic Publishers: Dordrecht, The Netherlands; Boston, MA, USA; London, UK, 1988; 666p.

10. Fetzer, E.J.; Teixeira, J.; Olsen, E.T.; Fishbein, E.F. Satellite remote sounding of atmospheric boundary layer temperature inversions over the subtropical eastern Pacific. *Geophys. Res. Lett.* **2004**, *31*. [CrossRef]

11. Wood, R.; Bretherton, C.S. Boundary Layer Depth, Entrainment, and Decoupling in the Cloud-Capped Subtropical and Tropical Marine Boundary Layer. *J. Clim.* **2004**, *17*, 3576–3588. [CrossRef]

12. Medeiros, B.; Hall, A.; Stevens, B. What Controls the Mean Depth of the PBL? *J. Clim.* **2005**, *18*, 3157–3172. [CrossRef]

13. Palm, S.P.; Spinhirne, J.; Benedetti, A. Validation of ECMWF global forecast model parameters using GLAS atmospheric channel measurements. *Geophys. Res. Lett.* **2005**, *32*, 109–127. [CrossRef]

14. Sokolovskiy, S.; Röcken, C.; Hunt, D.; Schreiner, W.; Johnson, J.; Masters, D.; Esterhuizen, S. GPS profiling of the lower troposphere from space: Inversion and demodulation of the open-loop radio occultation signals. *Geophys. Res. Lett.* **2006**, *33*. [CrossRef]

15. Beyrich, F. Mixing height estimation from sodar data—A critical discussion. *Atmos. Environ.* **1997**, *31*, 3941–3953. [CrossRef]

16. Dupont, E.; Menut, L.; Carissimo, B.; Pelon, J.; Flamant, P. Comparison between the atmospheric boundary layer in Paris and its rural suburbs during the ECLAP experiment. *Atmos. Environ.* **1999**, *33*, 979–994. [CrossRef]

17. Bianco, L.; Wilczak, J.M. Convective Boundary Layer Depth: Improved Measurement by Doppler Radar Wind Profiler Using Fuzzy Logic Methods. *J. Atmos. Ocean. Technol.* **2002**, *19*, 1745–1758. [CrossRef]

18. Holzworth, C.G. Estimates of mean maximum mixing depths in the contiguous United States. *Mon. Weather Rev.* **1964**, *92*, 235–242. [CrossRef]

19. Coulter, R.L. A Comparison of Three Methods for Measuring Mixing-Layer Height. *J. Appl. Meteorol.* **1979**, *18*, 1495–1499. [CrossRef]

20. Lokoshchenko, M.A. Long-Term Sodar Observations in Moscow and a New Approach to Potential Mixing Determination by Radiosonde Data. *J. Atmos. Ocean. Technol.* **2002**, *19*, 1151–1162. [CrossRef]

21. Emeis, S.; Münkel, C.; Vogt, S.; Müller, W.J.; Schäfer, K. Atmospheric boundary-layer structure from simultaneous SODAR, RASS, and ceilometer measurements. *Atmos. Environ.* **2004**, *38*, 273–286. [CrossRef]

22. Nielsen-Gammon, J.W.; Powell, C.L.; Mahoney, M.J.; Angevine, W.M.; Senff, C.; White, A.; Berkowitz, C.; Doran, C.; Knupp, K. Multisensor Estimation of Mixing Heights over a Coastal City. *J. Appl. Meteorol. Clim.* **2008**, *47*, 27–43. [CrossRef]

23. Guo, P.; Kuo, Y.-H.; Sokolovskiy, S.V.; Lenschow, D. Estimating Atmospheric Boundary Layer Depth Using COSMIC Radio Occultation Data. *J. Atmos. Sci.* **2011**, *68*, 1703–1713. [CrossRef]

24. Seidel, D.J.; Ao, C.O.; Li, K. Estimating climatological planetary boundary layer heights from radiosonde observations: Comparison of methods and uncertainty analysis. *J. Geophys. Res. Space Phys.* **2010**, *115*. [CrossRef]

25. Moss, M.S. Low-level turbulence structure in the vicinity of a hurricane. *Mon. Weather Rev.* **1978**, *106*, 841–849. [CrossRef]

26. French, J.R.; Drennan, W.M.; Zhang, J.A.; Black, P.G. Turbulent fluxes in the hurricane boundary layer. Part I: Momentum flux. *J. Atmos. Sci.* **2007**, *64*, 1089–1102. [CrossRef]

27. Zhang, J.A.; Drennan, W.M.; Black, P.G.; French, J.R. Turbulence Structure of the Hurricane Boundary Layer between the Outer Rainbands. *J. Atmos. Sci.* **2009**, *66*, 2455–2467. [CrossRef]

28. Zhang, J.A.; Drennan, W.M. An Observational Study of Vertical Eddy Diffusivity in the Hurricane Boundary Layer. *J. Atmos. Sci.* **2012**, *69*, 3223–3236. [CrossRef]

29. Guimond, S.R.; Tian, L.; Heymsfield, G.M.; Frasier, S.J. Wind Retrieval Algorithms for the IWRAP and HIWRAP Airborne Doppler Radars with Applications to Hurricanes. *J. Atmos. Ocean. Technol.* **2014**, *31*, 1189–1215. [CrossRef]

30. Guimond, S.R.; Zhang, J.A.; Sapp, J.W.; Frasier, S.J. Coherent turbulence in the boundary layer of Hurricane Rita (2005) during an eyewall replacement cycle. *J. Atmos. Sci.* **2018**, *75*, 3071–3093. [CrossRef]

31. Zhang, J.A.; Rogers, R.F.; Nolan, D.S.; Marks, F.D. On the Characteristic Height Scales of the Hurricane Boundary Layer. *Mon. Weather Rev.* **2011**, *139*, 2523–2535. [CrossRef]

32. Smith, R.K.; Montgomery, M.T.; Nguyen, S.V. Tropical cyclone spinup revisited. *Q. J. R. Meteor. Soc.* **2009**, *135*, 1321–1335. [CrossRef]

33. Ming, J.; Zhang, J.A.; Rogers, R.F. Typhoon kinematic and thermodynamic boundary layer structure from dropsonde composites. *J. Geophys. Res. Atmos.* **2015**, *120*, 3158–3172. [CrossRef]

34. Lee, C.-Y.; Chen, S.S. Symmetric and Asymmetric Structures of Hurricane Boundary Layer in Coupled Atmosphere–Wave–Ocean Models and Observations. *J. Atmos. Sci.* **2012**, *69*, 3576–3594. [CrossRef]

35. Wu, C.; Tu, W.; Pun, I.; Lin, I.; Peng, M.S. Tropical cyclone-ocean interaction in Typhoon Megi (2010)—A synergy study based on ITOP observations and atmosphere-ocean coupled model simulations. *J. Geophys. Res. Atmos.* **2016**, *121*, 153–167. [CrossRef]

36. Powell, M.D. Boundary layer structure and dynamics in outer hurricane rainbands. Part II: Downdraft modification and mixed layer recovery. *Mon. Weather Rev.* **1990**, *118*, 918–938. [CrossRef]

37. Kepert, J.D. Observed boundary layer wind structure and balance in the hurricane core. Part I: Hurricane Georges. *J. Atmos. Sci.* **2006**, *63*, 2169–2193. [CrossRef]

38. Kepert, J.D. Observed boundary layer wind structure and balance in the hurricane core. Part II: Hurricane Mitch. *J. Atmos. Sci.* **2006**, *63*, 2194–2211. [CrossRef]

39. Bell, M.M.; Montgomery, M.T. Observed structure, evolution, and potential intensity of category 5 Hurricane Isabel (2003) from 12 to 14 September. *Mon. Weather Rev.* **2008**, *136*, 2023–2046. [CrossRef]

40. Barnes, G.M. Atypical Thermodynamic Profiles in Hurricanes. *Mon. Weather Rev.* **2008**, *136*, 631–643. [CrossRef]

41. Lokoshchenko, M.T.; Zhang, J.A.; Smith, R.K. An analysis of the observed low-level structure of rapidly intensifying and mature hurricane *Earl* (2010). *Q. J. R. Meteorol. Soc.* **2014**, *140*, 2132–2146.

42. Zhang, J.A.; Nolan, D.S.; Rogers, R.F.; Tallapragada, V. Evaluating the impact of improvements in the boundary layer parameterization on hurricane intensity and structure forecasts in HWRF. *Mon. Weather Rev.* **2015**, *143*, 3136–3155. [CrossRef]

43. Hock, T.F.; Franklin, J.L. The NCAR GPS dropwindsonde. *Bull. Am. Meteor. Soc.* **1999**, *80*, 407–420. [CrossRef]

44. Vigh, J.L.; Dorst, N.M.; Williams, C.L.; Uhlhorn, E.W.; Klotz, B.W.; Martinez, J.; Willoughby, H.E.; Marks, F.D., Jr.; Chavas, D.R. *FLIGHT+: The Extended Flight Level Dataset for Tropical Cyclones (Version 1.0)*; Tropical Cyclone Data Project; National Center for Atmospheric Research, Research Applications Laboratory: Boulder, CO, USA, 2015. Available online: http://dx.doi.org/10.5065/D6WS8R93 (accessed on 21 January 2015).

45. Kepert, J.D. The Dynamics of Boundary Layer Jets within the Tropical Cyclone Core. Part I: Linear Theory. *J. Atmos. Sci.* **2001**, *58*, 2485–2501. [CrossRef]

46. Kepert, J.D.; Wang, Y. The dynamics of boundary layer jets within the tropical cyclone core. Part II: Nonlinear enhancement. *J. Atmos. Sci.* **2001**, *58*, 2485–2501. [CrossRef]

47. Shapiro, L.J. The Asymmetric Boundary layer Flow Under a Translating Hurricane. *J. Atmos. Sci.* **1983**, *40*, 1984–1998. [CrossRef]

48. Thomsen, G.L.; Smith, R.K.; Montgomery, M.T. Tropical cyclone flow asymmetries induced by a uniform flow revisited. *J. Adv. Model. Earth Syst.* **2015**, *7*, 1265–1284. [CrossRef]

49. Schwendike, J.; Kepert, J.D. The boundary layer winds in hurricanes Danielle (1998) and Isabel (2003). *Mon. Weather Rev.* **2008**, *136*, 3168–3192. [CrossRef]

50. Price, J.F. Upper Ocean Response to a Hurricane. *J. Phys. Oceanogr.* **1981**, *11*, 153–175. [CrossRef]

51. Shay, L.K.; Black, P.G.; Mariano, A.J.; Hawkins, J.D.; Elsberry, R.L. Upper ocean response to Hurricane Gilbert. *J. Geophys. Res. Space Phys.* **1992**, *97*, 20227. [CrossRef]

52. Lee, C.-Y.; Chen, S.S. Stable Boundary Layer and Its Impact on Tropical Cyclone Structure in a Coupled Atmosphere–Ocean Model. *Mon. Weather Rev.* **2014**, *142*, 1927–1944. [CrossRef]

53. Cione, J.J.; Kalina, E.A.; Zhang, J.A.; Uhlhorn, E.W. Observations of Air–Sea Interaction and Intensity Change in Hurricanes. *Mon. Weather Rev.* **2013**, *141*, 2368–2382. [CrossRef]

54. Zhang, J.A.; Rogers, R.F.; Reasor, P.D.; Uhlhorn, E.W.; Marks, F.D. Asymmetric Hurricane Boundary Layer Structure from Dropsonde Composites in Relation to the Environmental Vertical Wind Shear. *Mon. Weather Rev.* **2013**, *141*, 3968–3984. [CrossRef]

55. Ooyama, K.V. Numerical simulation of the life cycle of tropical cyclones. *J. Atmos. Sci.* **1969**, *26*, 3–40. [CrossRef]

56. Emanuel, K.A. An Air-Sea Interaction Theory for Tropical Cyclones. Part I: Steady-State Maintenance. *J. Atmos. Sci.* **1986**, *43*, 585–605. [CrossRef]

57. Emanuel, K.A. Sensitivity of Tropical Cyclones to Surface Exchange Coefficients and a Revised Steady-State Model incorporating Eye Dynamics. *J. Atmos. Sci.* **1995**, *52*, 3969–3976. [CrossRef]

58. Bryan, G.H.; Rotunno, R. The maximum intensity of tropical cyclones in axisymmetry numerical model simulations. *Mon. Weather Rev.* **2009**, *137*, 1770–1789. [CrossRef]

59. Foster, R.C. Boundary-Layer Similarity Under an Axisymmetric, Gradient Wind Vortex. *Bound. Layer Meteorol.* **2009**, *131*, 321–344. [CrossRef]

60. Montgomery, M.T.; Smith, R.K. Recent Developments in the Fluid Dynamics of Tropical Cyclones. *Annu. Rev. Fluid Mech.* **2017**, *49*, 541–574. [CrossRef]

61. Eliassen, A. On the Ekman Layer in a circular Vortex. *J. Meteorol. Soc. Jpn.* **1971**, *49*, 784–789. [CrossRef]

62. Carrier, G.F. Swirling flow boundary layers. *J. Fluid Mech.* **1971**, *49*, 133–144. [CrossRef]

63. Montgomery, M.T.; Snell, H.D.; Yang, Z. Axisymmetric Spindown Dynamics of Hurricane-like Vortices. *J. Atmos. Sci.* **2001**, *58*, 421–435. [CrossRef]

64. Nolan, D.S. Instabilities in hurricane-like boundary layers. *Dyn. Atmos. Oceans* **2005**, *40*, 209–236. [CrossRef]

65. Kepert, J.D. Slab- and height-resolving models of the tropical cyclone boundary layer. Part I: Comparing the simulations. *Q. J. R. Meteorol. Soc.* **2010**, *136*, 1686–1699. [CrossRef]

66. Kepert, J.D.; Schwendike, J.; Ramsay, H. Why is the tropical cyclone boundary layer not "well mixed"? *J. Atmos. Sci.* **2016**, *73*, 957–973. [CrossRef]

67. Uhlhorn, E.W.; Klotz, B.W.; Vukićević, T.; Reasor, P.D.; Rogers, R.F. Observed Hurricane Wind Speed Asymmetries and Relationships to Motion and Environmental Shear. *Mon. Weather Rev.* **2014**, *142*, 1290–1311. [CrossRef]

68. Klotz, B.W.; Jiang, H. Global composites of surface wind speeds in tropical cyclones based on a 12-year scatterometer database. *Geophys. Res. Lett.* **2016**, *43*, 10480–10488. [CrossRef]

69. Zhang, J.A.; Uhlhorn, E.W. Hurricane Sea Surface Inflow Angle and an Observation-Based Parametric Model. *Mon. Weather Rev.* **2012**, *140*, 3587–3605. [CrossRef]

70. Reasor, P.D.; Rogers, R.; Lorsolo, S. Environmental Flow Impacts on Tropical Cyclone Structure Diagnosed from Airborne Doppler Radar Composites. *Mon. Weather Rev.* **2013**, *141*, 2949–2969. [CrossRef]

71. Jorgensen, D.P.; Zipser, E.J.; LeMone, M.A. Vertical Motions in Intense Hurricanes. *J. Atmos. Sci.* **1985**, *42*, 839–856. [CrossRef]

72. Marks, F.D., Jr.; Houze, R.A., Jr.; Gamache, J.F. Dual-aircraft investigation of the inner core of Hurricane Norbert. Part I: Kinematic structure. *J. Atmos. Sci.* **1992**, *49*, 919–942. [CrossRef]

73. Marks, F.D. Evolution of the Structure of Precipitation in Hurricane Allen (1980). *Mon. Weather Rev.* **1985**, *113*, 909–930. [CrossRef]

74. Reasor, P.D.; Montgomery, M.T.; Marks, F.D.; Gamache, J.F. Low-Wavenumber Structure and Evolution of the Hurricane Inner Core Observed by Airborne Dual-Doppler Radar. *Mon. Weather Rev.* **2000**, *128*, 1653–1680. [CrossRef]

75. Franklin, J.L.; Lord, S.J.; Feuer, S.E.; Marks, F.D. The Kinematic Structure of Hurricane Gloria (1985) Determined from Nested Analyses of Dropwindsonde and Doppler Radar Data. *Mon. Weather Rev.* **1993**, *121*, 2433–2451. [CrossRef]

11

Seasonal and Interannual Variation Characteristics of Low-Cloud Fraction in Different North Pacific Regions

Qian Wang [1,2], Haiming Xu [1,2,*], Leying Zhang [3] and Jiechun Deng [1,2]

[1] Key Laboratory of Meteorological Disaster, Ministry of Education (KLME)/Joint International Research Laboratory of Climate and Environment Change (ILCEC)/Collaborative Innovation Center on Forecast and Evaluation of Meteorological Disasters (CIC-FEMD), Nanjing University of Information Science & Technology, Nanjing 210044, China; w_qian_qian@163.com (Q.W.); jcdeng@nuist.edu.cn (J.D.)

[2] School of Atmospheric Sciences, Nanjing University of Information Science & Technology, Nanjing 210044, China

[3] Joint Innovation Center for Modern Forestry Studies, College of Biology and Environment, Nanjing Forestry University, Nanjing 210037, China; zhangleyingzi@126.com

[*] Correspondence: hxu@nuist.edu.cn.

Abstract: In this study, we use the long-term satellite data to investigate seasonal and interannual variation of low-cloud fraction (LCF) and the associated controlling factors over the eastern and western North Pacific. On the seasonal time scale, the enhanced LCF over the eastern North Pacific in summer is actively coupled with strong estimated inversion strength (EIS) and 700-hPa relative humidity, and the LCF over the western North Pacific in winter is large and mainly caused by increased sensible heat flux and tropospheric low-level cold advection. On the interannual time scale, the increased LCF over the eastern North Pacific in summer is associated with increased EIS and decreased sea surface temperatures, in which the El Niño plays an important role; the enhanced LCF over the western North Pacific in spring and winter has a positive correlation with enhanced sensible heat flux (SHF) and tropospheric low-level cold advection, which can be partly explained by the subpolar frontal zone (SPFZ) intensity.

Keywords: low-level clouds; North Pacific; seasonal variation; interannual variation

---

1. Introduction

Low-level clouds play an important role in the global radiation balance, which includes longwave radiation emission, as well as absorption and reflection of solar shortwave radiation [1]. Garrett and Zhao found that where thin water clouds and pollution are coincident, there is an increase in cloud longwave emissivity resulting from elevated haze levels. This results in an estimated surface warming under cloudy skies [2]. They also found that in Alaska, the cloud radiative impact on the surface is a net warming effect between October and May and a net cooling in summer. During episodes of high surface haze aerosol concentrations and cloudy skies, both the net warming and net cooling are amplified. Thus the low cloud has an important influence on global climate change [3]. A small change in fractional coverage of low-level clouds can exert significant influences on weather and climate [4]. For example, the marine stratocumulus has a potential positive feedback to global warming [5]. However, since the formation of low-level clouds is governed by small-scale turbulent processes, the associated controlling factors of low-cloud fraction (LCF) are complex. Although Ma et al. has found a prognostic method of cloud-cover calculation (PROGCS) which has significant advantage over the conventional diagnostic one, the complex controlling factors still are the main sources of the uncertainty in state-of-the-art models [6]. For example, Fan et al. found that the aerosol errors have a certain contribution to cloud fraction biases in Coupled Model Intercomparison Project Phase 5 (CMIP5) simulations [7]. As a result, the variation of LCF has been poorly simulated in present climate projections [8,9]. Thus, it is very important to investigate the controlling factors of LCF for climate research [10,11].

Low-level clouds are frequently observed over the cool oceans where deep convection is unlikely to occur. Due to their significant potential impacts on the Earth's energy balance, low-level clouds have been intensively investigated at various time scales [12]. At the seasonal time scale, it is well known that the LCF is positively related to the inversion strength of lower-tropospheric temperature. In previous studies, the estimated inversion strength (EIS) was defined as a refinement of lower-tropospheric stability (LTS), and there is a linear relationship between LCF and EIS over the subtropical and mid-latitude oceans [13]. In fact, the EIS can only explain the seasonality of LCF over the eastern area of an ocean basin, which is located east of the western Pacific subtropical high that accompanies persistent mid-tropospheric subsidence and equatorward surface winds [14,15]. Besides the subtropical and mid-latitude oceans, large LCF also appears over the high-latitude and subpolar oceans. In these regions, one of the important factors that affect LCF is a prominent ocean front [16]. Over the ocean front, the sea surface temperature (SST) anomalies are controlled by surface temperature advection, resulting in a source of sensible heat flux (SHF) anomalies [17]. The upward SHF destabilizes the surface layer and facilitates shallow convection in the boundary layer to further increase LCF. In addition, the decreased relative humidity (RH) at 700-hPa acts to reduce cloudiness [18,19]. Besides, the seasonal variation of low clouds is also affected by aerosol and haze [20,21].

At the interannual time scale, the variation of LCF is considered to be associated with different environmental fields [22]. Previous studies focused on the variation of LCF and its relationship with SST anomalies (SSTA). Norris and Leovy [23] showed that LCF is negatively correlated with the SSTA in the eastern subtropical oceans, especially during summer [24]. They further noted that surface cold advection may play an important role in the interannual variation of LCF. The summertime interannual variation of LCF over the North Pacific is the largest in the central and western regions along 35° N and in the eastern region near 15° N. The LCF over these two regions are in good relationship with local SST and sea-level pressure (SLP) field [25]. Over the North Atlantic, the North Atlantic subtropical high (NASH) also plays an important role in the interannual variation of summertime LCF. A stronger NASH is often accompanied by increased LCF and cooler SSTs along the southeast of the NASH. The northeasterly surface wind anomalies associated with an intensified NASH tend to induce colder advection and stronger coastal upwelling in the LCF region, acting to decrease surface temperature. Meanwhile, the anomalous warm advection associated with the easterly wind anomalies from Africa leads to a warming at 700 hPa over the LCF region. Such warming and surface cooling increase atmospheric static stability, favoring the growth of LCF. The anomalous diabatic cooling associated with the growth of LCF dynamically excites an anomalous anticyclone to its north and enhances the NASH in turn. Besides the subtropical high, the El Niño-Southern Oscillation (ENSO) is a primary variability of interannual time scale in the Pacific Ocean, which has a positive relationship with summertime-enhanced LCF over the southeastern North Pacific [26].

The seasonal and interannual variations of LCF have been investigated in many studies. However, due to limited observation data, the data range used in previous studies is very short. In addition, previous studies showed that the EIS can only explain the variation of LCF over the eastern side of an ocean basin, while it is weakly related to the LCF over the western side [12]. Thus, different controlling factors of LCF between the eastern and western sides need to be studied. In this study, we use long-term satellite data to explore seasonal and interannual variations of low-level clouds over the North Pacific, where the associated controlling factors exhibit significant differences in the eastern and western regions.

The rest of this paper is organized as follows. In Section 2, we introduce the data and methods used in this study. In Section 3, we investigate the seasonal distribution of LCF over the North Pacific and its associated controlling factors. Multiple linear regression model analysis is used. Interannual variation of LCF is also explored in Section 3 and a conclusion is given in Section 4.

2. Data and Methods

2.1. Data

The cloud data used in this study are the collection 06 Level-3 monthly cloud product of the Moderate Resolution Imaging Spectroradiometer (MODIS), which has a horizontal resolution of $1° \times 1°$, and cover the period from January 2003 to December 2015 [27,28]. It is noteworthy that time representation errors exist for cloud fraction observed by MODIS, since it only observes clouds twice a day. The correlation coefficient between MODIS monthly cloud fraction (CF) and continuous day-and-night radar/lidar CF is 0.97. This small error will not affect our results [21,29]. MODIS cloud data include cloud fraction and cloud top pressure. Clouds with top pressure higher than 700 hPa are considered as low-level clouds [30]. Since the MODIS instruments cannot detect low-level clouds that are overlapped with mid- and high-level clouds, the random overlap assumption is used to reduce the influence of mid- and high-level clouds [31]. It is a reasonable assumption outside the areas of deep convection and landmass [32,33].

The meteorological fields used in this study are ERA-Interim global atmospheric reanalysis at $1° \times 1°$ grid from European Centre for Medium-Range Weather Forecasts (ECMWF) [34], including 700-hPa subsidence (W), 700-hPa RH, 700-hPa potential height (Z), 2-m surface air temperature (SAT), dew point temperature (T_d), 10-m surface wind, and SLP. All variables used cover the period from January 2003 to December 2015.

In addition, the SST at $1° \times 1°$ grid from the Hadley Center [35], the SHF at $1° \times 1°$ grid from the Woods Hole Oceanographic Institution (WHOI), and the Niño-3.4 index provided by the National Oceanic and Atmospheric Administration (NOAA) are used in this study [36,37]. For consistency, all variables used cover the period from January 2003 to December 2015.

2.2. Methods

The EIS defined by Wood and Bretherton [13] is used as a measure of inversion layer strength at the top of the boundary layer:

$$EIS = \left(\theta_{700} - \theta_{sfc}\right) - \gamma_m^{850}(Z_{700} - Z_{LCL}), \tag{1}$$

where θ_{700} and θ_{sfc} are the potential temperatures at 700 hPa and surface, respectively. Z_{700} is the 700-hPa height, Z_{LCL} is the lifting condensation level, and γ_m^{850} is the 850-hPa moist adiabatic lapse rate. Z_{LCL} is calculated by using SAT and T_d:

$$Z_{LCL} = 123 \times (SAT - T_d) \tag{2}$$

The near-surface temperature advection (advT) is calculated by using $-V_{sfc} \cdot \nabla SST$, where V_{sfc} represents surface zonal and meridional winds, and ∇SST are the zonal and meridional SST gradients [38–40].

We also define the subpolar frontal zone (SPFZ) to measure SST gradient strength in the subpolar North Pacific [41]. The SPFZ intensity index (Iint) is defined as the meridional SST gradient ($-\partial SST/\partial y$) averaged over the climatological SPFZ area (145°–170° E, 35°–47° N).

To quantify the relative importance of the associated controlling factors in seasonal variation of LCF, LCF dependence on these factors is derived using multiple linear regression. Although the multiple linear regression method cannot perfectly extract the impact of individual large-scale forcing, the derived local dependence is useful for quantifying their local controls on LCF [5].

In this study, spring refers to the period of March, April and May; summer refers to the period of June, July, August; autumn refers to the period of September, October, November; winter refers to the period of December, next January, next February.

3. Results

3.1. Climatological Distribution of LCF

Figure 1a,b displays the distributions of climatological LCF over the North Pacific in summer and winter, respectively. Winter is defined from December to the following February, and summer is defined from June to August. The LCF over the North Pacific is zonally inhomogeneous in both winter and summer, and exhibits obvious seasonal difference. In summer, the LCF over the eastern North Pacific is larger than that over the western North Pacific, with a local maximum around 20° N (Figure 1a). In contrast, the LCF maximum in winter appears over the western North Pacific (Figure 1b). Note that the LCF over the Bering Sea is large in both summer and winter.

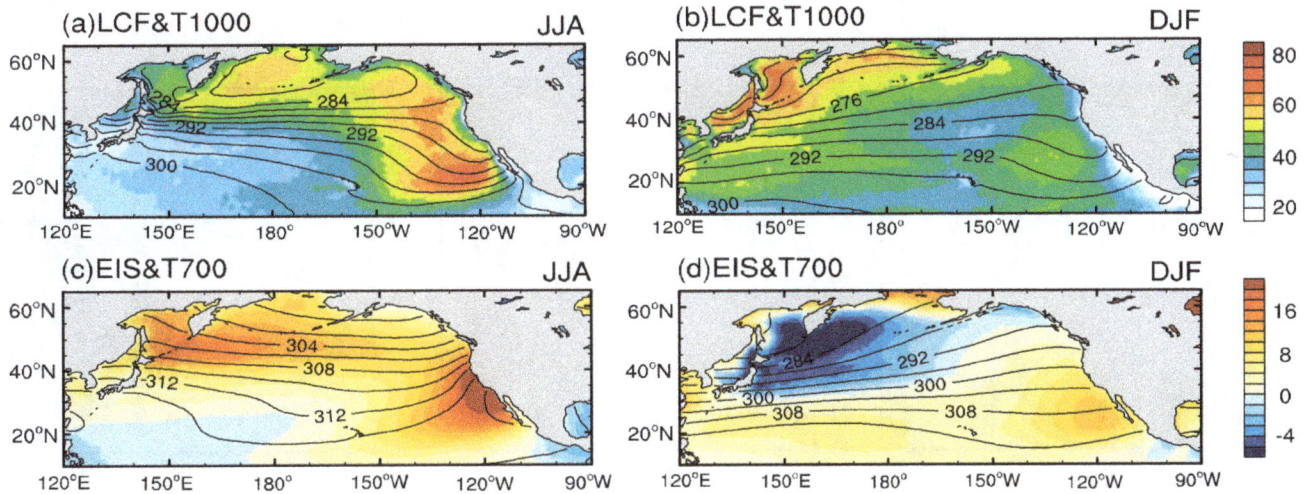

Figure 1. Climatological low-cloud fraction (LCF; shading; units: %) and surface potential temperature (contour interval: 2 K) in summer (**a**) and in winter (**b**); (**c,d**) are the same as (**a,b**), but for estimated inversion strength (EIS; shading; units: K) and 700-hPa potential temperature (contour interval: 4 K).

Previous studies indicated that the climatological distribution of LCF and the seasonality of LCF can be well explained by the EIS, whose enhancement acts to increase LCF [42,43]. Atmospheric circulation also make contribution to the variation of LCF, such as subtropical high, Hadley–Walker circulation, and mesoscale waves [12,44]. However, EIS was proved to be the dominating factor in the seasonal variation of the LCF in previous studies [12,31]. Thus, EIS is focused on in our study. Hence, the climatological EIS defined in Equation (1) is shown in Figure 1c,d for summer and winter, respectively. Across the summertime subtropical basin (Figure 1c), the EIS is maximal off the west coast of North America around 125° W, in good correspondence with the spatial pattern of LCF (Figure 1a). Compared to the summertime situation, the EIS in winter exhibits a zonal minimum (negative center) distribution over the mid-latitude western North Pacific (Figure 1d), where a maximum LCF dominates. This is in contrast to the well-known liner relationship between EIS and LCF [13]. The enhancement of EIS could maintain a strong temperature inversion at the top of the boundary layer, inhibiting cloud-top entrainment of dry air, further contributing to LCF increase. Overall, the LCF over the eastern North Pacific is positively associated with the EIS, while the relationship between LCF and EIS over the western North Pacific is negative. Thus, the EIS alone is not sufficient to explain the observed LCF. Next, we will discuss the possible factors dominating the seasonal cycles of LCF over the eastern and western North Pacific, respectively.

3.2. Seasonal Cycle of LCF

According to previous studies, the EIS, cool advT, SHF, 700-hPa subsidence, and 700-hPa RH are the main factors affecting the formation of low-level clouds over the oceans. Overall, the enhancements of these factors contribute to LCF increase [45]. The enhancement of EIS could maintain a strong temperature inversion at the top of the boundary layer, inhibiting cloud-top entrainment of dry air, further contributing to LCF increase. The cool advT could expand the difference value between SST and SAT, further increasing SHF. The increased SHF could destabilize the surface layer, and thereby facilitate shallow convection in the boundary layer, to further increase LCF. The enhanced 700-hPa W acts to warm the mid-troposphere, inhibiting cloud-top entrainment of dry air, further increasing LCF. The 700-hPa RH could contribute to the increase of LCF by increasing the water vapor content in the air. To discuss the seasonality of LCF and its distribution over the eastern and western North Pacific, longitude-time sections of climatological LCF and the associated controlling factors along 25° N and 45° N are shown in Figures 2 and 3, respectively.

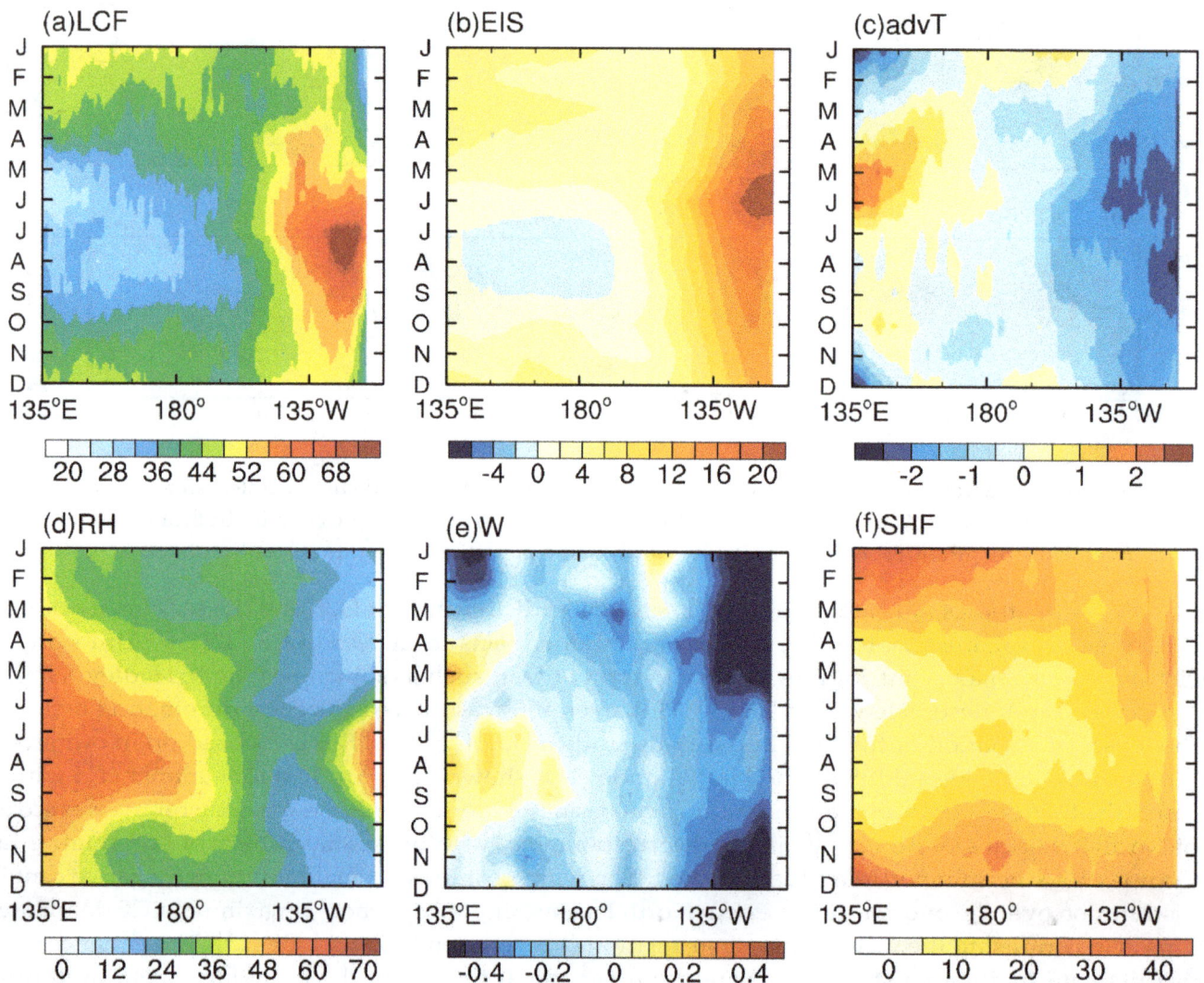

Figure 2. Time-longitude section of climatological (**a**) LCF (shading; units: %), (**b**) EIS (shading; units: K), (**c**) near-surface temperature advection (advT; shading; units: K/day), (**d**) 700-hPa relative humidity (RH; shading; units: %), (**e**) 700-hPa subsidence (W; shading; units: m/s), and (**f**) sensible heat flux (SHF; shading; units: W/m^2) along 25° N.

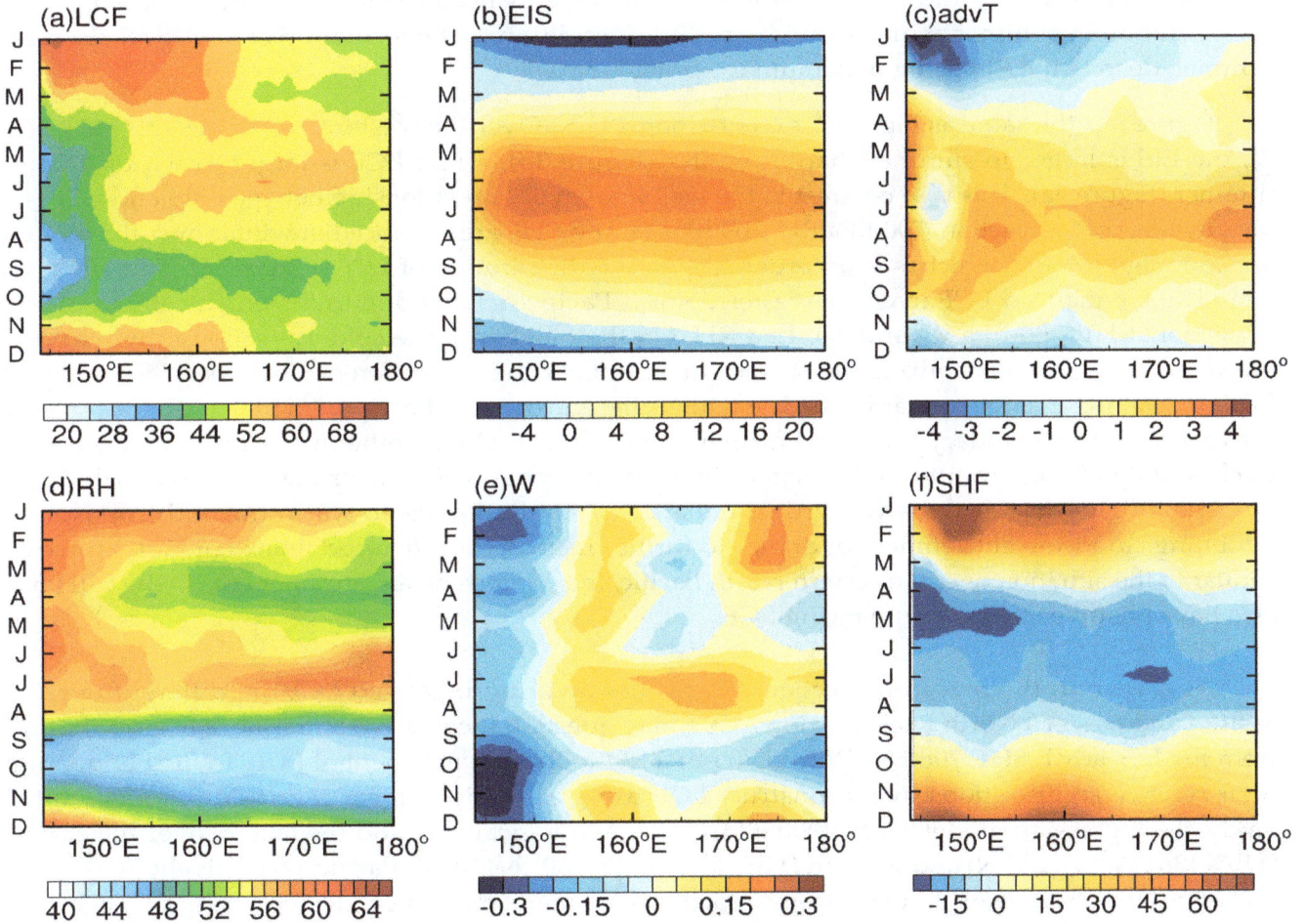

Figure 3. Time-longitude section of climatological (**a**) LCF (shading; units: %), (**b**) EIS (shading; units: K), (**c**) near-surface advT (shading; units: K/day), (**d**) 700-hPa RH (shading; units: %), (**e**) 700-hPa W (shading; units: m/s), and (**f**) SHF (shading; units: W/m²) along 45° N.

These figures reveal complex relationships of LCF with its controlling factors in the course of seasonal cycles over the North Pacific. At 25° N (Figure 2a), LCF is larger in summer than in winter over the eastern subtropics (115°–135° W), which is consistent with the winter-summer difference of the EIS shown in Figure 2b. As evident in Figure 2c,d, the distributions of cold temperature advection and 700-hPa RH are also in good accordance with LCF. However, 700-hPa W and SHF are relatively weaker in summer, which is in contrary to the seasonality of LCF (Figure 2e,f). Therefore, the EIS, cold advection, and 700-hPa RH have great contributions to the enhancement of LCF over the eastern North Pacific in summer. In summer, LCF prevails over the eastern portion of the subtropical North Pacific, which is located east of the surface subtropical high that accompanies persistent mid-troposphere subsidence and equatorward surface winds. The equatorward winds induce coastal upwelling (west coast of Mexico), upper-ocean mixing, and surface evaporation, acting to maintain relatively low SST. Meanwhile, the mid-troposphere subsidence associated with the subtropical high acts to warm the mid-troposphere. The combination of cool SST and warm mid-troposphere maintains a strong temperature inversion at the top of the boundary layer, inhibiting cloud-top entrainment of dry air, further increasing LCF. Thus, the EIS can explain the summertime enhancement of LCF to a certain extent. Note that the EIS reaches its maximum in spring, while the maximum of LCF is in summer (Figure 2a,b). This may be due to both EIS and RH being large in summer, while only the EIS is large in spring. Thus, LCF reaches its maximum in summer, resulting from the combined contribution of EIS and RH, rather than in spring, when only the EIS is the strongest, suggesting the essential influence of RH on the seasonal variability of LCF over the eastern North Pacific. RH may lead to a time-lag correlation between EIS and LCF in the subtropics. The enhanced RH indicates an increase in vapor concentration in summer, which provides positive condition to the formation of low-level clouds in summer, despite the EIS being maximum in spring [46].

Figure 3 is the same as Figure 2, except along 45° N. Over the western North Pacific (145°–160° E), the EIS is larger in summer than in winter (Figure 3b), while LCF is larger in winter than in summer (Figure 3a). This is in contrast to the well-known liner relationship between EIS and LCF [47]. Meanwhile, cold advection, 700-hPa RH, 700-hPa W, and SHF are enhanced in winter over the western portion (Figure 3c–f), which is in accordance with the distribution of LCF (Figure 3a). Therefore, the wintertime enhanced LCF over the western North Pacific may be due to the enhancement of cold advection, SHF, 700-hPa W, and 700-hPa RH. On the one hand, the cold advection in winter over the western region destabilizes the surface layer, increasing the difference between SST and SAT, further resulting in large upward SHF. The wintertime enhancement of upward SHF facilitates shallow convection in the boundary layer, and further increases LCF. On the other hand, the enhanced storm track activity also contributes to the wintertime enhancement of SHF over the western North Pacific (not shown). In this area, the wintertime enhanced 700-hPa W acts to warm the mid-troposphere, inhibiting cloud-top entrainment of dry air to further increase LCF to a certain extent [20]. Moreover, similar to the situation over the eastern North Pacific, the wintertime enhancement of 700-hPa RH may also have positive effects on enhancing LCF.

As shown in the preceding section, the factors favoring increased LCF are different over the eastern and western North Pacific regions. One may question the relative importance of contributions from the EIS, advT, SHF, 700-hPa W, and 700-hPa RH to the enhancement of LCF. To further quantify their relative contributions, we reconstruct LCF using a multiple linear regression model. In this study, the regression model is constructed from climatological LCF and the factors over the eastern (115°–136° W, 15°–28° N) and western (140°–155° E, 47°–60° N) North Pacific, respectively. To describe the relative importance of cloud controlling factors, the annual mean has been removed. The regression slope of LCF variation against each predictor is given in Table 1.

Table 1. Regression slope for each predictor (EIS: estimated inversion strength; advT: surface temperature advection; SHF: sensible heat flux; W: 700-hPa subsidence; RH: 700-hPa relative humidity).

Area	$\partial LCF/\partial EIS$ (% K^{-1})	$\partial LCF/\partial advT$ (%(Kday^{-2})$^{-1}$)	$\partial LCF/\partial SHF$ (%(Wm^{-2})$^{-1}$)	$\partial LCF/\partial W$ (%(m^{-1})$^{-1}$)	$\partial LCF/\partial RH$ (%%$^{-1}$)
Western North Pacific	−0.08	−0.19	0.12	48.6	−0.14
Eastern North Pacific	0.47	−2.86	0.16	58.89	0.27

Figures 4 and 5 show the longitude-time distributions of the predicted climatological seasonal cycles# of LCF along 25° N and 45° N, respectively. Besides, the corresponding LCF predicted by the multiple linear regression model is also shown in Figures 4b and 5b, named "Total". At 25° N, the multiple linear regression model explains 71% of the total variance of LCF regionality and its seasonal cycle, whereas the root mean square error (RMSE) between observed and predicted LCF is 8%. The model reproduces the summertime LCF maximum from July to September over the eastern subtropics (Figure 4a,b). The reconstruction indicates that the EIS (Figure 4c) and RH (Figure 4e) make the greatest contributions to the summertime enhancement of LCF, and the contribution from cold advection is also important (Figure 4d). In contrast, the 700-hPa W and SHF act to reduce LCF in summer (Figure 4f,g).

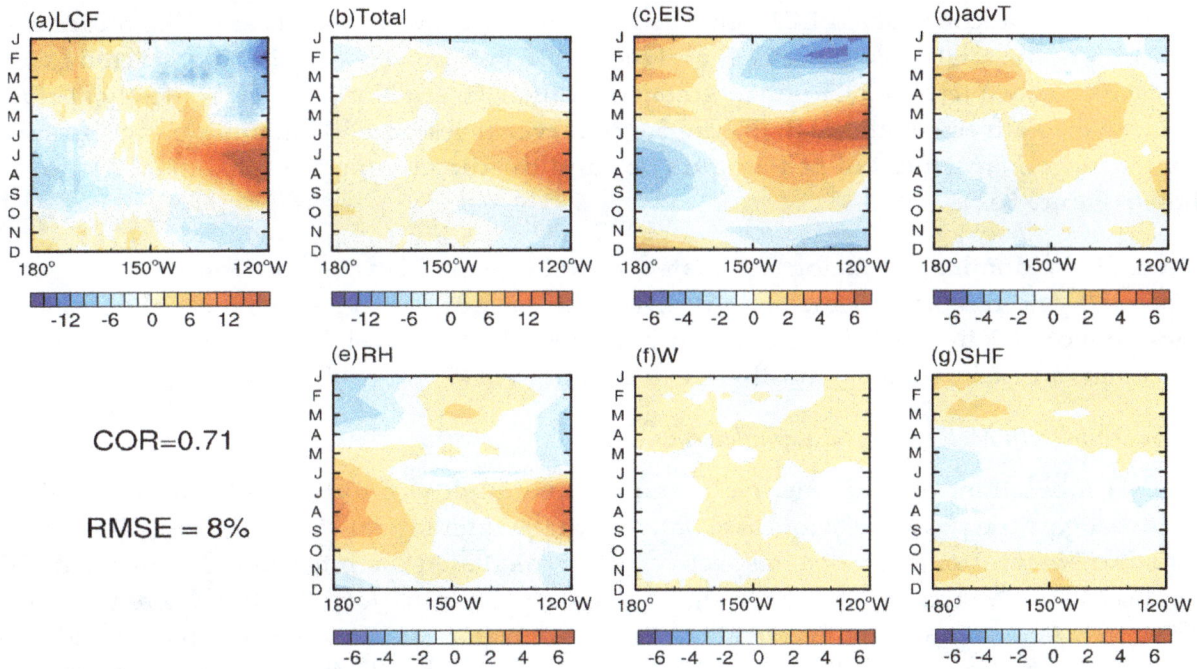

Figure 4. (**a**) Same as Figure 2a, except that the annual-mean LCF (%) has been removed. (**b**) Same as (**a**), except for the corresponding LCF (%) predicted by the multiple linear regression model. The correlation and root mean square error (RMSE) between (**a,b**) are shown below (**a**). (**c–g**) Same as (**b**), except for individual contributions from EIS, advT, 700-hPa RH, 700-hPa subsidence, and SHF to (**b**), respectively.

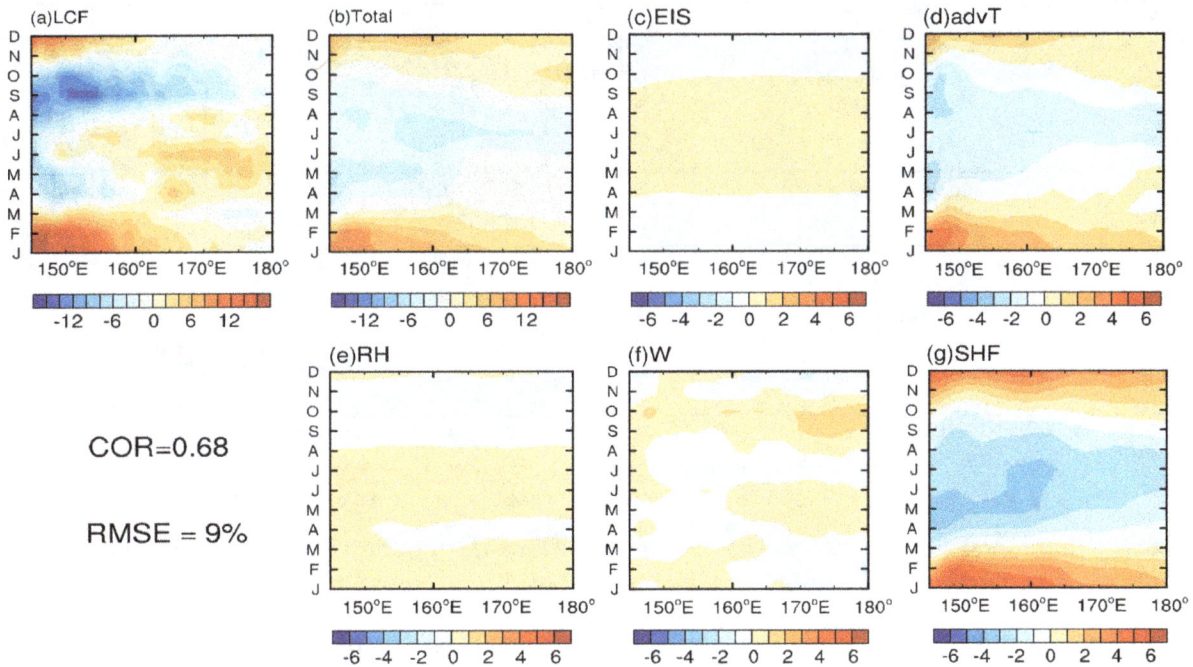

Figure 5. (**a**) Same as Figure 3a, except that the annual-mean LCF (%) has been removed. (**b**) Same as (**a**), except for the corresponding LCF (%) predicted by the multiple linear regression model. The correlation and root mean square error (RMSE) between (**a,b**) are shown below (**a**). (**c–g**) Same as (**b**), except for individual contributions from EIS, advT, 700-hPa RH, 700-hPa subsidence, and SHF to (**b**), respectively.

At 45° N, the reconstructed LCF can also reproduce LCF well. In this area, the reconstruction explains 68% of the total variance of LCF, and the RMSE between observed and predicted LCF is 9% (Figure 5a,b). The wintertime LCF enhancement over the western North Pacific is mostly attributable to enhanced cold advection and SHF (Figure 5d,g). Nevertheless, the EIS in this area acts to suppress the wintertime enhancement of LCF (Figure 5c), and the direct impacts of 700-hPa RH and W are negligible (Figure 5e,f).

Overall, the dominating factors associated with the seasonal cycle of LCF are different over the eastern and western North Pacific regions. Over the eastern North Pacific, the EIS dominates the enhancement of LCF in summer, together with the 700-hPa RH. Over the western North Pacific, the enhancement of LCF in winter is mostly due to enhanced SHF and cold advection.

3.3. Association with Meteorological Parameters

In this subsection, we examine the interannual variability of EIS, 700-hPa W, advT, SHF, and SST to investigate the possible factors associated with the interannual variability of LCF over eastern and western North Pacific regions, respectively. We calculated the interannual variance of LCF in different seasons. We first choose three regions over the eastern North Pacific where the variances are large. However, in order to investigate the different factors of LCF between eastern and western North Pacific, we also choose three regions over the western North Pacific where the variances of LCF are also large (Figure 6). The six different regions are defined as follows: the Okhotsk Sea (OS; 140°–155° E, 47°–60° N), the Kuroshio Extension (KE; 142°–180° E, 37°–44° N), the south basin of Japan (SJ; 124°–145° E, 25°–31° N), the center of the North Pacific (CE; 136°–165° W, 26°–37° N), the southeastern North Pacific (SE; 115°–136° W, 15°–28° N), and the northeastern North Pacific (NE; 124°–148° W, 26°–43° N).

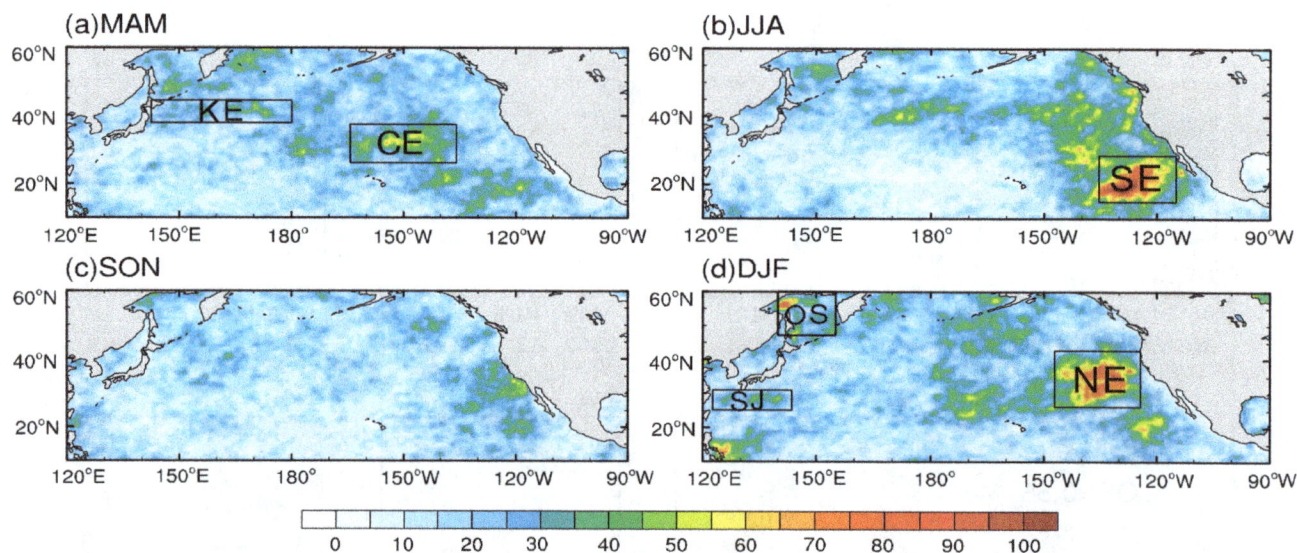

Figure 6. Interannual LCF variance (shading; units: %²) in (a) spring, (b) summer, (c) autumn, and (d) winter. Thick solid rectangles designate different regions discussed in Section 4.

First, we calculated the correlation coefficients between regional-mean LCF and corresponding meteorological variables for each region. Table 2 lists these correlation coefficients. For the correlation coefficients of different regions and seasons, we use the method of significance test of correlation coefficients [48]. According to the significance test table of correlation coefficient, we can see that when the degree of freedom $n = 11$, the correlation coefficient (COR) which is larger than 0.553 (COR < −0.553) exceeds a 95% confidence level; the COR which is less than 0.476 (−0.476 < COR < 0) cannot exceed a 90% confidence level.

Table 2. Correlation coefficients of regionally-averaged interannual anomalies of various meteorological parameters with those of LCF. The bold type exceed a 95% confidence level, while the parentheses indicate that the correlation coefficient is not statistically significant. (SE: the southeastern North Pacific; NE: the northeastern North Pacific; CE: the center of the North Pacific; OS: the Okhotsk Sea; KE: the Kuroshio Extension; SJ: the south basin of Japan).

Area	Season	Estimated Inversion Strength (EIS)	700-hPa Subsidence (W)	Surface Temperature Advection (advT)	Sensible Heat Flux (SHF)	Sea Surface Temperature (SST)
SE	Spring	0.49	(−0.25)	(0.04)	(−0.24)	(−0.41)
	Summer	**0.57**	**−0.64**	(−0.11)	(−0.07)	**−0.57**
	Autumn	(0.36)	(0.04)	(−0.07)	**−0.61**	**−0.72**
	Winter	**0.55**	**−0.65**	(0.12)	(0.21)	(−0.43)
NE	Spring	(0.25)	**−0.51**	**−0.73**	(0.46)	**−0.64**
	Summer	**0.73**	**−0.60**	(−0.38)	(−0.39)	**−0.74**
	Autumn	**0.56**	(−0.45)	**−0.66**	(−0.02)	(−0.12)
	Winter	**0.57**	**−0.81**	**−0.85**	**0.54**	**−0.57**
CE	Spring	**0.71**	**−0.91**	**−0.81**	**0.68**	(−0.37)
	Summer	**0.71**	(−0.41)	(−0.37)	(0.42)	**−0.59**
	Autumn	**0.59**	**−0.81**	(−0.29)	(0.001)	(−0.25)
	Winter	**0.60**	**−0.84**	**−0.48**	(0.40)	(0.35)
OS	Spring	(−0.34)	(−0.41)	(0.05)	(0.15)	(0.07)
	Summer	(0.07)	(−0.35)	(0.16)	**0.51**	(−0.24)
	Autumn	**−0.48**	(−0.35)	(−0.35)	(0.39)	(−0.37)
	Winter	(−0.36)	**−0.58**	**−0.65**	**0.61**	(0.02)
KE	Spring	(−0.16)	**−0.59**	**−0.56**	**0.61**	(−0.08)
	Summer	(0.21)	(−0.42)	**−0.62**	**0.67**	(−0.05)
	Autumn	(−0.28)	(−0.32)	**−0.48**	(0.27)	(0.02)
	Winter	(−0.30)	**−0.62**	**−0.60**	(0.45)	(−0.43)
SJ	Spring	(−0.01)	(−0.22)	(−0.17)	(0.38)	(−0.10)
	Summer	(−0.24)	**−0.55**	(0.13)	(0.31)	(0.21)
	Autumn	(0.37)	**−0.47**	(−0.09)	**0.56**	(−0.21)
	Winter	(−0.18)	**−0.80**	**−0.7**	**0.65**	**−0.85**

Over the eastern portions (SE, NE, and CE), the LCF is positively correlated with the EIS and negatively correlated with the SST, especially in summer and winter; the 700-hpa W also plays a positive role in the interannual variation of LCF over the eastern portions. In NE and CE, there is a positive correlation between LCF and cold advection. Over the eastern portions, the correlation between LCF and SHF is not obvious.

In contrast to the eastern regions, the correlations between LCF and meteorological parameters over the western regions (OS, KE, and SJ) present radically different features. Over the western regions, the LCF is positively connected with cold advection and SHF, especially in winter and spring. Note that there is no relationship between LCF and EIS, or between LCF and SST. Over the western regions, the 700-hPa W also has a positive correlation with LCF, similar to that in the eastern regions.

These correlation coefficients reflect regional contrast at the interannual time scale. The LCF over the eastern North Pacific is positively correlated with both EIS and SST; the LCF over the western regions coincides well with SHF and cold advection. The 700-hPa W has positive relation with LCF over both western and eastern regions.

3.4. Association with El Niño and SPFZ

What causes the characteristic interannual variations in SST, EIS, advT, and SHF, which are closely linked with the interannual variation of LCF? In this subsection, we first discuss the possible relationship of LCF with the El Niño, which is the primary variability in the North Pacific at the interannual time scale. Figure 7 shows the interannual anomalies of LCF, SST, and EIS over the North

Pacific regressed onto the synchronization Niño-3.4 index in different seasons. Usually, the El Niño occurs in the eastern equatorial Pacific. Warm SST anomalies simultaneously occur in the surrounding regions (Figure 7a–d), which are well-known as the typical SST anomaly distribution associated with the El Niño [49,50]. Meanwhile, the warm SST benefits warm surface potential temperature, corresponding to the overlying negative EIS. This is consistent with the negative correlation between EIS and El Niño over the eastern North Pacific. The negative EIS over the eastern North Pacific favors negative LCF. Thus, El Niño-related SSTs have negative feedback on the LCF over the eastern North Pacific, in which the EIS plays an important role. This is consistent with the results of previous studies [51].

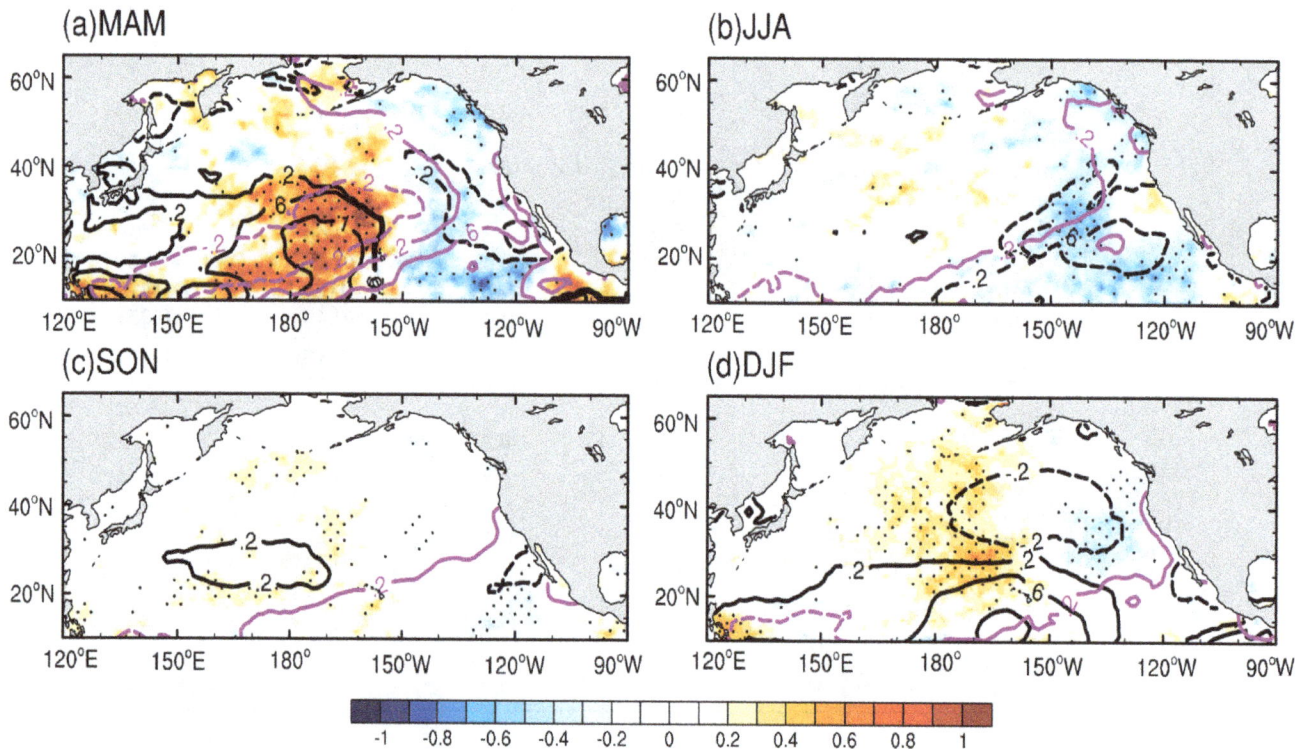

Figure 7. Coefficients of interannual LCF (shading; units: %), SST (purple contour; units: K), and EIS (black contour; units: K) onto the synchronization Niño-3.4 index for (**a**) spring, (**b**) summer, (**c**) autumn, and (**d**) winter. Stippling indicates the 90% confidence level using the t test. Contours indicate statistically significant positive (solid) and negative (dashed) differences.

The LCF over the western North Pacific does not have a clear relationship with the El Niño. This begs the question: Is there any possible link with other climate modes? From Table 2, we can see that the LCF over the western North Pacific is positively related to cold advection and SHF. Thus, in Figure 8, we show the maps of LCF, advT, and SHF anomalies regressed onto the normalized subpolar frontal zone (SPFZ) index, similar to Figure 7. The interannual LCF anomalies are positively correlated with the intensity of the SPFZ over the western North Pacific (140°–170° W, 40°–60° N), especially in winter and spring (Figure 8a,d). At the same time, both cold advection and SHF present positive correlations with the SPFZ (Figure 8a,d). In spring and winter, the enhancement of the SPFZ over the western North Pacific can strengthen the cold advection through northerly wind, rendering SAT lower than the SST underneath, further resulting in large upward SHF [52]. The wintertime enhancement of SHF destabilizes the surface layer and facilitates shallow convection in the boundary layer, thus increasing the convective LCF. This implies that the SPFZ-related cold advection has a positive effect on the enhanced LCF by interacting with the SHF over the western North Pacific.

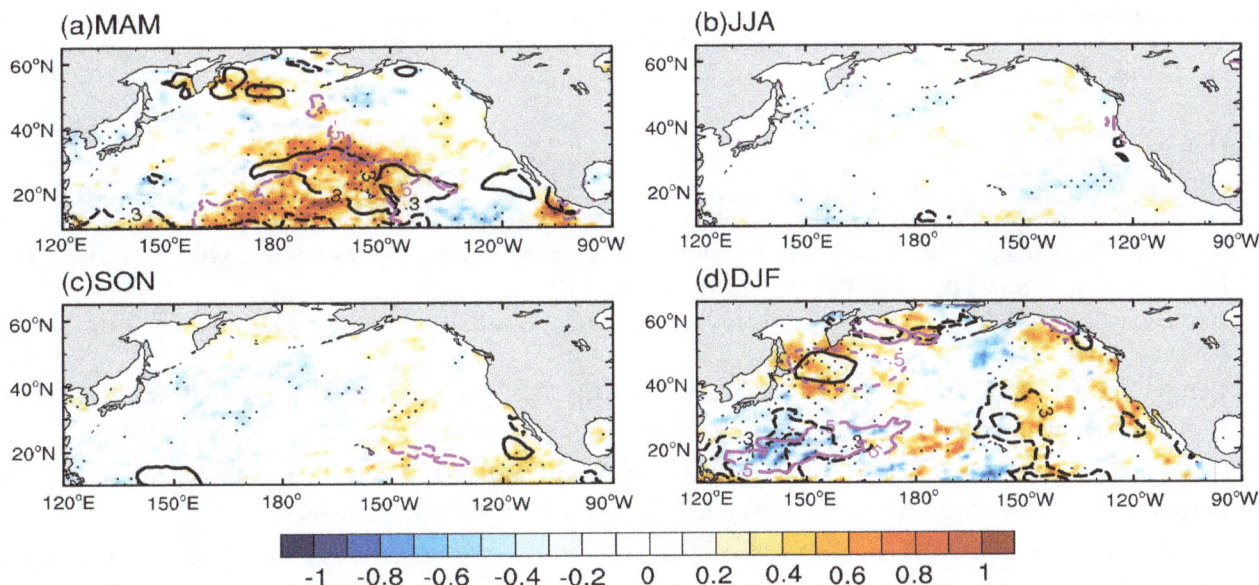

Figure 8. Regression coefficients of interannual LCF (shading; units: %), advT (purple contour; units: K/day), and SHF (black contour; units: W/m2) onto the SPFZ index for (**a**) spring, (**b**) summer, (**c**) autumn, and (**d**) winter. Stippling indicates the 90% confidence level using the t test. Contours indicate statistically significant positive (solid) and negative (dashed) differences at the 90% confidence level based on the t test.

Overall, the interannual variation of LCF is different over the eastern and western North Pacific regions, due to different controlling factors and large-scale atmospheric processes. The increased LCF over the eastern North Pacific is associated with increased EIS and decreased SST, especially in summer. Our regression analysis signifies the El Niño contribution. Over the western North Pacific, the springtime and wintertime enhanced LCF has a positive correlation with enhanced SHF and cold advection, which can be partly explained by SPFZ intensity.

4. Discussion and Concluding

In this study, we investigate the seasonal and interannual variability of LCF and associated controlling factors over the eastern and western North Pacific. On the seasonal time scale, LCF shows obvious regional contrast. Over the eastern North Pacific, the EIS dominates the enhancement of LCF in summer, together with the 700-hPa RH. Over the western North Pacific, the enhancement of LCF in winter is mostly due to enhanced SHF and surface cold advection.

In terms of interannual variation, the increased LCF over the eastern North Pacific is associated with increased EIS and decreased SSTs, especially in summer. Our regression analysis indicates that El Niño contributes most. Over the western North Pacific, the springtime and wintertime enhanced LCF has a positive correlation with enhanced SHF and cold advection, which can be partly explained by SPFZ intensity.

This paper only discusses the factors that influence LCF variability in previous work. However, other factors, such as ocean or atmospheric circulation, may also play an important role in LCF changes, which need to be further studied in the future. Cloud microphysical processes and aerosol properties are also important for the formation of low-level clouds [53]. For example, oceanic aerosol productivity plays an important role in determining cloud condensation nuclei. Zhao et al. have found that if liquid water content (LWC) is high and aerosol amount is not too large, both cloud droplet number concentration (N) and effective radius (re) increase with increasing aerosols; if LWC is low or if LWC is high but aerosol amount is too large, cloud N increases but re decreases with increasing aerosols [54]. These aspects will be explored in our future study.

Author Contributions: Conceptualization, H.X. Formal analysis, Q.W. Methodology, Q.W., H.X. and L.Z. Software, Q.W. Supervision, H.X. Validation, Q.W. and L.Z. Writing—original draft, Q.W. and L.Z. Writing—review and editing, Q.W., H.X., L.Z. and J.D.

References

1. Wood, R. Stratocumulus Clouds. *Mon. Weather Rev.* **2012**, *140*, 2373–2423. [CrossRef]
2. Garrett, T.J.; Zhao, C. Increased Arctic cloud longwave emissivity associated with pollution from mid-latitudes. *Nature* **2006**, *440*, 787. [CrossRef] [PubMed]
3. Zhao, C.; Garrett, T.J. Effects of Arctic haze on surface cloud radiative forcing. *Geophys. Res. Lett.* **2015**, *42*, 557–564. [CrossRef]
4. Randall, D.A.; Coakley, J.A., Jr.; Fairall, C.W.; Kropfli, R.A.; Lenschow, D.H. Outlook for research on subtropical marine stratiform clouds. *Bull. Am. Meteorol. Soc.* **1984**, *65*, 1290–1301. [CrossRef]
5. Oreopoulos, L.; Davies, R. Statistical dependence of albedo and cloud cover on sea surface temperature for two tropical marine stratocumulus regions. *J. Clim.* **1993**, *6*, 2434–2447. [CrossRef]
6. Ma, Z.; Liu, Q.; Zhao, C.; Shen, X.; Wang, Y.; Jiang, J.H.; Li, Z.; Yung, Y. Application and Evaluation of an Explicit Prognostic Cloud-Cover Scheme in GRAPES Global Forecast System. *J. Adv. Model. Earth Syst.* **2018**, *10*, 652–667. [CrossRef]
7. Fan, T.; Zhao, C.; Dong, X.; Liu, X.; Yang, X.; Zhang, F.; Shi, C.; Wang, Y.; Wu, F. Quantify contribution of aerosol errors to cloud fraction biases in CMIP5 Atmospheric Model Intercomparison Project simulations. *Int. J. Climatol.* **2018**, *38*, 3140–3156. [CrossRef]
8. Bony, S.; Dufresne, J.L. Marine boundary layer clouds at the heart of tropical cloud feedback uncertainties in climate models. *Geophys. Res. Lett.* **2005**, *32*, 023851. [CrossRef]
9. Zhai, C.; Jiang, J.H.; Su, H. Long-term cloud change imprinted in seasonal cloud variation: More evidence of high climate sensitivity. *Geophys. Res. Lett.* **2015**, *42*, 8729–8737. [CrossRef]
10. Trenberth, K.E.; Fasullo, J.T. Simulation of present-day and twenty-first-century energy budgets of the southern oceans. *J. Clim.* **2010**, *23*, 440–454. [CrossRef]
11. Grise, K.M.; Polvani, L.M.; Fasullo, J.T. Re-examining the relationship between climate sensitivity and the southern hemisphere radiation budget in cmip models. *J. Clim.* **2015**, *28*. [CrossRef]
12. Koshiro, T.; Yukimoto, S.; Shiotani, M. Interannual variability in low stratiform cloud amount over the summertime North Pacific in terms of cloud types. *J. Clim.* **2017**, *30*, 6107–6121. [CrossRef]
13. Wood, R.; Bretherton, C.S. On the relationship between stratiform low cloud cover and lower-tropospheric stability. *J. Clim.* **2006**, *19*, 6425–6432. [CrossRef]
14. Klein, S.A.; Hartmann, D.L. The seasonal cycle of low stratiform clouds. *J. Clim.* **1993**, *6*, 1587–1606. [CrossRef]
15. Myers, T.A.; Norris, J.R. Observational evidence that enhanced subsidence reduces subtropical marine boundary layer cloudiness. *J. Clim.* **2013**, *26*, 7507–7524. [CrossRef]
16. Miyasaka, T.; Nakamura, H. Structure and formation mechanisms of the northern hemisphere summertime subtropical highs. *J. Clim.* **2005**, *18*, 5046–5065. [CrossRef]
17. Deser, C.; Wahl, S.; Bates, J.J. The influence of sea surface temperature gradients on stratiform cloudiness along the equatorial front in the Pacific Ocean. *J. Clim.* **1993**, *6*, 1172–1180. [CrossRef]
18. Dong, S.; Gille, S.T.; Sprintall, J. An assessment of the Southern Ocean mixed layer heat budget. *J. Clim.* **2007**, *20*, 4425–4442. [CrossRef]
19. Bretherton, C.S.; Blossey, P.N.; Jones, C.R. Mechanisms of marine low cloud sensitivity to idealized climate perturbations: A single-les exploration extending the cgils cases. *J. Adv. Model. Earth Syst.* **2013**, *5*, 316–337. [CrossRef]
20. Garrett, T.J.; Zhao, C.; Dong, X.; Mace, G.G.; Hobbs, P.V. Effects of varying aerosol regimes on low-level Arctic stratus. *Geophys. Res. Lett.* **2004**, 31. [CrossRef]
21. Zhao, C.; Chen, Y.; Li, J.; Letu, H.; Su, Y.; Chen, T.; Wu, X. Fifteen-year statistical analysis of cloud characteristics over China using Terra and Aqua Moderate Resolution Imaging Spectroradiometer observations. *Int. J. Climatol.* **2018**. [CrossRef]

22. Seethala, C.; Norris, J.R.; Myers, T.A. How has subtropical stratocumulus and associated meteorology changed since the 1980s? *J. Clim.* **2015**, *28*. [CrossRef]

23. Tselioudis, G.; Rossow, W.B.; Rind, D. Global patterns of cloud optical thickness variation with temperature. *J. Clim.* **1992**, *5*, 1484–1495. [CrossRef]

24. Van der Dussen, J.J.; De Roode, S.R.; Dal Gesso, S.; Siebesma, A.P. An les model study of the influence of the free tropospheric thermodynamic conditions on the stratocumulus response to a climate perturbation. *J. Adv. Model. Earth Syst.* **2015**, *7*, 670–691. [CrossRef]

25. Norris, J.R.; Leovy, C.B. Interannual variability in stratiform cloudiness and sea surface temperature. *J. Clim.* **1994**, *7*, 1915–1925. [CrossRef]

26. Norris, J.R. On trends and possible artifacts in global ocean cloud cover between 1952 and 1995. *J. Clim.* **1999**, *12*, 1864–1870. [CrossRef]

27. Norris, J.R. Interannual and interdecadal variability in the storm track, cloudiness, and sea surface temperature over the summertime North Pacific. *J. Clim.* **2000**, *13*, 422–430. [CrossRef]

28. Remer, L.A.; Kaufman, Y.J.; Tanré, D.; Mattoo, S.; Chu, D.A.; Martins, J.V. The modis aerosol algorithm, products, and validation. *J. Atmos. Sci.* **2005**, *62*, 947–973. [CrossRef]

29. Wang, Y.; Zhao, C. Can MODIS cloud fraction fully represent the diurnal and seasonal variations at DOE ARM SGP and Manus sites? *J. Geophys. Res. Atmos.* **2017**, *122*, 329–343. [CrossRef]

30. Ackerman, S.A.; Strabala, K.I.; Menzel, W.P.; Frey, R.A.; Moeller, C.C.; Gumley, L.E. Discriminating clear sky from clouds with modis. *J. Geophys. Res.* **1998**, *103*, 32141. [CrossRef]

31. Miyamoto, A.; Nakamura, H.; Miyasaka, T. Influence of the subtropical high and storm track on low-cloud fraction and its seasonality over the south Indian Ocean. *J. Clim.* **2018**, *31*, 4017–4039. [CrossRef]

32. Weare, B.C. Near-global observations of low clouds. *J. Clim.* **2000**, *13*, 1255–1268. [CrossRef]

33. McCoy, D.T.; Burrows, S.M.; Wood, R.; Grosvenor, D.P.; Elliott, S.M.; Ma, P.L.; Hartmann, D.L. Natural aerosols explain seasonal and spatial patterns of Southern Ocean cloud albedo. *Sci. Adv.* **2015**, *1*, e1500157. [CrossRef] [PubMed]

34. Li, J.; Huang, J.; Stamnes, K.; Wang, T.; Lv, Q.; Jin, H. A global survey of cloud overlap based on calipso and cloudsat measurements. *Atmos. Chem. Phys.* **2015**, *15*, 519–536. [CrossRef]

35. Dee, D.P.; Uppala, S.M.; Simmons, A.J.; Berrisford, P.; Poli, P.; Kobayashi, S.; Bechtold, P. The ERA-Interim reanalysis: Configuration and performance of the data assimilation system. *Q. J. R. Meteorol. Soc.* **2011**, *137*, 553–597. [CrossRef]

36. Wei, W.; Li, W.; Deng, Y.; Yang, S.; Liu, W.T. Dynamical and thermodynamical coupling between the north atlantic subtropical high and the marine boundary layer clouds in boreal summer. *Clim. Dyn.* **2017**, *19*, 1–13. [CrossRef]

37. Yu, L.; Weller, R.A. Objectively analyzed air–sea heat fluxes for the global ice-free oceans (1981–2005). *Bull. Am. Meteorol. Soc.* **2007**, *88*, 527–539. [CrossRef]

38. Bunge, L.; Clarke, A.J. A verified estimation of the El Niño index Niño-3.4 since 1877. *J. Clim.* **2009**, *22*, 3979–3992. [CrossRef]

39. Klein, S.A.; Hartmann, D.L.; Norris, J.R. On the relationships among low-cloud structure, sea surface temperature, and atmospheric circulation in the summertime northeast pacific. *J. Clim.* **1995**, *8*, 2063–2078. [CrossRef]

40. Norris, J.R.; Iacobellis, S.F. North pacific cloud feedbacks inferred from synoptic-scale dynamic and thermodynamic relationships. *J. Clim.* **2005**, *18*, 4862–4878. [CrossRef]

41. Mansbach, D.K.; Norris, J.R. Low-level cloud variability over the equatorial cold tongue in observations and models. *J. Clim.* **2007**, *20*, 1555–1570. [CrossRef]

42. Yao, Y.; Zhong, Z.; Yang, X.Q. Impacts of the subarctic frontal zone on the North Pacific storm track in the cold season: An observational study. *Int. J. Climatol.* **2018**, *38*, 2554–2564. [CrossRef]

43. Qu, X.; Hall, A.; Klein, S.A.; Deangelis, A.M. Positive tropical marine low-cloud cover feedback inferred from cloud-controlling factors. *Geophys. Res. Lett.* **2015**, *42*, 7767–7775. [CrossRef]

44. Bony, S.; Stevens, B.; Frierson, D.M.; Jakob, C.; Kageyama, M.; Pincus, R.; Shepherd, G. Clouds, circulation and climate sensitivity. *Nat. Geosci.* **2015**, *8*, 261–268. [CrossRef]

45. Koshiro, T.; Shiotani, M. Relationship between low stratiform cloud amount and estimated inversion strength in the lower troposphere over the global ocean in terms of clouds. *J. Meteorol. Soc. Jpn. Ser. II* **2014**, *92*, 107–120. [CrossRef]

46. Eastman, R.; Warren, S.G.; Hahn, C.J. Variations in cloud cover and cloud types over the ocean from surface observations, 1954–2008. *J. Clim.* **2011**, *24*, 5914–5934. [CrossRef]

47. Xu, H.; Xie, S.P.; Wang, Y. Subseasonal variability of the southeast Pacific stratus cloud deck. *J. Clim.* **2005**, *18*, 131–142. [CrossRef]

48. Van der Dussen, J.J.; De Roode, S.R.; Ackerman, A.S.; Blossey, P.N.; Bretherton, C.S.; Kurowski, M.J.; Lock, A.P. The gass/euclipse model intercomparison of the stratocumulus transition as observed during astex: Les results. *J. Adv. Model. Earth Syst.* **2013**, *5*, 483–499. [CrossRef]

49. Lau, N.C.; Nath, M.J. Impact of ENSO on SST variability in the north pacific and north atlantic: Seasonal dependence and role of extratropical sea–air coupling. *J. Clim.* **2001**, *14*, 2846–2866. [CrossRef]

50. Zhu, P.; Hack, J.J.; Kiehl, J.T.; Bretherton, C.S. Climate sensitivity of tropical and subtropical marine low cloud amount to ENSO and global warming due to doubled CO_2. *J. Geophys. Res. Atmos.* **2007**, *112*. [CrossRef]

51. Park, S.; Leovy, C.B. Marine low-cloud anomalies associated with ENSO. *J. Clim.* **2004**, *17*, 3448–3469. [CrossRef]

52. Lana, A.; Bell, T.G.; Simó, R.; Vallina, S.M.; Ballabrera-Poy, J.; Kettle, A.J.; Johnson, J.E. An updated climatology of surface dimethlysulfide concentrations and emission fluxes in the global ocean. *Glob. Biogeochem. Cycles* **2011**, *25*. [CrossRef]

53. Zhao, C.; Qiu, Y.; Dong, X.; Wang, Z.; Peng, Y.; Li, B.; Wu, Z.; Wang, Y. Negative Aerosol-Cloud re Relationship from Aircraft Observations Over Hebei, China. *Earth Space Sci.* **2018**, *5*, 19–29. [CrossRef]

54. Yang, Y.; Zhao, C.; Dong, X.; Fan, G.; Zhou, Y.; Wang, Y.; Zhao, L.; Lv, F.; Yan, F. Toward understanding the process-level impacts of aerosols on microphysical properties of shallow cumulus cloud using aircraft observations. *Atmos. Res.* **2019**. [CrossRef]

12

Diurnal Variations of Different Cloud Types and the Relationship between the Diurnal Variations of Clouds and Precipitation in Central and East China

Cuicui Gao [1,2], Yunying Li [1,*] and Haowei Chen [2]

[1]   College of Meteorology and Oceanography, National University of Defense Technology, Nanjing 211101, China; gaocc2016@foxmail.com
[2]   Shaoguan Meteorological Office, Shaoguan, Guangdong 512000, China; chen_haowei@foxmail.com
*   Correspondence: ghlyy@mail.iap.ac.cn

Abstract: In this paper, the diurnal variations of various clouds are analyzed using hourly cloud observations at weather stations in China from 1985 to 2011. In combination with merged hourly precipitation data, the relationship between the diurnal variations of clouds and precipitation in the summers from 2008 to 2011 are studied. The results show that the occurrence frequencies of total cloud and various cloud types exhibit significant diurnal variations. The diurnal variations of the occurrence frequencies of altocumulus and stratocumulus show a bimodal pattern, with peaks appearing in the early morning and late afternoon. The early morning peaks of altocumulus and stratocumulus appear earlier in the summer than in the other seasons, while the late afternoon maxima show an opposite trend. The occurrence frequency of nimbostratus peaks in the morning between 07 and 12 LST (local solar time), and the peak value lags 2 to 3 h from west to east along the Yangtze River valley; meanwhile, the diurnal variation shows no clear differences caused by changes in the latitude or seasons. Cumulus shows an afternoon (14 LST) maximum, while cumulonimbus peaks in the late afternoon during 16–20 LST, and both of them present a great diurnal range. Cirrus usually reaches its peak at 17–18 LST, and it differs by 1 to 2 h with a change in the latitude. The results of the study first show that the diurnal variations of precipitation among different regions are dominated by different clouds. The upper reaches of the Yangtze River valley present a midnight precipitation maximum that is mainly dominated by cumulonimbus. For the middle reaches of the Yangtze River valley impacted by nimbostratus, the precipitation peaks in the early morning. In South and Northeast China, the precipitation peaks in the afternoon and is determined by the diurnal variations of convective clouds. In the region between the Yangtze River valley and Yellow River valley, the precipitation peaks in the early morning and afternoon; the early morning peak is mainly determined by stratiform clouds, while the afternoon peak is closely related to convective clouds.

Keywords: clouds; precipitation; diurnal variation

1. Introduction

Diurnal variations are the most basic period of change in the Earth's climate system, the most important driving force of which is solar radiation. Various meteorological parameters, such as the temperature, wind, pressure and precipitation, all show significant diurnal variations. Rutledge noted that stratiform and convective precipitation are produced by different cloud microphysical processes [1]. The formation mechanism of precipitation is complex, but it ultimately originates from clouds. Clouds are the external manifestation of dynamic and thermodynamic processes in the atmosphere, and there are many differences in the nature of precipitation sourced from different types of clouds. The diurnal variation of clouds reflects the changes in the internal movements throughout the atmosphere.

However, the diurnal cycle of clouds also plays a feedback role in the movements of the atmosphere in addition to the radiation process and the water cycle [2]. Furthermore, the diurnal variation of clouds also reflects the atmospheric stability and changes in the weather. Therefore, studying the diurnal variation of clouds is of significance for understanding the diurnal cycle of the internal movements throughout the atmosphere. In addition, understanding the relationship between the diurnal variations of clouds and precipitation is beneficial to provide a physical basis and observe relevant facts for cloud parameterization schemes in weather and climate models, which is helpful for improving the ability to simulate weather and climate models.

Precipitation exhibits significant diurnal variation characteristics. Consequently, many scholars have carried out a great deal of research [3–9]. Yu et al. noted that the summer precipitation over contiguous China has large diurnal variations with considerable regional features [10]. Over southern inland China and northeastern China, the summer precipitation peaks in the late afternoon, while the precipitation in some regions peaks between midnight and the early hours of the morning. Zhou et al. verified the above results by comparing satellite to rain gauge observations, showing that the diurnal phase of rainfall frequency and intensity were similar to those of the rainfall amount in southern China [11]. A study by Li et al. showed the clear differences in the diurnal cycle of precipitation between southwestern China and southeastern China [12]. The diurnal cycle of the annual mean precipitation in southwestern China tends to reach a maximum either at midnight or during the early morning, while the precipitation in eastern China peaks in the late afternoon. Yuan et al. analyzed the sub-seasonal characteristics of the diurnal variation of summer monsoon rainfall over eastern central China and found that the early-morning diurnal peaks experience sub-seasonal movements similar to those of the monsoon rain belt [13]. The aforementioned scholars have obtained many meaningful conclusions about the diurnal variation of precipitation. Moreover, since precipitation originates from clouds, the diurnal variation of precipitation is closely related to the diurnal variations of different types of clouds, and therefore, studying the relationships between the diurnal variations of clouds and precipitation could be helpful for further understanding the mechanism of precipitation.

William et al. presents observational evidence in support of the existence of a large diurnal cycle of oceanic, tropical, deep cumulus convections [14]. Zheng et al. used satellite infrared temperature of black body (TBB) data to explore the climatological characteristics of deep convection over southern China and the adjacent seas during the June-August periods of 1966–2007 [15]. They found that sea-land and mountain-valley breezes accounted for the propagation of deep convection from sea to land in the afternoon and from land to sea after midnight. Based on hourly infrared brightness temperature data acquired by the FY-2C satellite to classify clouds into cold clouds, middle clouds and warm clouds according to the cloud top temperature, Chen et al. analyzed the diurnal variations of three types of clouds in southern China during the summer periods from 2005 to 2008 and the seasonal changes in the diurnal variations of those clouds [16]. Fujinami et al. studied the diurnal variations of high clouds and precipitation over the Tibetan Plateau, where observation stations are sparse, using satellite observations. He noted that the cloud cover frequency for high clouds increased in the afternoon along the ridges and reached a maximum near 18 LST, after which high cloud cover frequencies moved over the valley and persisted until early morning [17]. In addition, a similar evolution occurred in the frequency of rainfall. Zhao et al. based on MODIS observations, showed the statistical characteristics of cloud properties, along with the difference between morning and afternoon [18]. These conclusions have promoted the understanding of the diurnal variations of clouds; however, the corresponding research is relatively fragmented, and the relationship between the diurnal variations of clouds and precipitation has not yet been established. Therefore, this paper aims to analyze the climatic characteristics of the diurnal changes of clouds over China in addition to the relationship between the diurnal variations of clouds and precipitation to deepen the existing understanding of the correlation between clouds and precipitation and to provide the basis for a simulation of the diurnal variations of clouds and precipitation in weather and climate models.

2. Data and Analysis Method

2.1. Data

In this study, hourly cloud data during 1985–2011 are acquired from 114 surface observation stations after a quality-control procedure is performed. Since the distribution of stations to the west of 105° E is sparse, this study mainly analyzes the diurnal variations of clouds over eastern and central China. The dataset includes the information of total cloud (TC), and 10 cloud genera (cirrus (Ci), cirrocumulus (Cc), cirrostratus (Cs), altocumulus (Ac), altostratus (As), nimbostratus (Ns), stratocumulus (Sc), stratus (St), cumulus (Cu), and cumulonimbus (Cb)). Due to the light intensity and visual reasons, artificial observations on the ground will produce some random observation errors, however, many years of climatological normal results can significantly reduce the instabilities of artificial observations. In addition, the reliability of the dataset has been tested and verified [19]. From the perspective of surface observations, the diurnal variation of low clouds is relatively accurate. High clouds and middle clouds are sometimes blocked by low clouds, and the article therefore analyzes the diurnal variation of a certain type of high (middle) cloud when it appears.

Hourly merged precipitation data during 2008–2011 were obtained from the National Meteorological Information Center of the China Meteorological Administration with a spatial resolution of 0.1° × 0.1°. These hourly precipitation data were merged with more than 30,000 automatic weather stations over China in addition to global Climate Prediction Center morphing technique (CMORPH) precipitation estimates with a temporal interval of 30 min and a resolution of 8 km provided by the American Climate Prediction Center. The merged precipitation product effectively combines the advantages of surface precipitation observations with satellite retrieved precipitation products. The overall error of the product is less than 10%, which is better than the same type of product worldwide [20]. The quality of the merged precipitation products in the sparse station areas still needs improvement. However, the stations in eastern and central China studied in this paper are densely and effectively distributed, and the precipitation values are accurate since station data were utilized as the main data source in the fusion process. Since 56.5 percent of the precipitation falls during the summertime throughout most of China [21], this paper mainly analyzes the relationship between the diurnal variations of clouds and precipitation during the summer.

The diurnal variation of clouds is the average of the data from the stations from 1985 to 2011. Since the precipitation data range from 2008 to 2011, to match the precipitation data, the cloud data are truncated from 2008 to 2011, when the relationship between the diurnal variations of clouds and precipitation is analyzed.

2.2. Analysis Method

The diurnal variations of the meteorological parameters are the changes in the meteorological parameters with the local solar time (LST) [5,22]. Therefore, statistical work on the diurnal variations of clouds and precipitation should be conducted according to the local solar time. The time system for the hourly cloud data used in this paper corresponds to China Standard Time (CST), and the method used to convert CST into LST is as follows:

$$T1 = \left[\frac{\text{lon} \pm 7.5}{15}\right] - 8 + T \tag{1}$$

$$LST = \begin{cases} T1 & 0 < T1 \leq 24 \\ 24 + T1 & T1 \leq 0 \\ T1 - 24 & T1 > 24 \end{cases} \tag{2}$$

where lon represents the longitude of the station, and the unit of lon is degree, T denotes CST, LST is the local solar time, and [] indicates an operation to change the number into an integer.

The time system for the precipitation data used in this paper corresponds to Greenwich Mean Time (GMT), and the method used to convert GMT into LST is similar to CST, and it just don't need to subtract the time zone in the first step. The time used hereafter in this study is LST.

The occurrence frequency of a cloud is defined as $F = (rec_{cld}/rec) \times 100\%$, where F represents the occurrence frequency of a cloud at a certain time, rec_{cld} represents the number of cloud occurrences at a certain time within the analysis period, rec is the number of observations at a certain time, and the definition of the occurrence frequency is suitable for all types of cloud, including TC. The maximum occurrence frequency of a cloud from 01 to 24 LST in a day is the daily peak. The amplitude of the diurnal variation of clouds reflects the magnitude of the maximum and minimum occurrence frequency of cloud in 24 h and is defined as $A = cld_{max}/cld_{avg}$, where A is the amplitude, cld_{max} is the daily peak of the cloud occurrence frequency, and cld_{avg} is the daily occurrence frequency of the cloud. When analyzing the diurnal variation of clouds in different seasons, to eliminate the differences in the occurrence frequencies of cloud among the different seasons, this paper calculates the anomalies in the occurrence frequencies of clouds in each month from 01 to 24 LST.

3. Climatic Characteristic of Diurnal Variation of Clouds

3.1. Diurnal Variation of Different Types of Clouds

To obtain an overall understanding of the diurnal variations of the total clouds (TC) and of the various types of clouds over central and eastern China, this paper first analyzes the diurnal variations of the average occurrence frequencies of the TC and various clouds at 114 stations over central and eastern China. The diurnal variation of the frequency of TC shows a bimodal pattern; the main peak appears in the late afternoon, and the second peak appears in the early morning (Figure 1a). During the day, the occurrence frequency of the TC is significantly greater than that of nocturnal clouds. After 04 LST, the occurrence frequency of the TC increases rapidly and reaches a peak at 07 LST. The frequency of the TC changes slightly during 07–09 LST, and it begins to increase again after 09 LST and reaches the main peak at 14–15 LST. Ac also shows a diurnal variation with a bimodal pattern with peaks at 07 LST and 18 LST, and the morning peak is larger. As occurs at a frequency of less than 3.5% but at a relatively high frequency in the afternoon.

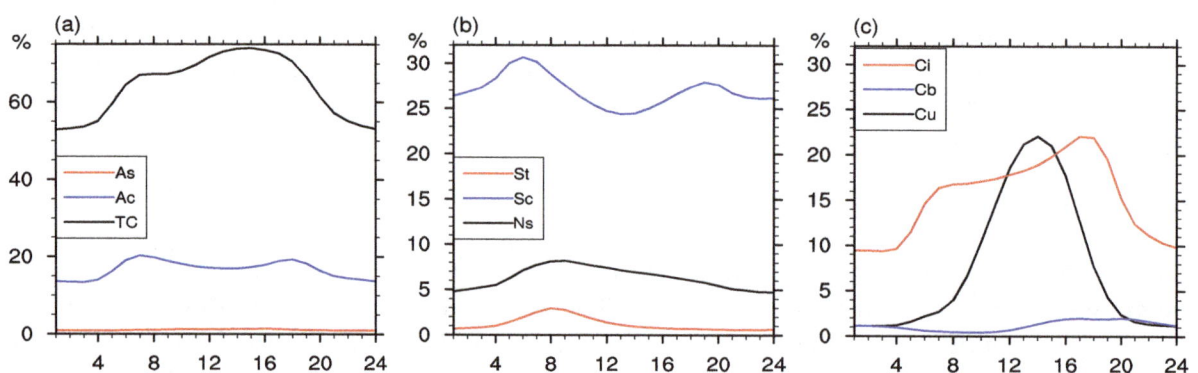

Figure 1. The diurnal variation of the occurrence frequencies (%) of various cloud types in central and east China during 1981–2011: (a) TC, Ac, As, (b) Ns, Sc, St, (c) Cu, Cb, Ci(the horizontal axis represents the local solar time, and the vertical axis represents the occurrence frequency of clouds).

The diurnal cycle of the occurrence frequency of Ns is unimodal, and it occurs at approximately 08 LST. Ns is a component of frontal clouds, which are mainly formed as deep, moist air rises along the front surface; the relative humidity of the air is larger in the morning and is saturated more easily and condensed, and thus, Ns appears more frequently in the early morning. The diurnal variation of St is consistent with that of Ns, and its diurnal peak appears at 08 LST. St, whose formation mechanism is the same as that of fog, is generally formed by fog rising. Fog usually appears during the night until

08 LST, and it especially occurs throughout most of the morning; St also shows such a pattern in the morning. The diurnal variation of the occurrence frequency of Sc presents a bimodal pattern with a main diurnal peak at 06 LST and a second peak at 17 LST. As a typical boundary layer cloud, the diurnal variation of Sc reflects the diurnal variation of the boundary layer. The low temperature in the boundary layer and the stable atmospheric stratification in the morning are beneficial for water vapor below the top of the boundary layer to accumulate and form Sc. When the atmosphere stability deteriorates, the conditions are no longer favorable to the formation of stratiform clouds and the occurrence of Sc. As the solar radiation weakens in the evening and the atmosphere stratification tends to become stable, the frequency of Sc starts to increase again and reaches the second peak at approximately 19 LST.

The diurnal variation of Cu is also unimodal (Figure 1c). During the daytime, as the solar radiation gradually enhances, the atmosphere temperature around and above the ground surface gradually rises while the instability of the atmospheric stratification increases, and thus, the occurrence frequency of Cu begins to increase beginning in the early morning and reaches a peak at 14 LST, after which it decreases. The diurnal variation of the occurrence frequency of Cb shows a unimodal pattern as well. The occurrence frequency of Cb is high at 16–20 LST, and its main diurnal peak appears at 16–17 LST, which is 4–6 h behind the time when Cu reaches its diurnal peak, indicating that it takes some time for shallow convection clouds to develop into deep clouds. The frequency of Ci increases from 04 to 05 LST continuously, and it reaches a maximum at 17 LST, after which it rapidly decreases after 18 LST. Sometimes, Ci acts as an anvil to the development of Cb to a mature stage, and thus, there is a lag in the time required for Cu, Cb to Ci to reach their diurnal peaks.

Figure 2 shows the times of the diurnal peaks of different clouds and the spatial distribution of the amplitudes of their diurnal variations over central and eastern China. The figure shows the time when the occurrence frequency peaks at each station in the form of a clock, and the number indicates the amplitude of the diurnal variation of the cloud occurrence frequency. Figure 2a shows that the diurnal peak of the TC appears from the afternoon to the late afternoon, but it appears in the early morning at some stations. For example, at most of the stations to the north of 32° N in China, the diurnal peak appears at 15–18 LST; meanwhile, over the Liaodong Peninsula and Shandong Peninsula, the diurnal peak appears a little earlier at 13–14 LST. However, to the south of 32° N, the diurnal peak appears in two periods at 14–18 LST and 06–07 LST. For instance, the diurnal peak of the TC in the southeastern coastal areas of China mainly occurs in the early morning at 06–07 LST. The amplitude of the occurrence frequency of the TC is large in northern China (approximately 1.2–1.3), while that at most stations in southern China is approximately 1.1, indicating that the diurnal variations of the occurrence frequencies of clouds in northern China are greater than those in southern China.

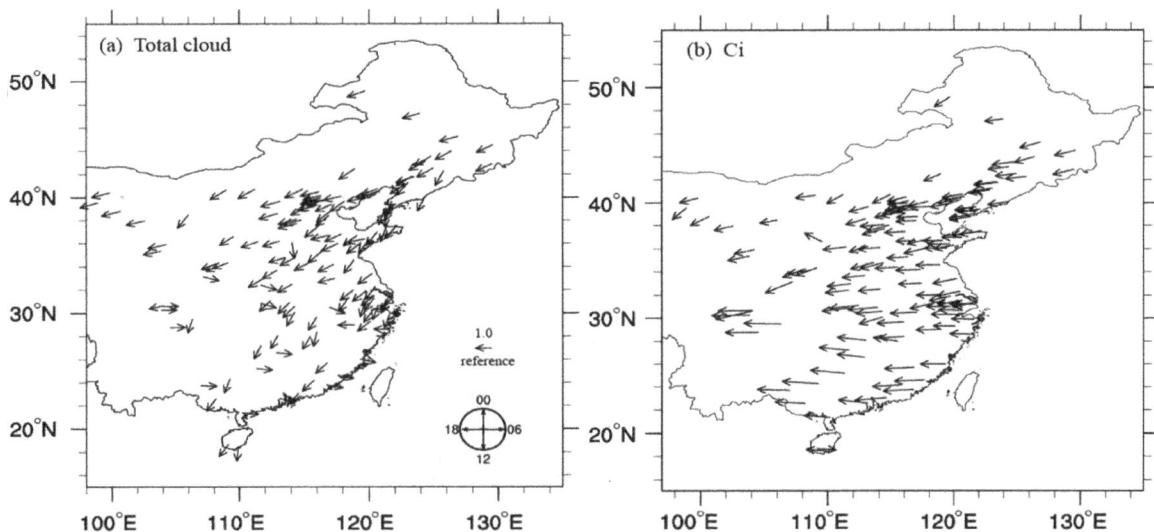

Figure 2. *Cont.*

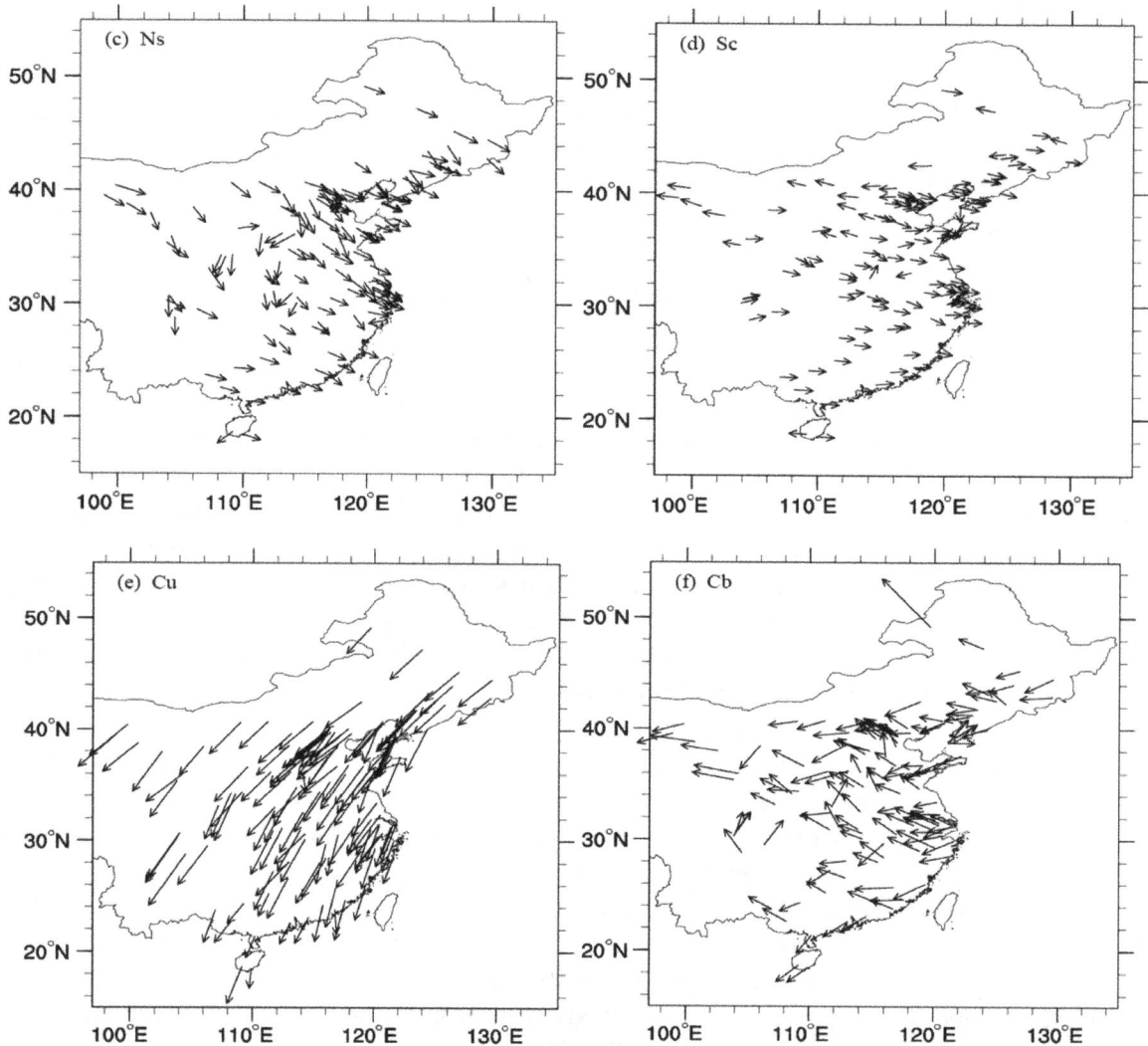

Figure 2. Spatial distributions of the phases and amplitudes of the diurnal cycles of various clouds: (**a**) TC, (**b**) Ci, (**c**) Ns, (**d**) Sc, (**e**) Cu, and (**f**) Cb. The directions of the vectors denote the local solar time (LST) of the maximum cloud occurrence frequency (phase clock in (**b**)), while the length of the vectors represent the amplitude of the diurnal cycle, the reference arrow represents the amplitude of diurnal cycle is 1.0.

According to this study, the diurnal peak of Ci appears at 17–18 LST consistently at each station (Figure 2b), and the diurnal variation amplitude is larger in South China and in the Sichuan Basin.

The diurnal peak of Ns occurs in the morning (08–09 LST) (Figure 2c), but those of some stations in Shanxi Province, western Henan Province, and Hubei Province and in the Sichuan basin occur around noon (11–14 LST). The Meiyu front near the Yangtze River valley is also a frequent area of Ns formation, yet the diurnal variation amplitude of Ns in this area is small. Sc is the cloud with the highest occurrence frequency over China. Its diurnal peak over most stations occurs at 06–07 LST (Figure 2d), but those in western Inner Mongolia and North China usually occur at 18–19 LST.

The diurnal peak of Cu over all stations appears in the afternoon (12–14 LST) (Figure 2e). The diurnal variation of Cu in most areas throughout China is large, and its amplitudes are basically above 2.9, while the diurnal variation of Cu in South China is relatively small with amplitudes between 1.9 and 3.1. The frequency of Cb usually reaches a peak value at 16–20 LST (Figure 2f); however, at very few stations, the peak occurs at 02–04 LST. In addition, the diurnal variation of Cb is large with the second-largest amplitude (approximately 1.5 to 2.7) following that of Cu.

The analysis above shows that there are regional differences among the diurnal peaks of Ac and Sc that are mainly due to the frequent occurrences of both cloud types in the early morning and evening with different peak times at different stations. As Ac and Sc occur more frequently, they have a great impact on the diurnal variation of the TC occurrence frequency, and thus, the TC shows a similar diurnal changing pattern with these two types of clouds. The diurnal peak of Ns mainly appears in the morning (07–12 LST), and the time difference in the diurnal peaks between different stations does not exceed 5 h. The diurnal peak of Ci usually appears in the evening (17–18 LST) while that of Cu appears in the afternoon but shows no clear regional difference. Finally, the diurnal peak of Cb mainly appears at 16–20 LST with slight differences among different regions. The diurnal variations of Cu and Cb are both relatively enormous.

3.2. Diurnal Variation Cycles of Clouds with the Latitude and Longitude

Figure 3 shows the diurnal variations of the TC and of various cloud types with the latitude; the zonal average of 110–120° E represents the occurrence frequency of cloud types at each latitude. Figure 3a illustrates that the diurnal peak of the TC to the south of 20° N appears in the afternoon (12–15 LST), which is closely related to the diurnal variation of convective clouds. Meanwhile, the diurnal peak of the TC at 20–26° N (i.e., southern China) appears in two periods during the early morning (06–08 LST) and afternoon (11–16 LST). This is probably due to the diurnal peaks of Ac and Sc in these areas that occur in the early morning and evening and have a great impact on the diurnal variation of the TC occurrence frequency. The diurnal peak of the TC at 26–36° N appears mainly in the afternoon. The occurrence frequency of the TC to the north of 36° N increases with the latitude with a maximum at 16 LST.

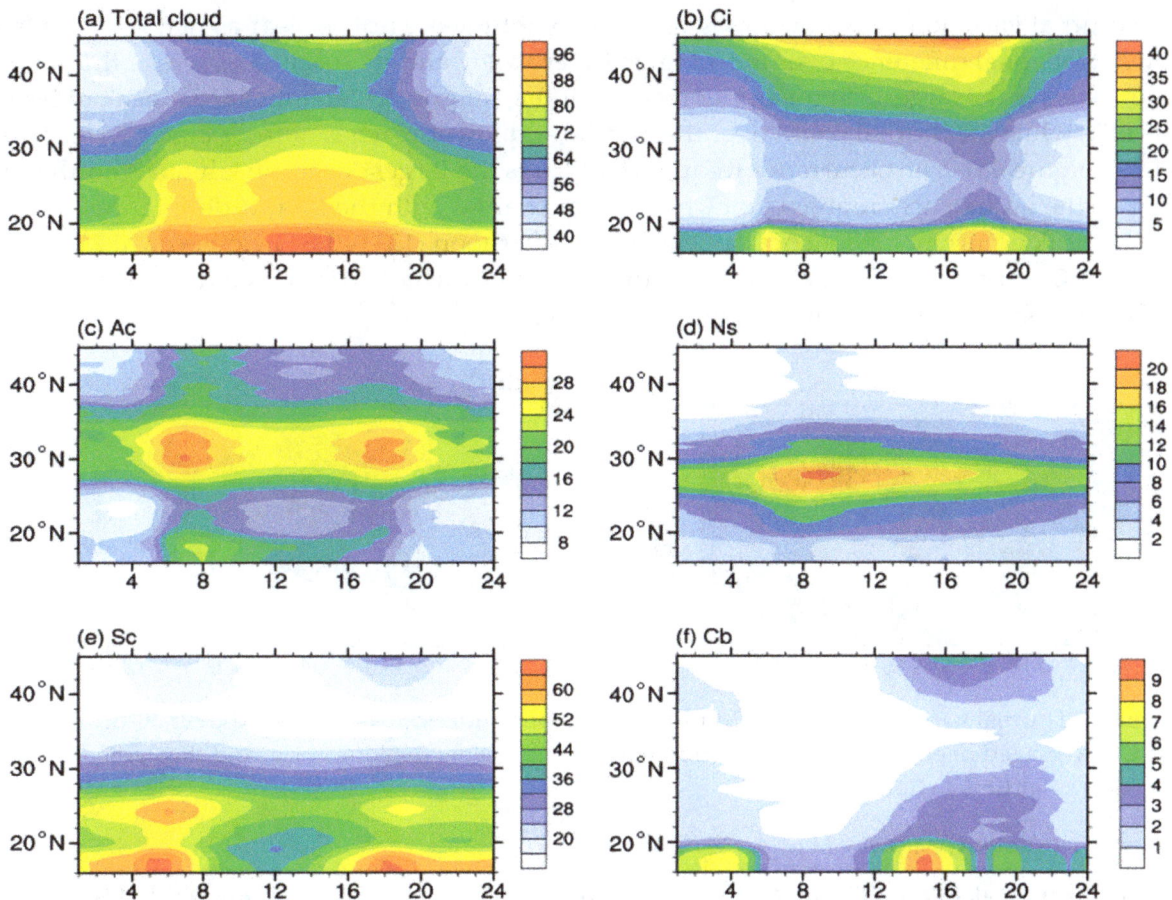

Figure 3. Diurnal variation cycles of occurrence frequency (%) of clouds with the latitude averaged over 110–120° E: (**a**) TC, (**b**) Ci, (**c**) Ac, (**d**) Ns, (**e**) Sc, and (**f**) Cb (the horizontal axis is the local solar time in hours, and the vertical axis represents the latitude).

The occurrence frequency of Ci at 20–32° N over China is small, but the occurrence frequencies over Hainan Province and to the north of 32° N are large, and the frequency of the latter area increases with the latitude (Figure 3b). Moreover, there is a slight difference in the diurnal peak of Ci between the two areas. The occurrence frequency of Ci over Hainan Province is large (approximately 35%), and the peak appears at 06 LST and 18 LST. The diurnal peak of Ci to the north of 32° N occurs at 16–17 LST, and the occurrence frequency of Ci near 45° N reaches 40% at 16 LST. The occurrence frequency of Ac increases with the latitude from south to north (Figure 3c); it reaches a maximum at 28–36° N and then decreases. At 28–36° N, the diurnal peak of Ac appears at 07 LST and 18 LST, while in other areas the diurnal peak of Ac appears at approximately 08 LST, indicating that its maximum daily peak appears at a difference of 1 to 2 h with a change in the latitude.

Figure 3d shows a high occurrence frequency of Ns within 26–29° N. After 04 LST, the high value of the frequency of Ns begins to expand to the south and to the north. In addition, a frequency of 10% expands to the south to 22° N at 08 LST and to the north to 32° N at 09 LST. After 09 LST, the frequency of Ns begins to shrink, as the maximum frequency of Ns occurs at 09 LST. The diurnal peak of Ns does not change significantly with the latitude. The diurnal peak of Sc occurs in two periods during the early morning (approximately 06 LST) and the late afternoon (18–19 LST) (Figure 3e). The diurnal peak of Sc over Hainan Province appears at 05 LST and 18 LST while that at 20–28° N over China occurs at 06 LST and 19 LST. Moreover, the diurnal peak of Sc at 18–25° N lags behind by approximately 1 h. The diurnal peak of Cb over Hainan Province is reached at 09 LST while that of Cb to the north of 20° N mainly occurs in the period of 16–20 LST, and its appearance time lags behind with the latitude (Figure 3f). There is no difference in the diurnal peak of Cu with the latitude, as it consistently occurs at 13–14 LST (figure omitted).

The diurnal variation cycles of clouds with the longitude are not as clear as that with the latitude. Ns shows clear zonal difference over the Yangtze River valley and its south side, where the occurrence frequency of Ns is the largest. In the upper reaches of the Yangtze River, the diurnal peak of Ns occurs at 06–09 LST, while the diurnal peaks over the middle and lower reaches of the Yangtze River appear at 07–13 LST (Figure 4a). The occurrence frequency of Cu is the largest in south China. And the diurnal variation of the occurrence frequency of Cu does not change with the longitude (Figure 4b) in south China. The peak occurrence frequency appears in the afternoon (13–14 LST); however, at 22–26° N, the frequency of Cu decreases clearly from west to east. The diurnal variations of the TC and the other types of clouds show no significant change with the longitude.

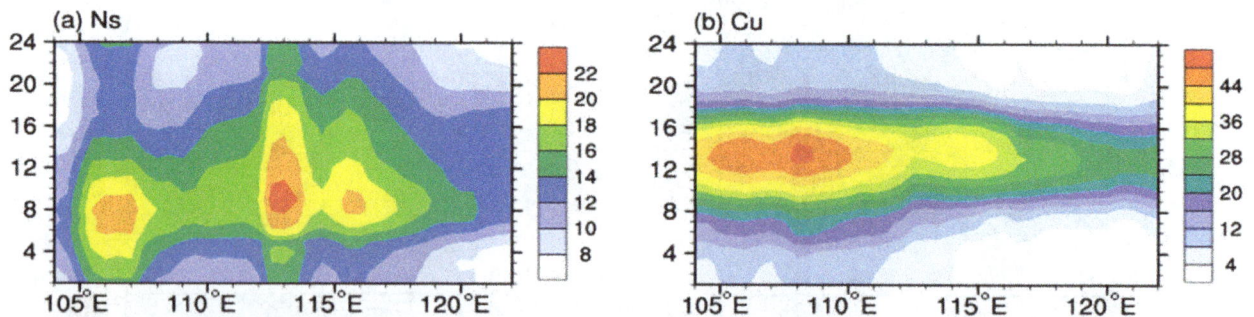

Figure 4. Diurnal variation cycles of clouds with the longitude: **(a)** Ns (averaged over 26–30° N) and **(b)** Cu (averaged over 22–26° N) (the horizontal axis represents the longitude, and the vertical axis is the local solar time in hours).

3.3. Diurnal Variation Cycles of Clouds with the Season

The diurnal peak of the TC over China during the spring occurs at approximately 16 LST (Figure 5a). In addition, there is a clear diurnal variation of the TC in the summer and autumn (i.e., from June to November). The maximum occurrence frequency occurs at 14 LST while the minimum occurs at 23 LST, and the difference between them reaches 28%. In addition, the amplitude of the diurnal variation of the TC in the winter is small; at 13–15 LST, however, the occurrence frequency is relatively large. In contrast, the occurrence frequency of the TC during the summer reaches a peak time 1–2 h earlier than that during the spring, which may occur because the mean temperature increases more rapidly in the summer than in the spring, and thus, the conditions that are unstable for cloud formation are reached earlier.

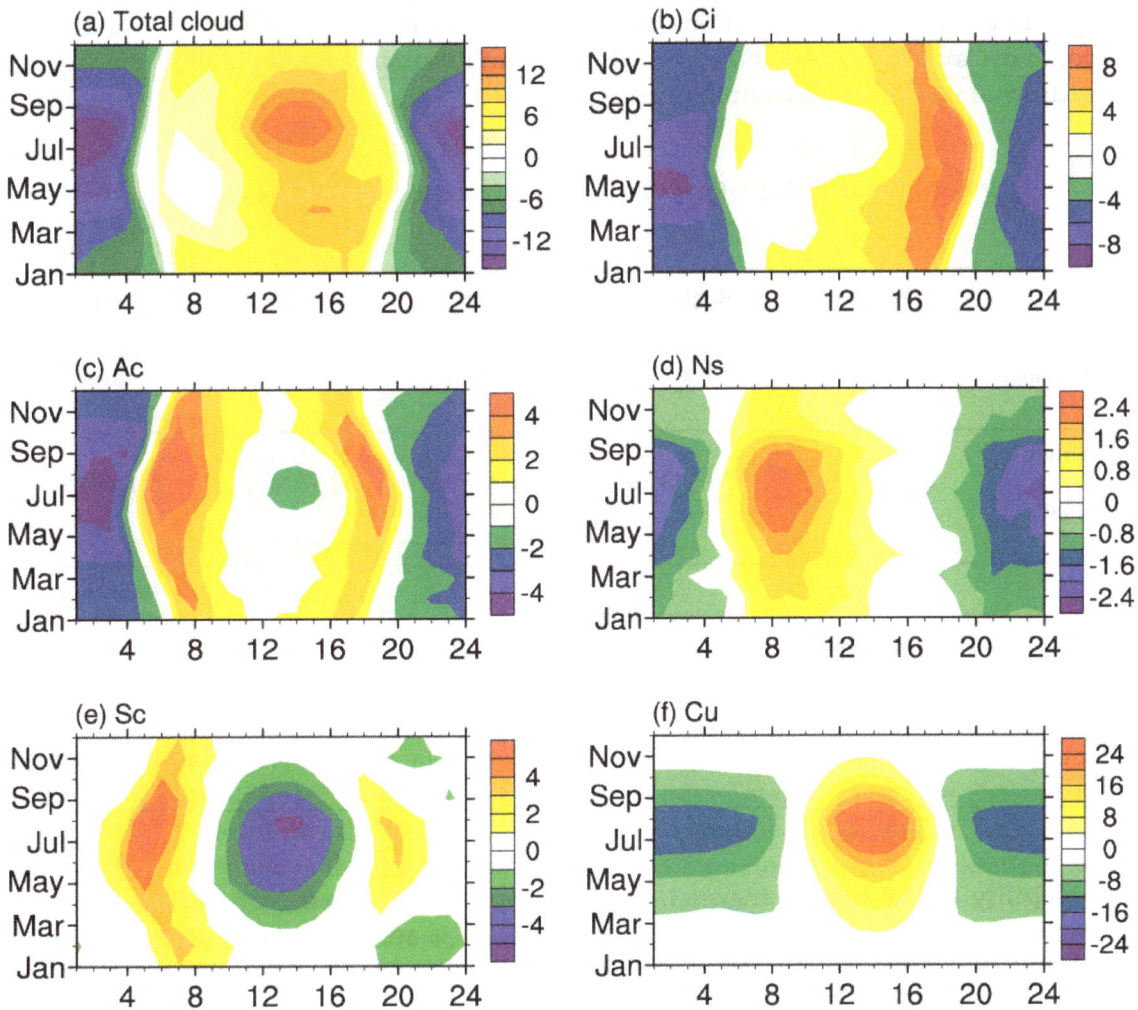

Figure 5. Seasonal variations in the diurnal cycle of anomalous values of the occurrence frequencies of clouds: (**a**) TC, (**b**) Ci, (**c**) Ac, (**d**) Ns, (**e**) Sc, (**f**) Cu (the horizontal axis is the local solar time in hours, and the vertical axis represents the month).

The diurnal peak of Ci also changes with the season (Figure 5b). Ci reaches its maximum at 16–17 LST in December and at 18–19 LST in July, thereby lagging 1–2 h between the winter and summer. The changes in the peak times of Ac are the most clear from June to September (Figure 5c). The peak times of the two peaks appear from early winter to summer, while the early morning peak appears earlier and the evening peak appears later in the summertime. The diurnal peak times of Ns and Cu do not change with the seasons (Figure 5d,f), but both of the amplitudes of their diurnal variations are relatively large in the summer. The diurnal variation of Sc with the seasons is somewhat similar to that of Ac (Figure 5e), but the minimum occurrence frequency always appears at 14 LST and does not change with the seasons. The occurrence frequency of Cb in the winter is very small, and its seasonal variation is also not clear and therefore is not discussed here.

Yangtze River relies upon the diurnal variations of Cb and Sc. Moreover, the correspondence between the diurnal variations of Cb and precipitation is better, and Cb may evolve into Sc after precipitation falls; thus, the diurnal peak of Sc lags behind that of precipitation by approximately 2 h.

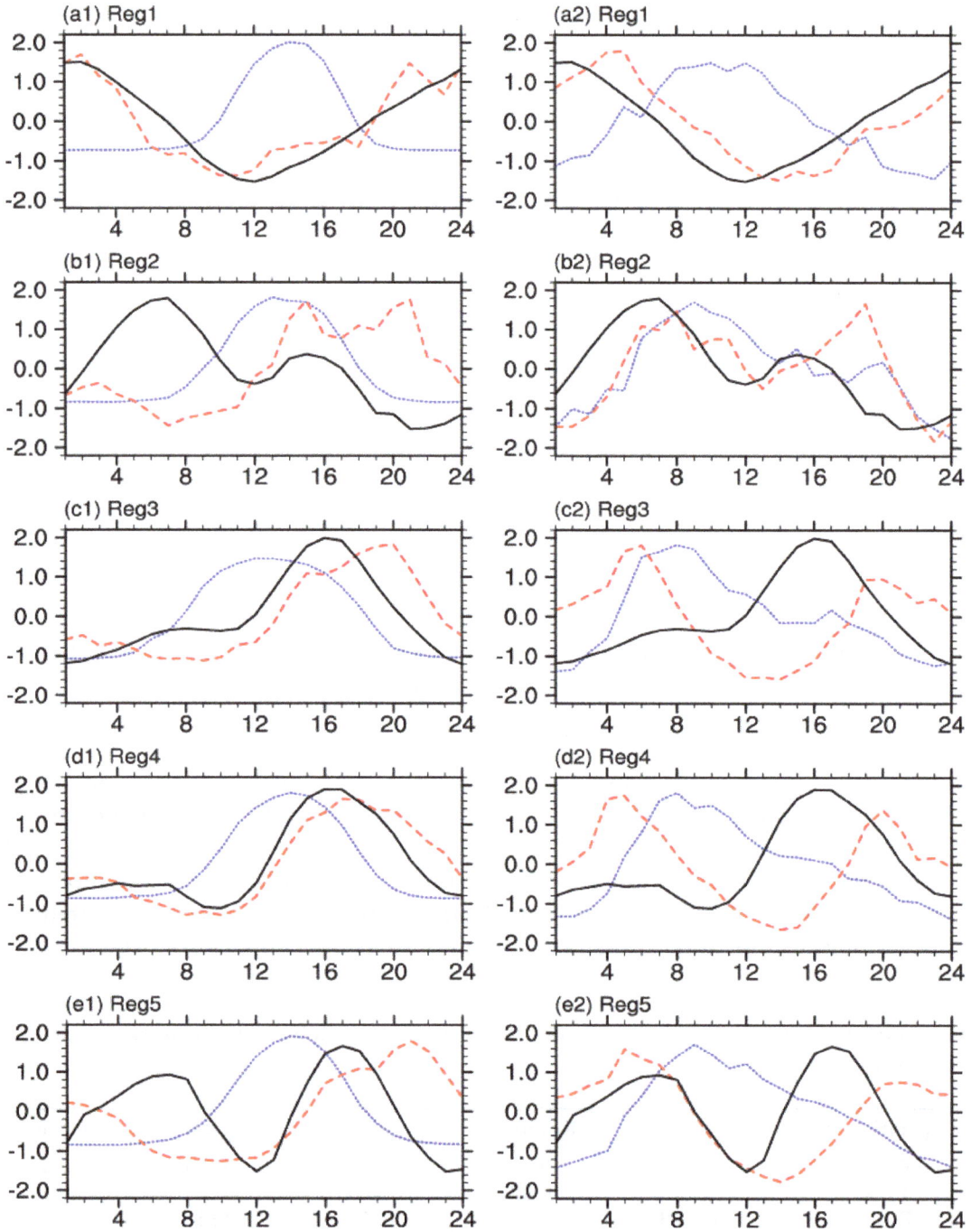

Figure 7. Diurnal variations of the standardized occurrence frequencies of Cu, Cb and precipitation (in **a1–e1**) and of the standardized occurrence frequencies of Ns, Sc and precipitation (in **a2–e2**) (the black solid line corresponds to precipitation, while the blue dotted line and red dashed line represent the occurrence frequencies of Cu and Cb, respectively, in **a1–e1**; meanwhile, the blue dotted line and red dashed line represent the occurrence frequencies of Ns and Sc, respectively in **a2–e2**).

The precipitation in the middle reaches of the Yangtze River mainly occurs in the early morning (Figure 7b1). There is a sub-peak at approximately 15 LST, the occurrence of which corresponds to the sub-peaks of both Cu and Cb, indicating that the afternoon rainfall in the middle reaches of the

Yangtze River mainly originates from Cu and Cb. Figure 7b2 shows that the corresponding relationship between the diurnal variations of precipitation and stratiform clouds in the middle reaches of the Yangtze River is good, and there is a notable positive correlation between precipitation and Ns with a correlation coefficient of 0.51. These results show that the peak of early morning rainfall in the middle reaches of the Yangtze River is mainly caused by stratiform clouds, while the sub-peak of afternoon rainfall mainly depends upon Cu and Cb.

The diurnal peaks of Cu, precipitation and Cb appear sequentially over southern China (Figure 7c1). The peak frequency of Cu occurs at 12 LST, while those of precipitation and Cb occur at 16 LST and 20 LST, respectively. In addition, the occurrence frequency of Cu decreases during 12–16 LST, but those of precipitation and Cb rapidly increase during this period, implying that Cu begins to produce precipitation during this period and that a part of Cu develops vigorously into Cb under the favorable conditions. However, the corresponding relationship between the diurnal variations of stratiform clouds and precipitation is poor over South China (Figure 7c2). In conclusion, both Cu and Cb over southern China show a good correspondence with the diurnal variation of precipitation, indicating that the diurnal variation of precipitation in southern China is mainly attributed to the diurnal variations of convective clouds.

There is a significant positive correlation between the diurnal cycles of Cu and precipitation in Northeast China with a correlation coefficient of 0.55. Cb shows a better correspondence with the diurnal variation of precipitation (Figure 7d1), with a significant positive correlation of 0.88. For example, the diurnal peaks of Cb and precipitation are both almost synchronous at 09–16 LST, although the diurnal peak of Cb that occurs during 01–09 LST is a little earlier than that of precipitation, and the occurrence frequency of Cb peaks one hour later than that of precipitation. In addition, the correspondence between the diurnal variations of stratiform clouds and precipitation is also poor in these areas (Figure 7d2). This shows that the diurnal variation of precipitation in Northeast China is primarily dominated by convective clouds, especially Cb.

There are two peaks in the diurnal variation of precipitation in the region between the Yangtze River and the Yellow River (region 5). In the afternoon, Cu, precipitation and Cb reach their peaks sequentially (Figure 7e1). Cu peaks at 14 LST, while the occurrence frequencies of precipitation and Cb increase rapidly after 12 LST. Cb and precipitation both peak at 17 LST, indicating that the afternoon peak of precipitation is determined by convective clouds. Figure 7e2 shows that the diurnal peaks of Sc and Ns occur at 05 LST and 09 LST, respectively, while the precipitation peaks at 07 LST, meaning that the morning precipitation peak in the area is dominated by stratiform clouds. Therefore, the results show that the diurnal variation of precipitation in the area is comparatively complicated and is closely related to both convective and stratiform clouds. The morning precipitation is closely related to stratiform clouds, while the afternoon precipitation is dominated by convective clouds.

All of the above analysis shows that the precipitation peak in central and eastern China usually occurs in the early morning or afternoon. The precipitation peak in early morning is dominated by stratiform clouds, while the peak in the afternoon is dominated by convective clouds.

5. Discussions and Conclusions

Based on hourly cloud observations at weather stations in combination with merged hourly precipitation data, the diurnal climatic features of different types of clouds in central and eastern China are revealed and the relationship between the diurnal variations of clouds and precipitation are analyzed. The primary conclusions are as follows:

The TC and the occurrence frequencies of all types of clouds show significant diurnal variations. The diurnal variation of the TC shows a bimodal pattern with a main peak in the late afternoon. Ac, Sc and Ci contribute the most to the diurnal variation of the TC. The diurnal cycles of the occurrence frequencies of Ac and Sc also show a bimodal pattern with peaks appearing in the early morning and late afternoon. The occurrence frequency of nimbostratus peaks in the morning between 07 and 12 LST, and the daily difference is weaker. The diurnal peak of Ci usually appears at 17–18 LST in the late afternoon, Cu peaks in the afternoon, and Cb peaks in the late afternoon during 16–20 LST, and there are slight differences among different regions. In addition, the diurnal variation amplitudes of Cu and Cb are larger, indicating that they both have significant diurnal variations.

The diurnal variations of clouds vary with both the latitude and the longitude. There is a slight difference in the time when the occurrence frequency of the TC reaches its peak value at different latitudes, while there is a difference of 1–2 h with a change in the latitude for both Ci and Ac. The peak times of Ns, Sc, Cu and Cb do not change with the latitude. In addition, the peak appearance time of the occurrence frequency of Ns in the Yangtze River valley lags by 1–2 h from west to east.

The diurnal features of clouds change with the seasons. All of the types of clouds show the largest diurnal amplitudes in the summer and autumn, and the time when the TC and Ci reach their maximum values tend to lag by 1–2 h from the spring to the summer. Ac and Sc present bimodal patterns with early morning peaks that appear earlier in the summer, while their late afternoon peaks appear later in the summer than in the other seasons. The maximum peak appearance times of Ns and Cu do not change with the seasons.

The diurnal variations of precipitation among the different regions are dominated by convective or stratiform clouds. The upper reaches of the Yangtze River valley present a midnight precipitation maximum that is mainly dominated by Cb. Both precipitation and Cb reach a peak around midnight with a significant positive correlation coefficient of 0.90. In other regions, the precipitation usually peaks in the early morning or in the afternoon, and there is a notable positive correlation between the early morning peak of precipitation and stratiform clouds, while the afternoon peak of precipitation shows positive relationship with convective clouds. The diurnal peak in the Yangtze River valley and the region between the Yangtze River valley and the Yellow River valley is mainly dominated by stratiform clouds, while it is mainly determined by convective clouds in South and Northeast China. The peak appearing in the afternoon over the Yangtze River valley is also dominated by convective clouds.

Nesbitt and Zipser suggested that the nocturnal rain is often caused by meso-scale convective systems (MCS) rather than isolated convection, and the MCSs are the strongest systems after midnight [24]. Lin et al. indicated that the nocturnal maximum rainfall is enhanced by instability due to nocturnal radiative cooling at cloud top [25]. This explains that the midnight precipitation maximum is mainly dominated by Cb in the upper reaches of the Yangtze River valley. The afternoon peaks of convective clouds and precipitation can be explained by the highest surface air temperature and the strongest instability in the lower troposphere occurring in the afternoon, which are beneficial to the trigger of the convection [22].

The study points out the relationship of diurnal variations of clouds and precipitation among different regions for the first time. However, the research on the physical mechanism of the diurnal variations of clouds and precipitation is not deep. We will study the diurnal variation of cloud microphysical characteristics, and analyze the influence of terrain, atmospheric stability and atmospheric circulation on the diurnal variation of clouds and precipitation in the next step. This is helpful for understanding the mechanism of clouds and precipitation, and for improving the ability to simulate weather and climate models.

Author Contributions: Conceptualization, C.G.; Data curation, C.G. and H.C.; Formal analysis, C.G.; Funding acquisition, Y.L.; Project administration, Y.L.; Validation, H.C.; Visualization, C.G. and H.C.; Writing – original draft, C.G.; Writing – review and editing, Y.L.

References

1. Rutledge, S.A.; Houze, R.A. A diagnostic modeling study of the trailing stratiform region of a midlatitude squall line. *J. Atmos. Sci.* **1987**, *44*, 2640–2656. [CrossRef]

2. Bao, S.H.; Letu, H.; Zhao, C.F.; Tana, G.; Shang, H.; Wang, T.; Lige, B.; Bao, Y.; Purevjav, G.; He, J.; et al. Spatiotemporal distributions of cloud parameters and the temperature response over the Mongolian Plateau during 2006-2015 based on MODIS data. *IEEE J. Sel. Top. Appl. Earth Obs. Remote Sens.* **2018**, *12*, 549–558. [CrossRef]

3. Wallace, J.M. Diurnal variations in precipitation and thunderstorm frequency over the conterminous United States. *Mon. Weather Rev.* **1975**, *103*, 406–419. [CrossRef]

4. Higgins, R.W.; Janowiak, J.E.; Yao, Y.P. A gridded hourly precipitation data base for the United States (1963–1993). *NCEP Clim. Predict. Center ATLAS* **1996**, *1*, 47.

5. Dai, A.G. Global precipitation and thunderstorm frequencies. Part II: Diurnal variations. *J. Climate* **2010**, *14*, 1112–1128. [CrossRef]

6. Sorooshian, S.; Gao, X.; Maddox, R.A.; Hong, Y.; Imam, B. Diurnal variability of tropical precipitation retrieved from combined GOES and TRMM satellite information. *J. Clim.* **2002**, *15*, 983–1001. [CrossRef]

7. Liang, X.Z.; Li, L.; Dai, A.G. Regional climate model simulation of summer precipitation diurnal cycle over the United States. *Geophys. Res. Lett.* **2004**, *31*, L24208. [CrossRef]

8. Lolli, S.; Di Girolamo, P.; Demoz, B.; Li, X.; Welton, E.J. Rain evaporation rate estimates from dual-wavelength lidar measurements and intercomparison against a model analytical solution. *J. Atmos. Ocean. Technol.* **2017**, *34*, 829–839. [CrossRef]

9. Lolli, S.; D'Adderio, L.; Campbell, J.; Sicard, M.; Welton, E.; Binci, A.; Rea, A.; Tokay, A.; Comeron, A.; Barragan, R.; et al. Vertically resolved precipitation intensity retrieved through a synergy between the ground-based NASA MPLNET lidar network measurements, surface disdrometer datasets and an analytical model solution. *Remote Sens.* **2018**, *10*, 1102. [CrossRef]

10. Yu, R.C.; Zhou, T.J.; Xiong, A.Y.; Zhu, Y.J.; Li, J.M. Diurnal variations of summer precipitation over contiguous China. *Geophys. Res. Lett.* **2007**, *34*, L01704. [CrossRef]

11. Zhou, T.J.; Yu, R.C.; Chen, H.M. Summer precipitation frequency, intensity, and diurnal cycle over China: A comparison of satellite data with rain gauge observations. *J. Climate* **2008**, *21*, 3997–4010. [CrossRef]

12. Li, J.; Yu, R.C.; Wang, J.J. Seasonal variation of the diurnal cycle of rainfall in the southern contiguous China. *J. Climate* **2008**, *21*, 6036–6043. [CrossRef]

13. Yuan, W.H.; Yu, R.C.; Chen, H.M.; Li, J.; Zhang, M.H. Subseasonal characteristics of diurnal variation in summer monsoon rainfall over central eastern China. *J. Climate* **2010**, *23*, 6684–6695. [CrossRef]

14. Gray, W.M.; Jacobson, R.W., Jr. Diurnal variation of deep cumulus convection. *Mon. Weather Rev.* **1977**, *105*, 1171–1188. [CrossRef]

15. Zheng, Y.G.; Chen, J. A climatology of deep convection over south China and adjacent seas during summer. *J. Trop. Meteor.* **2011**, *27*, 495–508.

16. Chen, H.M.; Yu, R.C.; Wu, B.Y. FY-2C derived diurnal features of clouds in the southern contiguous China. *J. Geophys. Res.* **2012**, *117*, D18101. [CrossRef]

17. Fujinami, H.; Nomura, S.; Yasunari, T. Characteristics of diurnal variations in convection and precipitation over the southern Tibetan Plateau during summer. *SOLA* **2005**, *1*, 49–52. [CrossRef]

18. Zhao, C.F.; Chen, Y.Y.; Li, J.M.; Letu, H.; Su, Y.; Chen, T.; Wu, X. Fifteen-year statistical analysis of cloud characteristics over China using Terra and Aqua moderate resolution imaging spectroradiometer observations. *Int. J. Climatol.* **2018**, *39*, 2612–2629. [CrossRef]

19. Gao, C.C.; Fang, L.X.; Li, Y.Y.; Kou, X.W. Cloud occurrence frequency, duration time and accompanying rainfall probability in China during 1985–2011. *Torrential Rain Disaster* **2015**, *34*, 206–214.

20. Shen, Y.; Zhao, P.; Pan, Y.; Yu, J.J. A high spatiotemporal gauge-satellite merged precipitation analysis over China. *J. Geophys. Res.* **2014**, *119*, 3063–3075. [CrossRef]

21. Yao, S.B.; Jiang, D.B.; Fan, G.Z. Seasonality of precipitation over China. *Chin. J. Atmos. Sci.* **2017**, *41*, 1191–1203. [CrossRef]

22. Yu, R.C.; Xu, Y.P.; Zhou, T.J.; Li, J. Relation between rainfall duration and diurnal variation in the warm season precipitation over central eastern China. *Geophys. Res. Lett.* **2007**, *34*, L13703. [CrossRef]

23. Shen, Y.; Pan, Y.; Yu, J.J.; Zhao, P.; Zhou, Z.J. Quality assessment of hourly merged precipitation product over China. *Trans. Atmos. Sci.* **2013**, *36*, 37–46.

24. Nesbitt, S.W.; Zipser, E.J. The diurnal cycle of rainfall and convective intensity according to three years of TRMM measurements. *J. Climate* **2003**, *16*, 1456–1475. [CrossRef]

25. Lin, X.; Randall, D.A.; Fowler, L.D. Diurnal variability of the hydrologic cycle and radiative fluxes: Comparisons between observations and a GCM. *J. Climate* **2000**, *13*, 4159–4179. [CrossRef]

13

Diurnal Variations in Surface Wind over the Tibetan Plateau

Yufei Zhao [1],*, Jianping Li [2,3], Qiang Zhang [1], Xiaowei Jiang [1] and Aixia Feng [1]

[1]   National Meteorological Information Centre, China Meteorological Administration, Beijing 100081, China; zhangq@cma.gov.cn (Q.Z.); jiangxw@cma.gov.cn (X.J.); fengax@cma.gov.cn (A.F.)

[2]   College of Global Change and Earth System Science, Beijing Normal University, Beijing 100875, China; ljp@bnu.edu.cn

[3]   Laboratory for Regional Oceanography and Numerical Modeling, Qingdao National Laboratory for Marine Science and Technology, Qingdao 266237, China

*   Correspondence: zhaoyf@cma.gov.cn

Abstract: This study uses hourly surface wind direction and wind speed observations from 53 meteorological stations on the Tibetan Plateau (TP) (70–105° E, 25–45° N) between 1995 and 2017 to investigate diurnal variations in the surface wind. The results show large diurnal variations in surface wind on the TP. The minimum wind speed occurs in the morning and the maximum in the afternoon. In all four seasons, the prevailing meridional wind is a southerly, and this is typically evident for more than two-thirds of each day. However, in the mornings during December–February and September–November, this southerly wind is replaced by a northerly, but remains southerly in the afternoon. The TP shows remarkable regional characteristics with respect to diurnal variations in wind speed. In the eastern region, the minimum and maximum daily wind speeds occur about 1 h later than in the west. Among the 53 meteorological stations, 79% observed that it took less time for the minimum speed to rise to the maximum speed than for the maximum to drop to the minimum. The blocking effect of the high surrounding terrain causes the diurnal variations seen in the surface winds at the three stations in the Qaidam Basin to differ significantly from those observed at the other stations elsewhere on the plateau. These Qaidam Basin stations recorded their maximum wind speeds around noon, with the minimum at dusk, which is around 1900 LST. The EOF1 (EOF = empirical orthogonal function) of the hourly wind speed on the TP indicates the key daily circulation feature of the region; i.e., the wind speed is high in the afternoon and low in the morning. The EOF2 reflects the regional differences in the diurnal variations of wind speed on the TP; i.e., the eastern region reaches the daily maximum and minimum wind speeds slightly later than the western region.

Keywords: Tibetan Plateau; surface wind; diurnal variation

1. Introduction

Research into diurnal variations in meteorological parameters is important in improving our understanding of weather and climate systems. There has been much research into diurnal variations in global or local meteorological parameters (e.g., precipitation and surface winds). However, the limited amount of available observational data meant that early studies of these diurnal variations focused mainly on tropical areas. Infrared satellite data have been used to investigate the diurnal variations in convective activity using cloud cover, cloud crest brightness temperature, and water vapor (e.g., [1–4]). Studies in tropical regions have shown that the daily cycle of convective activity is strongly dependent on geographical location, and that topography plays a crucial role in the daily cycle of weather patterns [1,5].

The Tibetan Plateau (TP), also known as the third pole, is the world's largest landform and has an average elevation of about 4500 m. The TP has a significant effect on its surroundings through thermal and dynamic processes [6,7]. Wind is an important indicator of atmospheric circulation, and changes in wind speed are an indication of circulation changes caused by natural or anthropogenic processes [8]. The wind is known to be distinctively turbulent and non-stationary. As a consequence, the wind velocity varies rather randomly on many different time scales [9]. Previous studies have examined the surface wind regime on the TP. For example, analysis of daily wind speed data from ground observation stations on the TP has shown seasonal differences in the surface wind, with the wind speed dropping most significantly during March–May [10]. Other research has found that changing wind speed is the most important meteorological control on trends in potential evapotranspiration on the TP [11].

There are limited observational data to study diurnal variations in meteorological elements over the TP. This has led to some studies having to rely on data with a low temporal resolution, or data obtained from indirect observations, to analyze such variations. For example, data with a temporal resolution of 3 h have been used to analyze diurnal variations in precipitation, thunderstorms [12], and surface winds [13]. Other studies have used satellite and radar data to explore the changing weather and climate on the TP (e.g., [14–17]).

Following the rapid deployment of automatic weather stations across China in recent years, hourly observational data are now available. However, these automatic weather stations are typically about 10 years old, so their records are short. Before the establishment of automatic weather stations, various self-recording instruments were widely used for hourly and even minute-by-minute observations, such as self-recording rain gauges and self-registering anemometers, but most of these datasets were recorded on paper, which is more difficult to collate and analyze than digital data, thereby restricting the application of these valuable data. In 2017, the China Meteorological Administration (CMA), after nearly 10 years of data processing work that included integrating the surface wind data observed by automatic weather stations, established the first hourly wind series from 2400 Chinese stations covering the period since 1951. The CMA also developed suitable quality control procedures, based on the characteristics of hourly wind data, to check the accuracy and completeness of the hour-by-hour wind series, and so established the Hourly Surface Wind dataset (HSW dataset) for mainland China [18].

The reliability of various reanalysis data is relatively low (e.g., [19,20]). Observational data recorded at meteorological stations are fundamental to the data processing and analysis that underpins climate research. The direction and speed of surface wind are major elements of China's meteorological observational dataset. However, few studies have used the long sequence of hourly wind data observed by the stations to analyze diurnal variations in surface winds on the TP. Accordingly, this study uses hourly wind direction and speed data from the TP in the HSW dataset to analyze diurnal variations in surface winds on the TP.

2. Data and Methodology

We used the HSW dataset from mainland China, which was developed, collated, and quality controlled by the National Meteorological Information Center (NMIC) of the CMA (Zhao et al., 2017 [18]). In this study, the TP region is defined as that bounded by 70–105° E and 25–45° N. Our study area in west China encompasses Qinghai Province, Tibet, and parts of neighboring Xinjiang, Gansu, Yunnan, and Sichuan provinces. To avoid biases introduced by missing data, we limited our analysis to the 53 stations on the TP that provide a complete hourly wind speed series from 1995 to 2017 and are located at altitudes above 2000 m (figure omitted).

In China, the standard time (Beijing time) is UTC + 8, and this is the time used for meteorological observations. However, because of its vast size, China is divided into five time zones, and the TP alone spans three time zones (UTC + 5 to UTC + 7). Therefore, using LST to analyze diurnal variations on the TP avoids the difference in diurnal variations in wind speed caused by the gap between LST and Beijing time. In the diurnal cycle of surface wind speed at each station, the hour when the wind speed reached its maximum is defined as the 'max-hour', and the hour when the wind speed reached its minimum is defined as the 'min-hour'. The difference between the maximum and minimum is the daily range of wind speed. The 'max-dir' is the wind direction in the hour when the wind speed reached its daily maximum, and the 'min-dir' is the wind direction in the hour when the wind speed reached its daily minimum. For our analysis, we divided the year into four equal periods: December–February (DJF), March–May (MAM), June–August (JJA), and September–November (SON).

3. Diurnal Variations in Wind Speed

Figure 1 shows the seasonal and annual mean wind speed, variances of wind speed, zonal wind speed, and meridional wind speed on the TP. The maximum wind speed occurs during MAM, with the average wind speeds at all hours being higher than those of the other three seasons. MAM also sees the largest range in wind speed of up to 2.7 m·s^{-1}, whereas JJA sees the smallest range of 1.8 m·s^{-1}. The minimum wind speeds on the TP in DJF, MAM, JJA, and SON were recorded at 0800, 0600, 0600, and 0600 LST, respectively, and the maximum wind speeds at 1500, 1600, 1600, and 1500 LST, respectively. It is suggested that increased downward turbulent mixing of momentum during the day could be one of the main causes for the early afternoon maximum of surface wind speed [13]. In DJF, the wind speed takes the shortest time to rise from the daily minimum to the daily maximum, of up to 7 h, meaning it takes 17 h to drop from the maximum to the minimum. In MAM and JJA, it takes 10 h for the wind speed to rise from the lowest in the morning to the highest in the afternoon, compared with 9 h in SON. Previous research has shown that in eastern China, wind speed drops to its minimum at 0500 LST and rises to its maximum at 1500 LST each day [21], with a variation of 1.2 m·s^{-1}. On the TP, however, the wind speed reaches its minimum at 0600 LST and maximum at 1500 LST, with an annual average diurnal variation of 2.2 m·s^{-1}. Therefore, the minimum wind speed in the daily cycle of wind speed on the TP arrives 1 h, on average, later than in the eastern part of China, whereas the maximum wind speed arrives at the same time. The amplitude of the variation in the former is nearly twice that in the latter.

The change in wind speed is large when the wind speed is high (Figure 1a,b). In addition, although in the daily cycle of wind speed the hourly wind speeds during SON are lower than those during MAM, the changes in SON wind speed are larger than those in MAM between 1200 and 1500 LST. The maximum hourly wind speed variation during MAM and SON arrives 1 h earlier than the maximum hourly wind speeds.

In the zonal wind diagram (Figure 1c), the prevailing wind alternates between an easterly and a westerly in JJA and SON. In most cases, a westerly wind prevails in JJA, and the westerly component reaches a maximum at 2000 LST of only up to 0.52 m·s^{-1} which is much smaller than the maxima in other seasons. The zonal wind speed is compensated in averaging on multiple station data, not weakened at each station (figure omitted). The effect of large terrain on airflow movement is mainly manifested in blocking and diverting. On the one hand, these effects of mountain ranges on airflow will lead to asymmetric distribution of surface pressure on windward and leeward slopes, which usually results in a gradient of pressure from the windward slope to the leeward slope. On the other hand, the southwest monsoon from the Indian Ocean can affect southeastern Tibet and southwestern Sichuan province in JJA. Therefore, the zonal wind at some of the 53 stations is westerly in JJA (Figure 1), and the weaker westerly wind will appear in most hours of a day by being compensated in averaging on multiple stations which are not affected by the southwest wind. An east wind prevails during SON, with the zonal wind speeds all above 0.5 m·s^{-1} between 1200 and 1700 LST, reaching a maximum at 1400 LST of up to 1.1 m·s^{-1}. During DJF and MAM, the average zonal wind for all hours on the TP is an easterly. During DJF, the zonal wind speeds between 1200 and 1700 LST exceed 1.0 m·s^{-1}, reaching a maximum at 1500 LST of up to 2.1 m·s^{-1}. During MAM, the zonal wind speeds between 1100 and 1700 LST exceed 1.0 m·s^{-1}, reaching a maximum at 1400 LST of up to 1.6 m·s^{-1}.

The diurnal variations in the average meridional wind speed during each season and the annual mean can be divided into two groups (Figure 1d): 2000–0800 and 0800–2000 LST. Overall, the prevailing meridional wind in each season is a southerly. To be specific, a southerly and northerly wind alternate during DJF and SON, with the southerly being replaced by a northerly at about 0800–0900 LST, returning to a southerly at 1500–1600 LST. Over the 24 h of each day, the north wind prevails for about 7 h in DJF and SON, while the south wind prevails for the remaining 17 h. During MAM and JJA, the south wind is replaced by a weak northerly for only the 2–3 h before noon, and the southerly prevails for the remaining 21–22 h of each day.

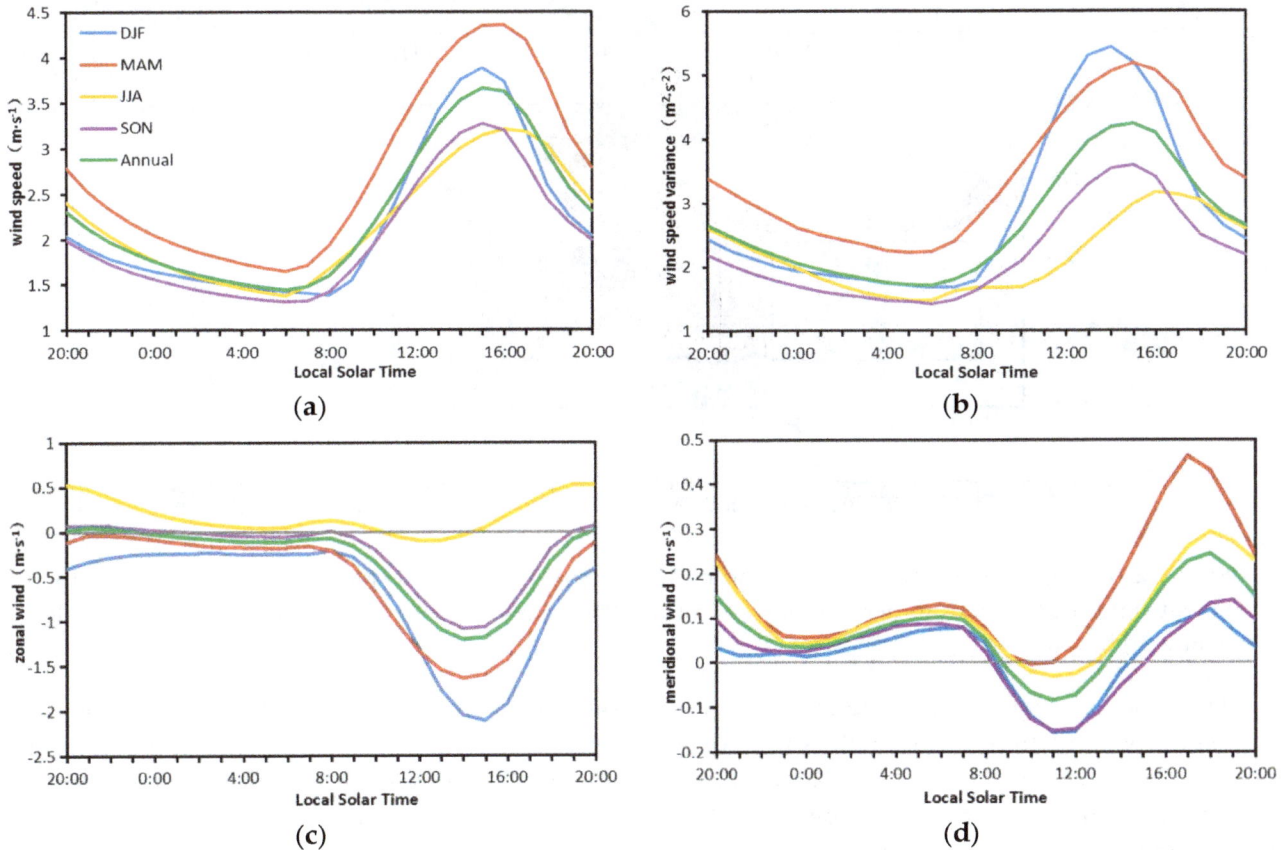

Figure 1. Diurnal variations in seasonal and annual mean wind speed (**a**), variances of wind speed (**b**), zonal wind speed (**c**), and meridional wind speed (**d**) averaged over the TP.

The max-hour and min-hour in the average diurnal wind speed cycle for each station on the TP are shown in Figure 2. The eastern region typically experiences the maximum wind speed about 1 h later than the western region. Among the 53 stations, 10, 15, and 15 stations reach their peak values at 1400, 1500, and 1600 LST, respectively, indicating that most stations (57%) reach the maximum wind speed between 1500 and 1600 LST each day. In contrast to most other stations, Xiaozaohuo (93.2° E, 36.9° N) in Qinghai Province reaches its peak wind speed at 0900 LST, the earliest among the stations, Nuomuhong (96.5° E, 36.4° N) in Qinghai Province reaches its peak at 1100 LST, while Huajialing (105.0° E, 35.4° N) in Gansu Province and Jianzha (102.0° E, 35.9° N) in Qinghai Province reach their peaks at 2100 and 1900 LST, respectively.

Similarly, according to Figure 2b, the eastern region attains a daily minimum wind speed about 1 h later than the western region. Thirty six stations (68% of the total) reach the daily valley at 0600–0700 LST. The Xiaozaohuo, Delingha (97.4° E, 37.4° N), and Nuomuhong stations (all in Qinghai Province) differ from the other stations, reaching their daily minimum wind speed at 1900 LST. Anduo (91.1° E, 32.4° N) in Tibet and Tianjun (99.0° E, 37.3° N) in Qinghai Province reach their daily minimum wind speed at 0100 LST.

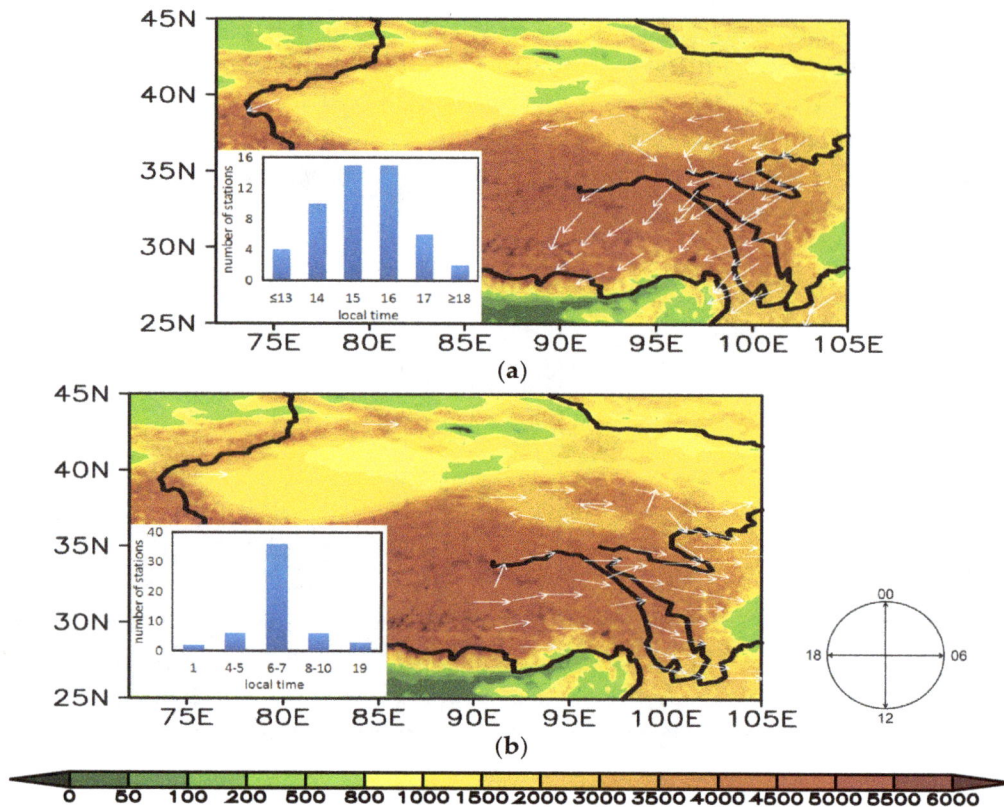

Figure 2. Timing of daily maximum (**a**) and minimum (**b**) surface wind speeds (vectors according to key at right, LST). Shading denotes elevation (m). The insets show histograms of the peak (**a**) and valley (**b**) hours.

The time interval between the min-hour and max-hour for the stations on the TP varies between 6 h (3 stations) and 19 h (2 stations). A total of 30 stations take 8–10 h to see the wind speed rise from the minimum to the maximum, and 42 stations need less than 12 h (Table 1). That is, 79% of the stations take a shorter time to see the wind speed rise from the minimum to the maximum than to see the wind speed drop from the maximum to the minimum. It takes 12 h for three stations to see their wind speed rise from the lowest to the highest, so 6% of the stations take an equal time to see the wind speed go from the lowest to the highest and from the highest to the lowest. A total of eight stations take more than 12 h to see their wind speed rise from the lowest to the highest, so 15% of the stations take longer to see their wind speed rise from the lowest to the highest than to see the wind speed drop from the highest to the lowest.

Table 1. Frequency distribution of the time (hours) required from the lowest to the highest daily wind speeds among the stations.

Number of hours	6	7	8	9	10	11	12	13	14	15	16	19
Number of stations	2	5	9	14	11	5	2	1	1	1	1	1
Percentage (%)	3.8	9.4	17.0	26.4	20.8	9.4	3.8	1.9	1.9	1.9	1.9	1.9

4. Diurnal Variations in Surface Wind Directions and Diurnal Range of Wind Speed

Figure 3 and Table 2 show that the peak wind directions at each station differ, being northeasterly, southeasterly, northwesterly, and min-hour southwesterly for 24, 22, 4, and 3 stations, respectively. For 87% of the stations, the max-dir is easterly and for 13% it is westerly (Figure 3). For 55% of the stations, the min-dir is easterly. In the min-hour, 10, 19, 15, and 9 stations recorded northeasterly, southeasterly, northwesterly, and southwesterly winds, respectively. For 74% of the stations, the average wind direction is easterly, and 17, 22, 6, and 8 stations have an average wind direction of northeast, southeast, northwest, and southwest, respectively.

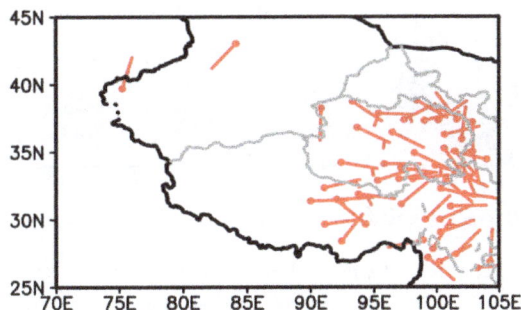

Figure 3. Surface wind direction and speed at the max-hour for each station.

Next, we consider the daily range of wind speed at the stations. Figure 4a,b shows the daily range of annual average wind speed and wind speed variance for all stations. The variation in hourly wind speed is greater in the areas with a larger daily range. Tuotuohe (92.6° E, 34.0° N) station in Qinghai Province has the largest daily range of wind speed, reaching 4.2 m·s^{-1}, and Delingha station in Qinghai Province in the northerly part of the TP has the smallest daily range of wind speed, at about 0.7 m·s^{-1}. The average daily range of wind speed across the TP is 2.4 m·s^{-1}. The daily range of wind speed for all stations on the TP varies significantly from one season to another (Figure 4c–f). The high-value zone with a daily range over 3.6 m·s^{-1} moves northwestward from DJF to MAM. From MAM to JJA, the high-value zone continues to move northwestward. The spatial distribution of the wind speed daily range in SON is similar to that in DJF, but with a smaller daily range.

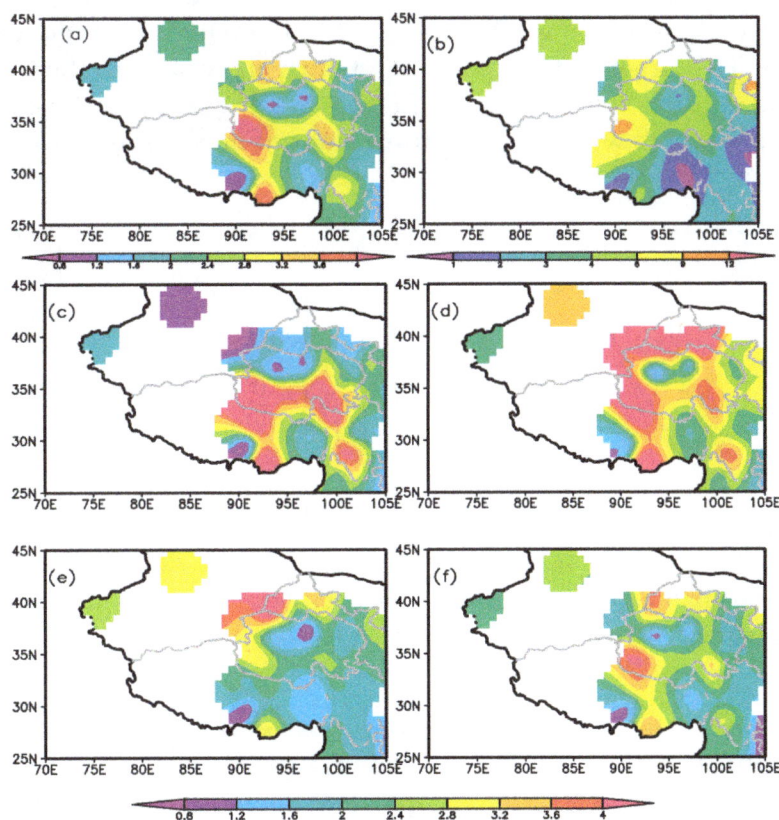

Figure 4. Daily range of annual average wind speed (**a**) (m·s^{-1}), hourly wind speed variance (**b**), and daily range of wind speed (m·s^{-1}) in December–February (DJF) (**c**), March–May (MAM) (**d**), June–August (JJA) (**e**), and September–November (SON) (**f**).

Table 2. The numbers of stations which recorded northeasterly, southeasterly, northwesterly, and southwesterly winds in max-dir, min-dir, and average wind direction.

Wind Direction	The Numbers of Stations		
	Max-Dir	Min-Dir	Average Wind Direction
Northeast	24	10	17
Southeast	22	19	22
Northwest	4	15	6
Southwest	3	9	8

5. Diurnal Cycle of Surface Winds in Different Zones

As mentioned above, the diurnal cycles of surface winds recorded in Xiaozaohuo, Delingha, and Nuomuhong are strikingly different from the other zones of the TP, and these three stations are all located in the Qaidam Basin. In this section, we analyze these diurnal variations by treating the three stations as a 'basin area' and the remaining 50 stations as a 'plateau area'. Figure 5 shows the annual average diurnal wind speed variations for the basin area and the plateau area. The diurnal wind speed variations of the two areas are obviously different. The wind speed in the basin area features a broad peak that reaches a maximum either side of noon and with a minimum at 1900 LST. In contrast, the plateau area sees a maximum wind speed at 1500 LST and a minimum wind speed at 0600 LST. At this time (0600 LST), the wind speed in the basin area is not at its minimum, but shows a small reduction compared with the wind speeds in the adjacent periods. Yu et al. (2009) [21] found a similar reversed day–night phase in the surface wind speed between mountain regions and plain regions of China. Based on a single year of wind data from television towers, Crawford and Hudson (1973) [22] concluded that the wind speed of the lower (higher) layers reached a minimum around midnight (noon) and a maximum in the afternoon (midnight). Their research target was diurnal wind speed variations on large plains and high mountain stations. In the present study, most of the stations from which data were collected on the TP are located in large high-altitude terrain, other than the stations at Xiaozaohuo, Delingha, and Nuomuhong, which are located in the Qaidam Basin at lower altitudes than the surrounding stations. Consequently, we suggest that the clear differences in the diurnal variations of the surface winds at these three stations are caused by the blocking effect of the surrounding mountainous terrain.

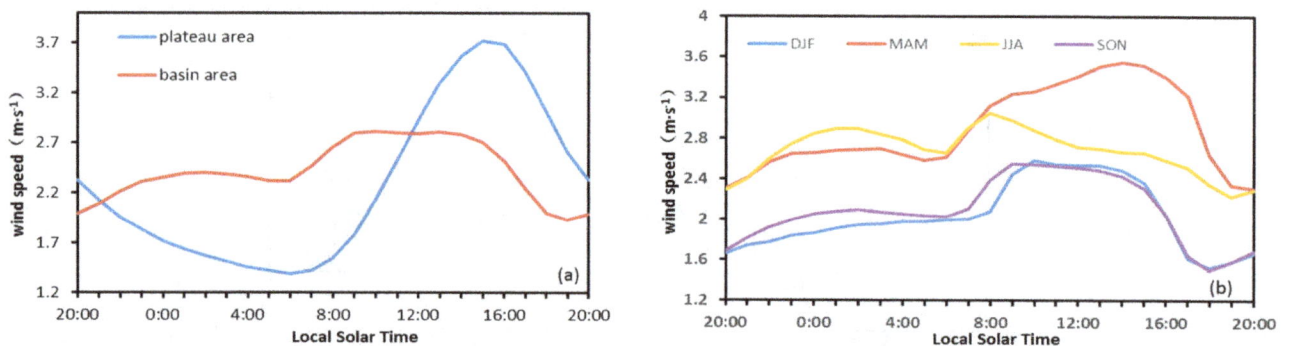

Figure 5. (**a**) Diurnal variations in annual mean wind speed averaged over the plateau area (blue line) and the basin area (red line). (**b**) Diurnal variations in seasonal mean wind speed averaged over the basin area.

In MAM, DJF, and SON the Qaidam Basin shows similar diurnal variations in wind speed to the annual mean (Figure 5b). The difference is that the wind speeds in MAM (SON and DJF) are higher (lower) than the annual mean wind speed. The diurnal variation in surface wind speed during JJA is different to that in the other three seasons. In JJA, the wind speed gradually decreases from 0800 to 1900 LST and reaches a daily minimum at 1900 LST. In the other three seasons, however, the diurnal variations in wind speed feature a gradual rise (in MAM) or continuous high speeds (in SON and DJF) from 0800 LST to about 1400 LST. In JJA afternoons on the TP the prevailing wind is a southwesterly.

In the afternoons of the other seasons, however, the prevailing wind is a southeasterly (Figure 1c,d). As the TP is higher in the northwest and lower in the southeast, and the Qaidam Basin is aligned northwest–southeast, the blocking effect of the surrounding terrain on the Qaidam Basin is more pronounced in JJA, when a southwest wind prevails, than in the other seasons when a southeasterly prevails. In other words, because the dominant wind direction is aligned with the basin except in JJA, the blocking effect would not play a major role in DJF, MAM, and SON like JJA.

We applied empirical orthogonal function (EOF) analysis to the hourly wind speed data of the TP (51 stations, excluding two remote stations in Xinjiang Province). The hourly wind speed series from each station was standardized prior to the EOF decomposition. The first and second leading EOF (EOF1 and EOF2) passed the North test [23]. EOF1 and EOF2 account for 83% and 13% of the total variance, respectively. EOF1 shows mainly positive values, with weakly negatives in the Qaidam Basin (Figure 6a). The time coefficient curve of EOF1 (Figure 6c) is similar to the series in Figure 1a, with both featuring a single peak. The peak in diurnal wind speed occurs between 1400 and 1500 LST, and the wind speed at night tends to decrease slowly. EOF1 reflects mainly the diurnal variation of wind speed in most areas of the TP, and the diurnal variations that differ between the Qaidam Basin and the much larger plateau area. EOF2 (Figure 6b) varies between positive and negative values with an east–west dipole pattern, while the corresponding time coefficient features a single peak (Figure 6d). EOF2 is negative in the Qilian Mountains in Gansu Province, eastern Qinghai Province, western Sichuan Province, and southeast Tibet (the eastern region), and is positive in mid-west Qinghai Province, eastern Sichuan Province, and central Tibet (the western region). This suggests that EOF2 indicates that the eastern region reaches the peak and valley of diurnal wind speed slightly later than the western region, which is consistent with the conclusion drawn above. The branching effect of the TP on airflow makes the surface wind of the eastern edge of the TP different from that of the western part of the TP. The northerly airflow generated by the circulation of the plateau forms a "leeward wake zone" on the eastern edge of the plateau, and the anticyclone vortex is very strong, while westerly winds prevail in the western part of the plateau [24]. The difference of atmospheric circulation pattern will inevitably lead to the difference of surface wind diurnal variation between the eastern and western regions.

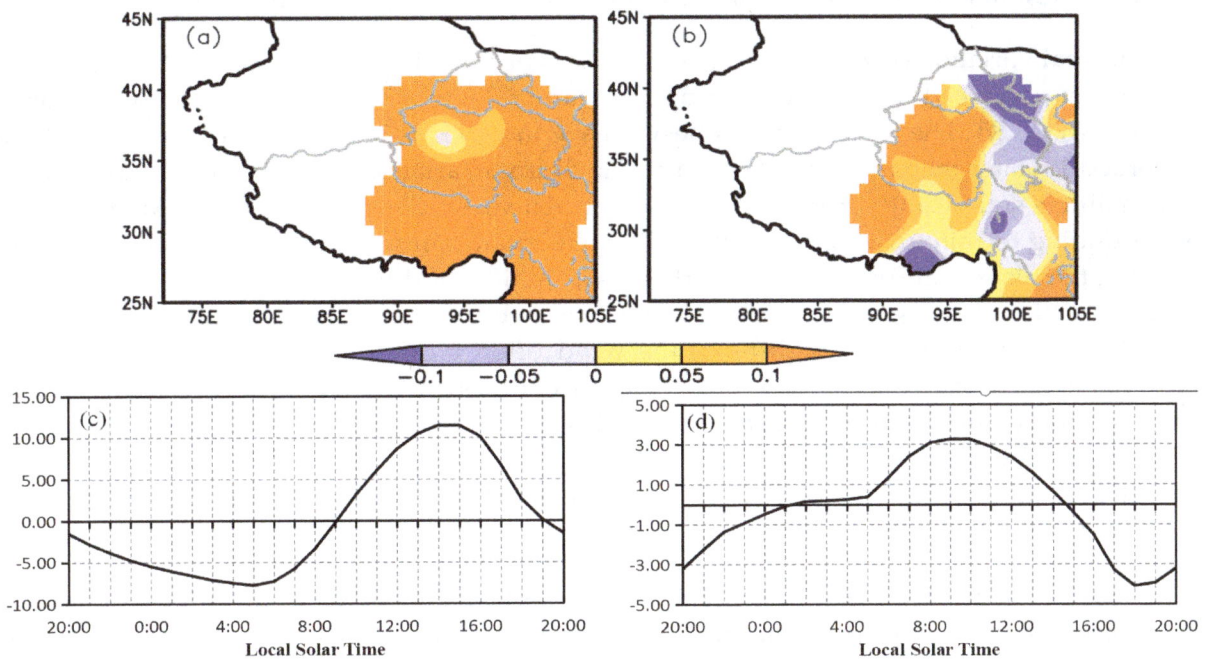

Figure 6. (a) The first leading empirical orthogonal function (EOF1) pattern of climatological hourly wind speeds on the Tibetan Plateau (TP). (b) As for (a), but for the second leading empirical orthogonal function (EOF2). (c) Time series corresponding to EOF1. (d) Time series corresponding to EOF2.

6. Summary and Discussion

The diurnal cycle of surface wind speed on the TP was analyzed using hourly wind observation data from 53 meteorological stations with 23 years (1995–2017) of complete data. Our results reveal some novel spatial and temporal characteristics. The main conclusions are summarized as follows.

(1)   The surface wind speed shows large diurnal variation. MAM has the highest wind speed and largest daily range of wind speed. On the TP, the minimum wind speed occurs in the early morning at 0600 LST and the maximum in the afternoon at 1500 LST. It is understandable that during daytime, as a response to surface solar heating, the downward vertical turbulent transport of momentum reaches its strongest in the afternoon. During night time, as a response to nocturnal cooling in the boundary layer, the eddy viscosity is reduced and less momentum is transported to the lower level. The surface wind slows down gradually due to the surface friction [21]. When the wind speed is high, the variation in wind speed is also large. On the TP the daily minimum wind speed arrives 1 h later than that recorded in eastern China by Yu et al. (2009) [21]. The two regions analyzed see the peak hourly wind speed in the same hour. The average daily range of wind speed on the TP is 2.4 m·s^{-1}, nearly twice that in eastern China.

(2)   The surface winds were decomposed into meridional and zonal winds, and their diurnal variation analyzed. The south wind dominated the meridional winds in each season, accounting for more than two-thirds of each day. In the mornings during DJF and SON, the south wind changes to a north wind, changing back to a south wind in the afternoon. In JJA and SON, east and west winds alternate as the prevailing wind direction. During DJF and MAM, the average wind direction on the TP is easterly throughout the day.

(3)   The seasonal variation in the daily wind speed range of each station on the TP is well defined. The area of high diurnal range moved to the northwest with seasonal changes from DJF to MAM and from MAM to JJA. The hourly wind speed at 57% of the stations takes 8–10 h to develop from the lowest to the highest speed. At 79% of the stations the wind speed rises from the lowest to the highest in less time than it takes to change from the highest to the lowest.

(4)   The diurnal variation of wind speed over the TP is noticeable [13]. At 0600–0700 LST, 68% of the stations experienced their lowest wind speed, and 57% recorded their highest wind speeds at 1500 or 1600 LST. The lowest and highest daily wind speeds in the eastern part of the TP were recorded 1 h later than in the western part. The diurnal variation in surface winds in the Qaidam Basin differs significantly from that at the other plateau stations due to the blocking effect of the surrounding terrain. In addition to the hilly area, mountain-valley breeze can also occur on the edge of the plateau and the basin, and there is a significant diurnal variation in wind speed and wind direction. During the daytime, the wind often blows from the valley to the mountainside and the mountaintop. At nighttime, the wind often blows from the mountaintop and the mountainside to the valley. Based on decomposing the observed wind into the system wind and the mountain-valley breeze, researchers could carry out research on the diurnal variation of the surface wind and the mountain-valley breeze. This may contribute to analyzing the causes of diurnal variation of surface wind on the TP, however, more detailed mechanisms are to be further explored in future studies.

(5)   The EOF of the hourly wind speeds reflects the high wind speeds in the afternoon and lower wind speeds in the morning over most of the TP (except for the Qaidam Basin). It also reflects that the wind speed in the eastern part of the TP reaches a diurnal maximum later than that in the western part. It may be reasonable that EOF2 may be related to the different atmospheric circulation systems [25], topographic characteristics, snow cover and vegetation cover in the eastern and western parts of the TP.

Author Contributions: Conceptualization, Y.Z. and J.L.; Data curation, Y.Z. and Q.Z.; Formal analysis, Y.Z.; Funding acquisition, Y.Z.; Investigation, Q.Z.; Methodology, J.L.; Project administration, Q.Z. and X.J.; Resources, Y.Z.; Software, X.J. and A.F.; Supervision, J.L.; Writing—Original Draft, Y.Z. and X.J.; Writing—Review & Editing, Y.Z. and J.L.

Acknowledgments: This work was supported by the Chinese National Natural Science Foundation 'Development of Data Sharing Platform of the Tibetan Plateau's Multi-Source Land-Atmosphere System Information' under grant number 91637313 and the National Natural Science Foundation of China (NSFC) Project (41530424).

References

1. Yang, G.-Y.; Slingo, J. The diurnal cycle in the tropics. *Mon. Weather Rev.* **2001**, *129*, 784–801. [CrossRef]
2. Nesbitt, S.-W.; Zipser, E.-J. The diurnal cycle of rainfall and convective intensity according to three years of TRMM measurements. *J. Clim.* **2003**, *16*, 1456–1475. [CrossRef]
3. Liu, C.-H.; Moncrieff, M.-W. A numerical study of the diurnal cycle of tropical oceanic convection. *J. Atmos. Sci.* **1998**, *55*, 2329–2344. [CrossRef]
4. Bowman, K.-P.; Collier, J.-C.; North, G.-R.; Wu, Q.; Ha, E.; Hardin, J. Diurnal cycle of tropical precipitation in Tropical Rainfall Measuring Mission (TRMM) satellite and ocean buoy rain gauge data. *J. Geophys. Res.* **2005**, *110*, 1–14. [CrossRef]
5. Zuidema, P. Convective clouds over the Bay of Bengal. *Mon. Weather Rev.* **2003**, *131*, 780–798. [CrossRef]
6. Xiao, D.; Zhao, P.; Wang, Y.; Tian, Q.-H.; Zhou, X.-J. Millennial-scale phase relationship between North Atlantic deep-level temperature and Qinghai-Tibet Plateau temperature and its evolution since the Last Interglaciation. *Chin. Sci. Bull.* **2014**, *59*, 75–81. [CrossRef]
7. Nan, S.-L.; Zhao, P.; Chen, J.-M. Variability of summertime Tibetan tropospheric temperature and associated precipitation anomalies over the central-eastern Sahel. *Clim. Dyn.* **2018**. [CrossRef]
8. Guo, H.; Xu, M.; Hu, Q. Changes in near-surface wind speed in China: 1969–2005. *Int. J. Climatol.* **2011**, *31*, 349–358. [CrossRef]
9. Fu, Z.-T.; Li, Q.-L.; Yuan, N.-M.; Yao, Z.-H. Multi-scale entropy analysis of vertical wind-speed series in atmospheric boundary-layer. *Commun. Nonlinear Sci.* **2014**, *19*, 83–91. [CrossRef]
10. Guo, X.-Y.; Wang, L.; Tian, L.-D.; Li, X.-P. Elevation-dependent reductions in wind speed over and around the Tibetan Plateau. *Int. J. Climatol.* **2017**, *37*, 1117–1126. [CrossRef]
11. Chen, S.-B.; Liu, Y.-F.; Thomas, A. Climatic change on the Tibetan Plateau: Potential evapotranspiration trends from 1961–2000. *Clim. Chang.* **2006**, *76*, 291–319.
12. Dai, A.-G. Global precipitation and thunderstorm frequencies. Part II: Diurnal variations. *J. Clim.* **2001**, *14*, 1112–1128. [CrossRef]
13. Dai, A.-G.; Deser, C. Diurnal and semidiurnal variations in global surface wind and divergence fields. *J. Geophys. Res.* **1999**, *104*, 109–126. [CrossRef]
14. Murakami, M. Analysis of the deep convective activity over the Western Pacific and Southeast Asia. Part I: Diurnal variation. *J. Meteorol. Soc. Jpn.* **1983**, *61*, 60–76. [CrossRef]
15. Bhatt, B.-C.; Nakamura, K. Characteristics of monsoon rainfall around the Himalayas revealed by TRMM precipitation Radar. *Mon. Weather Rev.* **2005**, *133*, 149–165. [CrossRef]
16. Liu, L.-P.; Feng, J.-M.; Chu, R.-Z.; Zhou, J.-J. The diurnal variation of precipitation in Monsoon season in the Tibetan Plateau. *Adv. Atmos. Sci.* **2002**, *19*, 365–378.
17. Bai, A.-J.; Liu, C.-H.; Liu, X.-D. Diurnal variation of summer rainfall over the Tibetan Plateau and its neighboring regions revealed by TRMM Multisatellite Precipitation Analysis. *Chin. J. Geophys.* **2008**, *51*, 704–714. (In Chinese) [CrossRef]
18. Zhao, Y.-F.; Zhang, Q.; Yu, Y.; Yang, G. Development of hourly wind speed dataset in China and application on Qinghai Tibetan Plateau. *Plat. Meteorol.* **2017**, *36*, 930–938.
19. Zhang, W.-J.; Jin, F.-F.; Zhao, J.-X.; Li, J.-P. On the bias in simulated ENSO SSTA meridional width of CMIP3 models. *J. Clim.* **2013**, *26*, 3173–3186. [CrossRef]
20. He, W.-P.; Zhao, S.-S. Assessment of the quality of NCEP-2 and CFSR reanalysis daily temperature in China based on long-range correlation. *Clim. Dyn.* **2018**, *50*, 493–505. [CrossRef]

21. Yu, R.-C.; Li, J.; Chen, H.-M. Diurnal variation of surface wind over central eastern China. *Clim. Dyn.* **2009**, *33*, 1089–1097. [CrossRef]

22. Crawford, K.-C.; Hudson, H.-R. The diurnal wind variation in the lowest 1500 ft in Central Oklahoma: June 1966–May 1967. *J. Appl. Meteorol.* **1973**, *12*, 127–132. [CrossRef]

23. North, G.-R.; Bell, T.-L.; Cahalan, R.-F.; Moeng, F.-J. Sampling errors in the estimation of empirical orthogonal functions. *Mon. Weather Rev.* **1982**, *110*, 699–706. [CrossRef]

24. Qin, Q.-C.; Shen, X.-S. An estimate of surface pressure drag of the Tibetan Plateau and its characteristic analysis. *Acta. Meteorol. Sin.* **2015**, *73*, 93–109. (In Chinese)

25. Yang, K.; Wu, H.; Qin, J.; Lin, C.-G.; Tang, W.-J.; Chen, Y.-Y. Recent climate changes over the Tibetan Plateau and their impacts on energy and water cycle: A review. *Glob. Planet Chang.* **2014**, *112*, 79–91. [CrossRef]

14

Air Mass Trajectories to Estimate the "Most Likely" Areas to be Affected by the Release of Hazardous Materials in the Atmosphere

Miguel Ángel Hernández-Ceballos * and Luca De Felice

European Commission, Joint Research Centre, 21027 Ispra, Italy; Luca.DE-FELICE@ec.europa.eu
* Correspondence: miguelhceballos@gmail.com.

Abstract: Countries continuously review and improve their Emergency Preparedness and Response (EP&R) arrangements and capabilities to take agile and rapid actions with the intent of minimizing health, environmental and economic impacts of potential harmful releases into the atmosphere. One of the specific topics within the EP&R field is the estimation of the areas that might be affected. A proposal is presented to estimate the spatial distribution of the released material. The methodology combines the computation of air mass trajectories and the elaboration of density maps from the corresponding end-point positions. To this purpose, density maps are created in a three-way procedure; first, forward trajectories are calculated from a certain location and for a long period of time, e.g., a decade; second, the selected end-point positions are aggregated in a density field by applying the kernel density estimation method, and then the density field is visualized. The final product reports the areas with the longest residence time of air masses, and hence, the areas "most likely" to be affected and where the deposit may be substantial. The usefulness of this method is evaluated taking as reference a ten-year period (2007–2016) and against two different radioactive release scenarios, such as the Chernobyl accident and the Algeciras release. While far from being fully comprehensive, as only meteorological data are used, the performance of this method is reasonably efficient, and hence, it is a desirable alternative to estimating those areas potentially affected by a substantial deposit following the releases of a harmful material in the atmosphere.

Keywords: Nuclear Emergency Preparedness and Response; air mass trajectory; density maps

1. Introduction

In the context of the effective management of crises and disasters, which is a global challenge, the risk of an accident releasing airborne harmful material to the atmosphere, like radioactive, chemical or bacteriological active substances, cannot be ruled out. To face these kinds of events, it is well recognized that good preparedness substantially improves the emergency response. In this line, efforts to promote and enhance global safety, and to limit and mitigate any consequences associated with these kinds of releases are continuously triggered at national and international levels [1].

The early phase of a harmful release to the atmosphere is a period usually characterized by large uncertainty in the source term release characteristics and the lack of field measurements. Within this period, to have available a prior estimation of the possible atmospheric transport and dispersion of the released material, and hence, the likely areas to be affected by the plume, is of importance to decision makers and emergency managers in an attempt at minimizing its health and environmental impacts.

Meteorological conditions determine the dispersion and transport of substances in the atmosphere. Rainfall, wind speed, wind direction and temperature are relevant parameters in establishing their temporal and spatial variability. Once in the atmosphere, the primary process favoring the transport and dispersion of substances is wind regimes, which is characterized by wind direction and wind speed, e.g., the greater the wind velocity is, then the greater the dispersion of the contaminants, and the lower their concentration is [2]. Many studies [3,4] have outlined the need of characterizing wind regimes to understand air pollution scenarios. This direct link makes the knowledge and characterization of wind regimes at a site become key information in estimating the atmospheric transport and dispersion of harmful substances and, therefore, in forecasting the likely areas affected by the plume passage.

Air trajectory analysis is a central scientific tool to characterize synoptic and regional wind regimes [5]. The basic methodological approach in this kind of analysis is to generate massive amounts of air mass trajectories with the purpose of considering a large number of transport and dispersion scenarios so that the statistical analysis leads to the extraction of representative information about flow patterns (e.g., residence time, range, distance, speed, etc.). When compiled and studied over multiple years [6,7], the outcome of this trajectory analysis is used to generate an estimation of future wind scenarios.

With this in mind, the aim of this paper is to describe and to evaluate a method to estimate the likely areas to be affected by a hypothetical release to the atmosphere. The present methodology, which is based on the influence that lower atmosphere meteorology has on the dispersion and transport of substances in the atmosphere, is based on the calculation of density maps from air mass trajectories by applying the kernel density estimation (KDE) method [8]. KDE's strength is its ability to provide an estimate of density at any location in the spatial frame. In this present framework, these density maps report the areas with the longest residence time of air masses, and hence, the areas where the deposit may be substantial. Residence time analysis [9] identifies the likelihood that an air mass will traverse a given region in its movement to or from the site of interest over a given time period [7,10]. The longer the residence time over certain areas is, the more favorable the situation for significant surface dry deposition from contaminant plumes is.

Considering that the calculation of air mass trajectories is purely based on meteorological fields, i.e., we are not considering any release to the atmosphere (source term) in the calculation, we can only address a qualitative comparison to evaluate the outcomes of this method, e.g., between areas of high residence time and high deposit. To estimate the bias of this meteorological approach, we have evaluated it against the period 2006–2017 and two radioactive dispersion events, such as Chernobyl accident and Algeciras release. The corresponding density maps for each case study, which identify those areas with the longest residence time of air masses, are then qualitatively compared with those areas affected by high deposits in each release scenario. Reasonable estimation of the areas affected by both releases would prove the usefulness of this method and the number of years selected in the estimation of likely areas to be affected by a hypothetical release.

While far from being fully comprehensive of the complexity behind the simulation and prediction of atmospheric dispersion, as only meteorological data are used, this method would provide useful guidance in estimating the transport and dispersion that might result if harmful material reaches the atmosphere, and hence, in increasing knowledge of crisis and disaster management for improving responsiveness. Being only based on the analysis of the wind field, this method can be used, for instance, for any kind of airborne material.

The article is structured as followed. In Section 2, we describe the methodology applied to obtain density maps from air mass end-point positions, while Section 3 is dedicated to showing and discussing the results obtained in the two study cases taken as reference. To finalize, the conclusions obtained from this work are shown in Section 4.

2. Materials and Methods

2.1. Trajectory Modelling

Air mass trajectory shows the pathway of an infinitesimal air parcel through a centerline of an advected air mass having vertical and horizontal dispersion [11]. Forward trajectory estimates the pathway to be followed by an air parcel downwind from the selected coordinates in due course of time.

The National Oceanic and Atmospheric Administration (NOAA) Air Resources Laboratory's (ARL) Hybrid Single-Particle Lagrangian Integrated Trajectory model (HYSPLIT 4.9) [12] is used to calculate forward kinematic three-dimensional trajectories. The computation of three-dimensional trajectories is based on the use of the vertical velocity field included in publicly available meteorological datafiles, such as the 3-hourly meteorological archive data from NCEP's GDAS (National Weather Service's National Centers for Environmental Prediction-Global Data Assimilation System) [13]. The GDAS covers from 2004 to the present, which is a big advantage in favor of using them in research studies, as they span 10 years or more. The GDAS is run 4 times a day, i.e., at 0000, 0600, 1200, and 1800 UTC. These data are archived and made available by the NOAA's ARL as a global and 1-degree latitude–longitude dataset on pressure surface.

Our method uses forward trajectories with duration of 120 hours (length of the trajectory). A complete description of all the equations and HYSPLIT calculation methods for trajectories can be seen in [14]. HYSPLIT provides the coordinates (longitude and latitude) and height (in meters above ground level, a.g.l) of every trajectory calculated at 1-h intervals. Therefore, each trajectory of 120 h is composed of 120 end-point positions. In our study, we work these 120 end-point positions, which are the position of the lagrangian particles, at a certain hour within the duration of 120 h.

2.2. Kernel Density Estimation

Kernel density estimation (KDE) [8] is an important method in visualizing and analyzing spatial data, with application in many fields (ecology, public health, etc.), with the objective of understanding and potentially predicting event patterns [15–17]. KDE is a non-parametric approach to the estimation of probability density functions using a finite number of cases [18]. This method compensates for in distance the influence of a case in its vicinity. To this purpose, it incorporates a decay function that assigns smaller values to locations which are still in the neighborhood, but further way from a case [15]. To achieve this, KDE fits a curved surface over each case such that the surface is highest above the center and zero at a specified distance from the case (the bandwidth, i.e., the distance around a case at which its influence is felt). The influence of a number of cases at a certain location, over a two-dimensional space, can be represented as the following definition [8,19]:

$$f(x, y) = \frac{1}{nh^2} \sum_{i=1}^{n} K\left(\frac{d_i}{h}\right),$$ (1)

where $f(x,y)$ is the density value at location (x,y), n is the total number of cases under concern, h is the bandwidth, d_i is the geographical distance between case i and the location and K is a density function (generally a radially symmetric unimodal probability density function) which integrates to one. Different density functions (K) can be used, e.g., Cauchy [20], Epanechnikov [21], Gaussian [22]. More information about different kernel-smoothing algorithms can be found in [23].

The KDE refers to high or low density of points, i.e., cases per unit area. Therefore, KDE method detects the highs and lows of cases densities of the pattern, and is useful for detecting hot spots. Figure 1 graphically shows an example of the procedure use in our method to determine the density value $f(x,y)$ (grey squared) over the raster data by considering two cases (red circles).

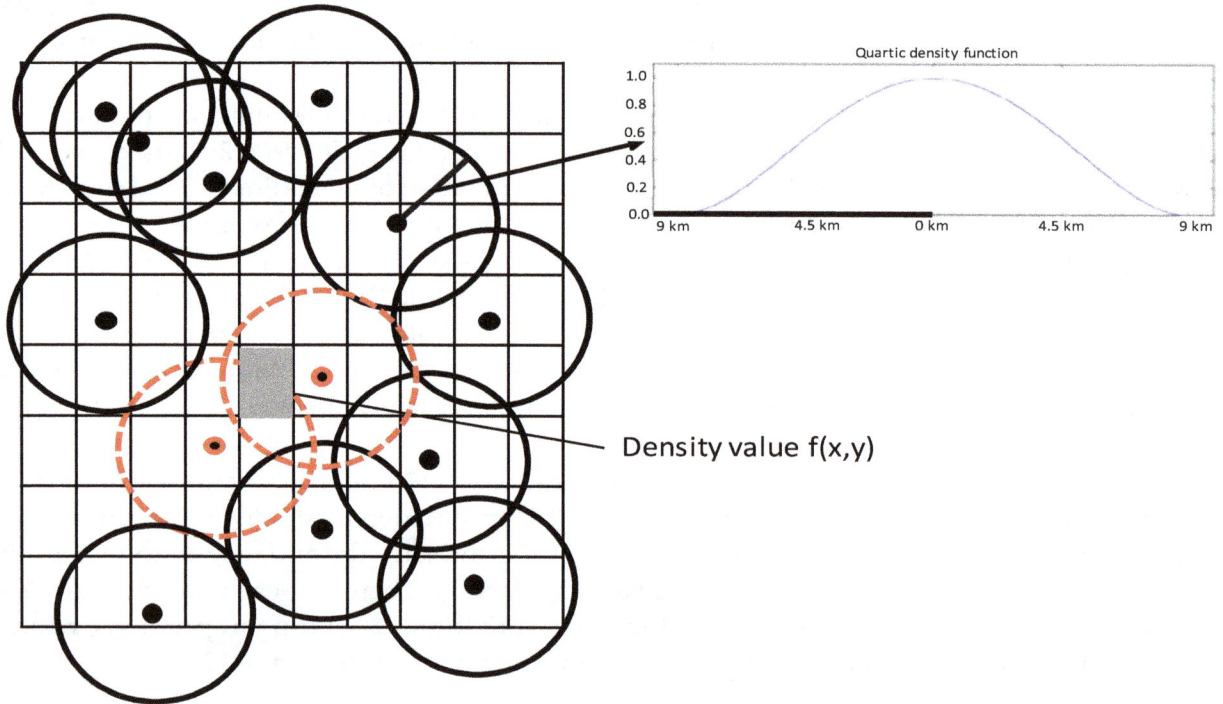

Figure 1. An illustration of the density value $f(x,y)$ calculations over raster data and the quartic kernel function used (K), with a bandwidth of 9 km.

2.3. Method

Figure 2 shows the procedure applied in our method to determine the corresponding density maps. Once calculated, the trajectories for the period 2007–2016 (Figure 2a), and the corresponding end-point positions (i.e., cases in Section 2.2), the subset of end-point positions within the analyzed period (hereafter, time window), are extracted (Figure 2b). We define the "time window" as the reference period for which end-point positions are collected and to which, as a result, the density map refers. This time window refers to one reference time (e.g., a specific day, such as the release date or the maximum concentration date). The definition of the time window is based on the fact of having evidences that cyclicality plays a significant role in the weather we experience, i.e., weather appears to be a regularly cycling pattern over short and long-range time. In the present method, we consider three different temporal cycling patterns of weather, e.g., two weeks, one month and two months. As an example, if the event date is 7 May, we work with all end-point positions within the period 2007–2016 from 1 May to 14 May (one week before and after the release date), from 24 April to 21 May (two weeks before and after), and from 7 April to 7 June (one month before and after).

Based on the end-point positions extracted, we provide a comprehensive analysis of the residence time of air masses. In order to accommodate as much as possible the residence time with hypothetical deposits from the plume, the present analysis is carried out by selecting the number of end-point positions below a vertical level, which is set to 100 m a.g.l., within the selected time window

Once the subset of end-point positions within the time window and with a height less than 100 m a.g.l are selected, these are then displayed as a density map using the KDE method indicated in Section 2.2 (Figure 2c,d). To make a KDE, the challenge is to select the kernel function, the bandwidth and the cell size. However, there are no rules and standards concerning this selection, and they are predominantly taken as the results of experimental studies [24]. After several analyses, in the present work, end-point positions are displayed on a map with a fixed grid cell size of 3 × 3 km and we have defined a bandwidth of 9 km, while the kernel function is based on the quartic kernel function described in [8], i.e., an inverse distance weighting (Figure 1). We have used the free and open source QGIS geographic information system [25].

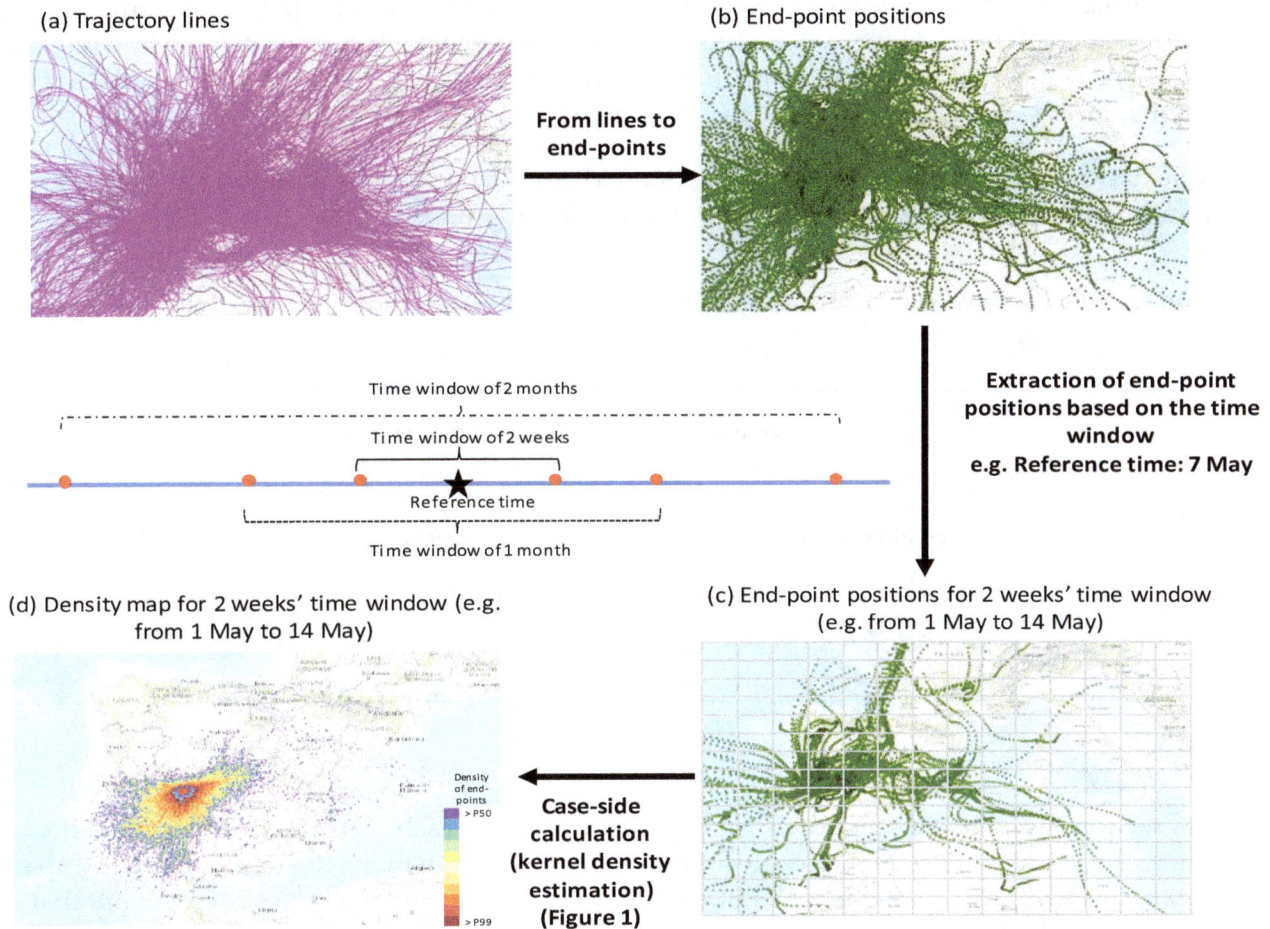

(a) Trajectory lines

(b) End-point positions

From lines to end-points

Time window of 2 months

Time window of 2 weeks

Reference time

Time window of 1 month

Extraction of end-point positions based on the time window
e.g. Reference time: 7 May

(d) Density map for 2 weeks' time window (e.g. from 1 May to 14 May)

(c) End-point positions for 2 weeks' time window (e.g. from 1 May to 14 May)

Density of end-points
> P50
> P99

Case-side calculation (kernel density estimation) (Figure 1)

Figure 2. Steps and information used to produce the density maps: (**a**) Set of trajectory lines, (**b**) end-point positions for the 2007–2016 period, (**c**) definition of time windows and extraction of end-point positions for the defined time window (2 weeks) and (**d**) example of density map, which expresses the density of end-point positions estimated above 50%.

3. Case Studies

Two case studies, the Chernobyl accident (25–26 April 1986) and the Algeciras release (30 May 1998), have been taken as reference in order to evaluate the use of trajectories and density maps as valuable information in the preparedness phase of an atmospheric event release.

An explosive accident took place at the Chernobyl nuclear power plant unit IV in Ukraine on 25 April 1986 at 2123 UTC. This accident led to a widespread dispersion of radioactive materials released at different times in the atmosphere at a continental scale. The meteorological conditions caused the various plumes to take different directions according to the release date [26], and therefore, contaminated clouds flew all around the world.

On 30 May, 1998, a ^{137}Cs medical source was accidentally melted in one of the furnaces at the Acerinox stainless steel production plant near Algeciras, southern Spain. An unknown amount of contaminated air was dispersed over the western coast of Spain and travelled all the way north, reaching the southern coast of France and northern Italy two to three days after the release. The International Atomic Energy Agency (IAEA) Emergency Response Centre reported the occurrence of a radiological accident on June 12.

Following the methodology explained in Section 2.3, we have calculated the forward trajectories and the corresponding end-point positions for the 10-year period (2007–2016) at Chernobyl (51°23′23.47″ N, 30°5′38.57″ E) and Algeciras (6°7′39″ N, 5°27′14″ W).

In accordance with the release characteristic of each release scenario [27,28], the initial height applied in the calculation of trajectories differs: Chernobyl at 1000 m a.g.l and Algeciras release at 100 m a.g.l. For each scenario, over 97% of 1-h trajectories in the period 2007–2016 are completed for all 120 h (an incomplete trajectory is due to missing data in the initialization fields).

Table 1 shows the three time windows defined for each case study, taking as reference the corresponding release date to calculate the associated density map.

Table 1. Time windows defined for Chernobyl and Algeciras case studies to extract the end-point positions in the period 2007–2016.

	Time Window of Two Weeks	Time Window of One Month	Time Window of Two Months
Chernobyl (reference date: 25 April)	From 18 April to 3 May	From 11 April to 9 May	From 25 March to 25 May
Algeciras (reference date: 30 May)	From 23 May to 6 June	From 16 May to 13 June	From 30 April to 30 June

4. Results and Discussion

This section presents the set of density maps calculated for each case study. To evaluate the bias of our meteorological approach, the results obtained in the density maps are qualitatively compared with bibliography regarding the depositional analysis of each case study. We need to point out that our method does not aim to estimate the deposit, but the areas affected.

4.1. Chernobyl Accident

Figure 3 shows the density maps produced applying each one of the three time windows (Table 1) for the period 2007–2016 from the Chernobyl site. The three maps display a large spatial dispersion from the release point. Our analysis finds high-density values (above the 90th percentile, P90) in central Europe and Scandinavia. This large dispersion provides an example of the differences in flow regimes, with areas being affected to the south (from southwest to southeast) and to the north of the release point. This spatial spread is in agreement with the variation in the meteorological conditions according to the release date. The emissions on 26 April were transported in the western and northern directions [29], while the plume released in the following days started to move in a south-westerly direction, towards central Europe [30], as well as towards Asia [31].

The highest density values (above the 99th Percentile (P99); the blue circles in Figure 3) in the three density maps are located to the southwest of the source region. This region covers areas of Romania and Bulgaria, such as the Western Romanian Carpathians, as well as the landmass formed inside the mountain circle, the Transylvanian Plateau and the Western Plain and Hills. The geographical distribution of elevated residence times in Figure 3 is in agreement with most of the areas identified in the "Atlas of cesium deposition on Europe after the Chernobyl accident" [32]. Figure 4 shows the high-resolution map of the total cumulative deposition of ^{137}Cs throughout Europe as a result of the Chernobyl accident, which is used in this paper to validate the areas identified in Figure 3. Figure 4 shows that high deposits occurred in Eastern Europe and the Balkan countries, following the dominant precipitation. These areas are well identified in our density maps, which report the highest residence time values on the northern and southern borders of Romania.

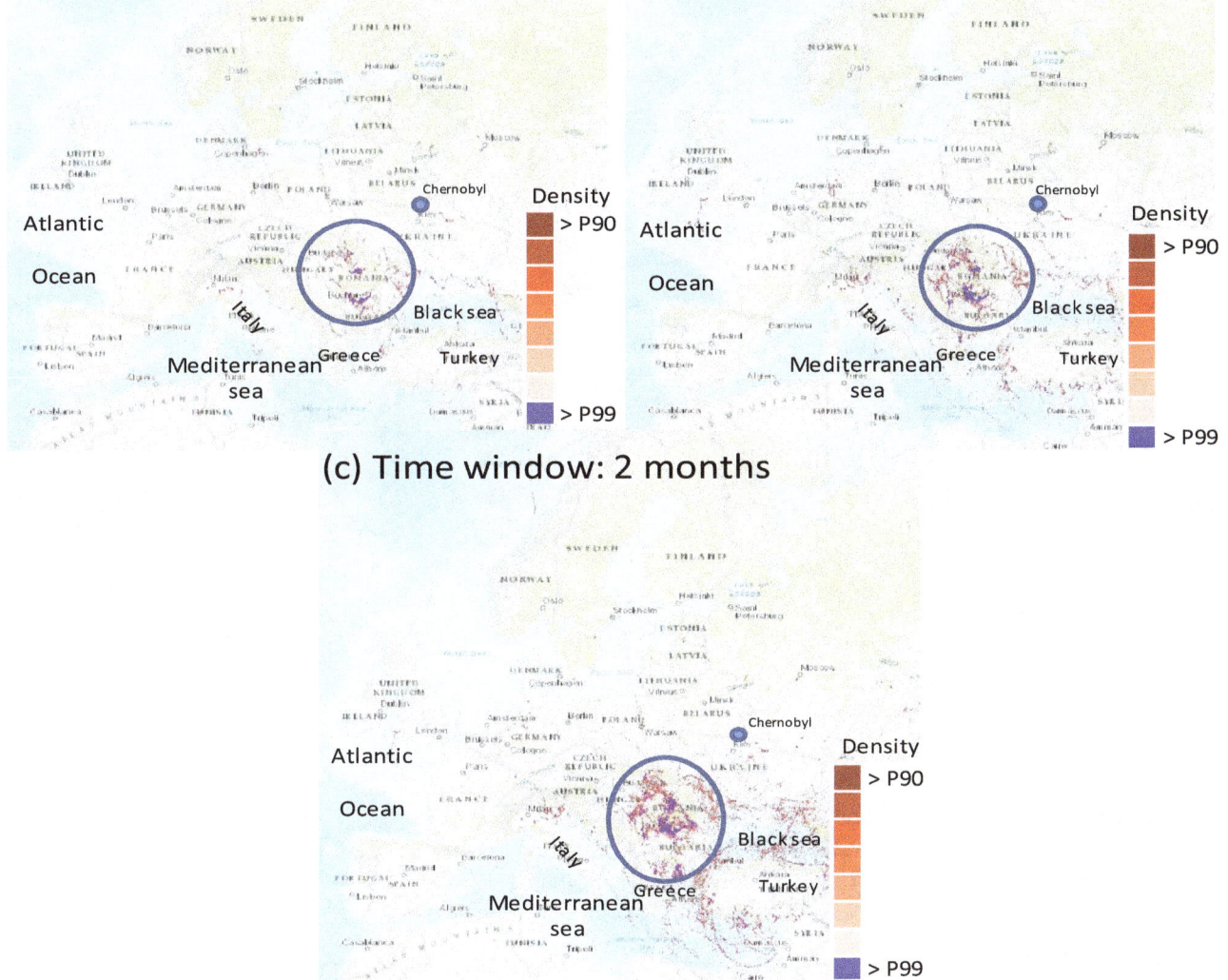

Figure 3. Density plots for 120-h trajectories starting at Chernobyl (blue point) and with a height below 100 m above ground level for (**a**) 2 weeks, (**b**) 1 month and (**c**) 2 months' time window (reference date of April 25) for the period 2007–2016 (Table 1). The density maps express the density of end-point positions estimated above 90%; quartic kernel function bandwidth 9 km. Blue circles express those areas with a density value above the 99th percentile.

Comparing the three maps (Figure 3), the area with the highest density values is better identified on increasing the time window from two weeks to two months, i.e., the number of end-point positions, and hence, the meteorological scenarios. The impact of the time window on results is well observed by comparing the greater spread of spots with values above P90 between Figure 3a (time window of two weeks) and Figure 3c (time window of two months). This impact is especially seen to the south of the release site, as high-density values show up in the Black and Aegean seas, and northern Italy–southern Austria. In addition, zones in northern Germany and in the United Kingdom are identified by using a longer time window. In contrast, the set of areas identified in northern countries does not vary with the time window. In this region, few spots can be identified over continental areas, such as southern and eastern coastal areas of Sweden. In contrast, Figure 4 displays a large spread of high ^{137}Cs in specific areas of Sweden and Finland. One of the reasons of this difference can be associated with the selection of density values up to P90.

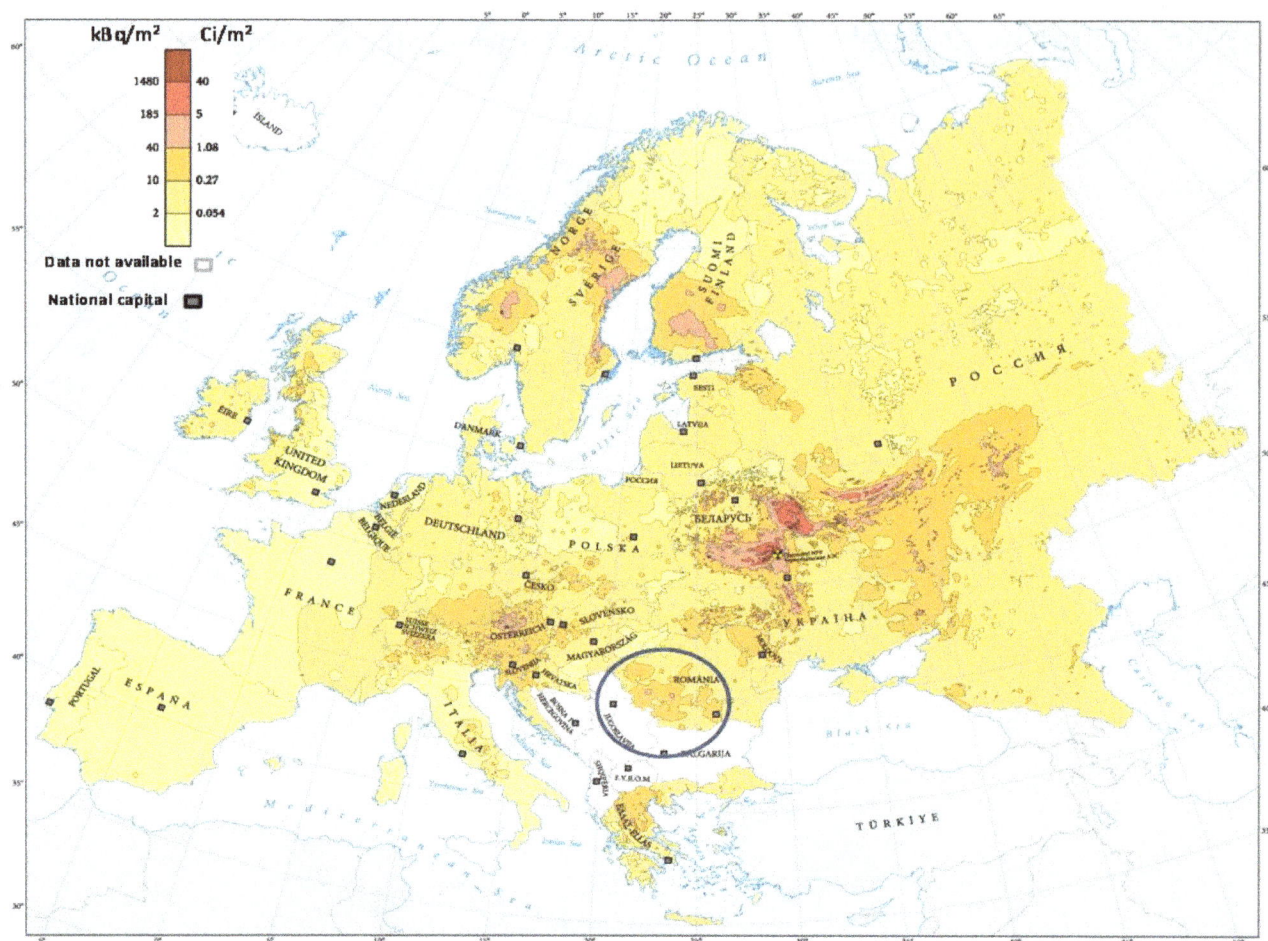

Figure 4. The Atlas map (using the Lambert azimuthal projection) depicting the total cumulative deposition of ^{137}Cs throughout Europe as a result of the Chernobyl accident from all available data of the REM database corrected for radioactive decay to 10 May 1986 [31]. Blue circle expresses the areas with a density value above the 99th Percentile obtained in Figure 3.

4.2. Algeciras Release

Figure 5 shows the corresponding density maps for the three time windows defined for the Algeciras case study. Density maps look similar for the three "time windows", with the increase in the most affected area (>P90) easily seen by increasing the time window considered. The trajectories have either an eastern or southwestern component, displaying two main branches of the possible dispersion of the plume released at Algeciras. While the branch to the east, over the Alboran Sea, turns to the northeast following the Spanish coastline, the one to the west veers to the southwest following the northwestern coast of Africa. In both cases, after the channeling effect created by the southern Iberian Peninsula and northern African mountain chains, there is a wide area over the Atlantic and Mediterranean covered by high-density values.

The three maps display a wider area corresponding to the highest density values (above P99) to the east of the release site, along the southern Mediterranean coast of the Iberian Peninsula. It is interesting to highlight how this Mediterranean branch also presents areas over central Italy and northern Africa with high-density values.

This distribution is completely in agreement with the wind dynamics in this area, which is clearly conditioned by the channeling effect of the strait of Gibraltar [33]. The highest density values are obtained to the east of the release site, over the Mediterranean Sea, being consistent with previous studies dealing with this event [34], which reported an easterly movement of the plume. According to the measurements of ^{137}Cs registered in southern France and northern Italy at the beginning of June, the radioactive plume continued its movement to the north-west over the Mediterranean Sea the following days. The density maps suggest this displacement, although the consideration of trajectories with a duration of 120 h limits the possible identification of continental areas affected.

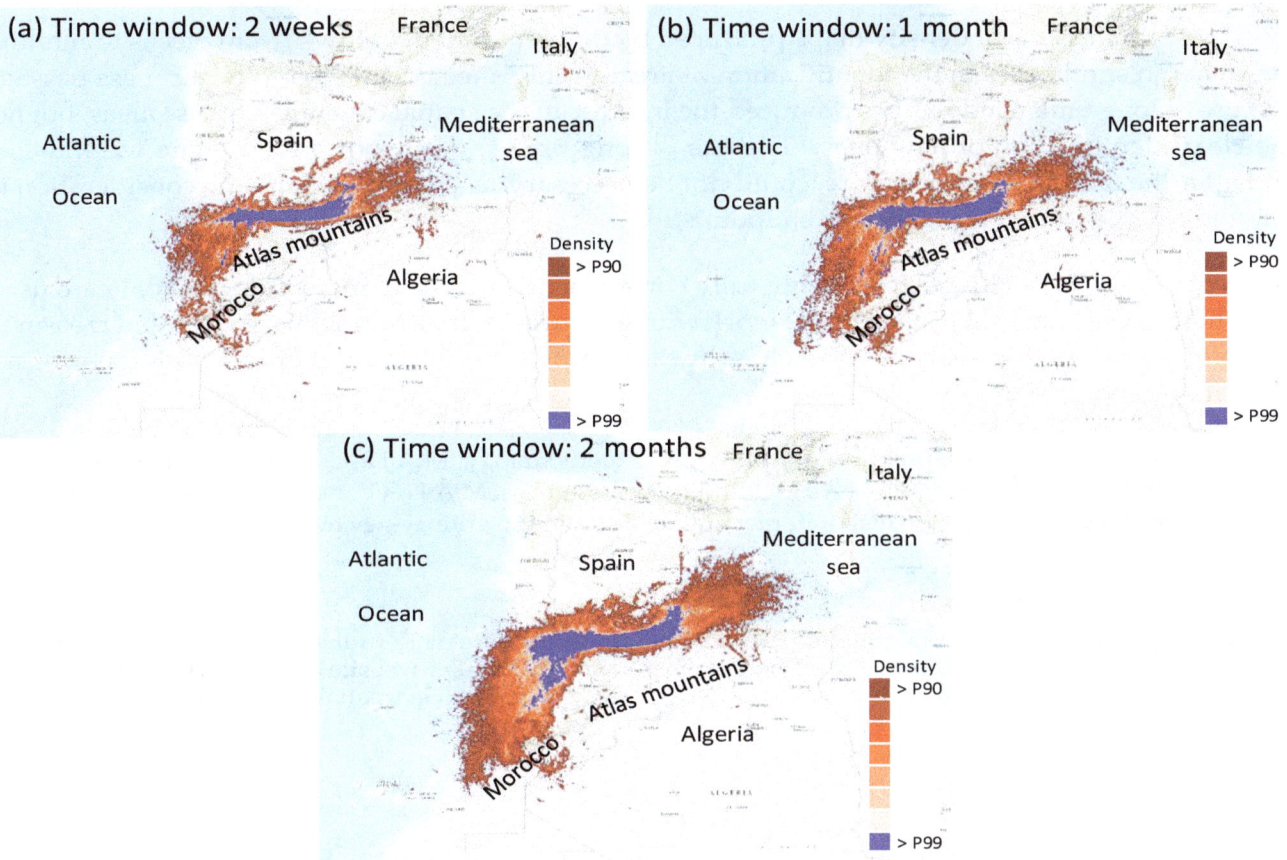

Figure 5. Density plots for 120-h trajectories starting at Algeciras and with a height below 100 m above ground level for (**a**) 2 weeks, (**b**) 1 month and (**c**) 2 months' time window (reference date of 30 May) for the period 2007–2016 (Table 1). The density maps express the density of end-point positions estimated above 90%; quartic kernel function bandwidth 9 km.

5. Conclusions

This paper addresses the method developed to estimate the areas that could be most affected by harmful airborne releases to the atmosphere. To the author's knowledge, this is the first study in which a long-term set of trajectories is used as part of an emergency preparedness infrastructure to early estimate areas likely to be affected by the release of harmful substances in the atmosphere. The bases for this approach are 5-days forward trajectories departing every 1 h at different heights above ground (in accordance with the study case) for the period 2007–2016.

We have qualitatively evaluated the method by comparing the results from density maps against the air concentration and deposition maps of ^{137}Cs from the Chernobyl and Algeciras releases. The comparison reports a reasonable agreement in the identification of both the atmospheric dispersion patterns and the areas most affected by ^{137}Cs activity concentrations.

Differences in the spatial coverage of the potential affected areas between these two events highlight the influence of the orography and of the release height in the transport and dispersion of substances in the atmosphere. While in Chernobyl (initial height of 1000 m a.g.l.), the transport occurs mainly above the boundary layer (1–2 km above surface) and hence, surface features affect the wind less (e.g., frictional drag); in Algeciras the wind is forced to move in a certain direction by the presence of mountains. In this sense, the Chernobyl map is a good example of the orographic blocking caused by the mountains, forcing the winds to slow down and/or change direction much more. The highest density values for the Chernobyl case study are found within the arc of the Carpathians and in the southern part of the country, in the Danube Plain, between the sub-Carpathians and Balkans. The main transport direction in the Algeciras case study is to the east, in agreement with previous studies.

The comparisons of density maps produced by different time windows (from weeks to months) have shown similarities in the identification of areas highly affected (>P99) by the air mass passage. The use of long time windows has produced the increase of the spatial coverage of these areas, but not the clear identification of new ones. The use of long time windows (more than one week before and after the release date), therefore, could not be necessary, as it could include the consideration of inhomogeneous and noisy trajectory endpoints.

While this method is far from being fully comprehensive, as only meteorological data are used, by building on the current results, future work with the purpose of including more years and measured concentrations into the analysis will help to further improve the present method, and to perform a quantitative evaluation.

Author Contributions: Conceptualization, M.Á.H.-C.; methodology, M.Á.H.-C. and L.D.F; formal analysis, M.Á.H.-C. and L.D.F; investigation, M.Á.H.-C. and L.D.F; resources, M.Á.H.-C. and L.D.F.; visualization, L.D.F.; data curation, L.D.F.; writing—original draft preparation, M.Á.H.-C.; writing—review and editing, M.Á.H.-C. and L.D.F; supervision, M.Á.H.-C.

Acknowledgments: The authors gratefully acknowledge the NOAA Air Resources Laboratory (ARL) for the provision of the HYSPLIT transport and dispersion model and/or READY website (http://www.ready.noaa.gov) used in this publication. The authors also acknowledge the Open Source Geospatial Foundation (OSGeo) for the provision of the QGIS project.

References

1. International Atomic Energy Agency (IAEA). *Preparedness and Response for a Nuclear or Radiological Emergency*; IAEA Safety Standards Series No. GSR Part 7; IAEA: Vienna, Austria, 2015.

2. Cooper, J.R.; Randle, K.; Sokhi, R.S. *Radioactive Releases in the Environment: Impact and Assessment*; John Wiley &Sons: Hoboken, NJ, USA, 2003; ISBN 978-0-471-89924-2.

3. Yang, X.; Wang, T.; Xi, M.; Gao, X.; Li, Q.; Zhang, N.; Gao, Y.; Lee, S.; Wang, X.; Xye, L.; et al. Abundance and origin of fine particulate chloride in continental China. *Sci. Total. Environ.* **2018**, *624*, 1041–4051. [CrossRef]

4. Wang, W.; Yu, J.; Cui, Y.; He, J.; Xue, P.; Cao, W.; Ying, H.; Gao, W.; Yan, Y.; Hu, B.; et al. Characteristics of fine particulate matter and its sources in an industrialized coastal city, Ningbo, Yangtze River Delta, China. *Atmos. Res.* **2018**, *203*, 105–117. [CrossRef]

5. Pérez, I.A.; Artuso, F.; Mahmud, M.; Kulshrestha, U.; Sánchez, M.L.; García, M.A. Applications of Air Mass Trajectories. *Adv. Meteorol.* **2015**. [CrossRef]

6. Toledano, C.; Cachorro, V.E.; De Frutos, A.M.; Torres, B.; Berjón, A.; Sorribas, M.; Stone, R.S. Airmass classification and analysis of aerosol types at El Arenosillo (Spain). *J. Appl. Meteorol. Climatol.* **2009**, *48*, 962–981. [CrossRef]

7. Hondula, D.M.; Sitka, L.; Davis, R.E.; Knight, D.B.; Gawtry, S.D.; Deaton, M.L.; Lee, T.R.; Normile, C.P.; Stenger, P.J. A back-trajectory and air mass climatology for the Northern Shenandoah Valley, USA. *Int. J. Climatol.* **2010**, *30*, 569–581. [CrossRef]

8. Silverman, B.W. *Density Estimation for Statistics and Data Analysis*; Chapman and Hall: New York, NY, USA, 1986.

9. Ashbaugh, L.L. A statistical trajectory technique for determining air pollution source regions. *APCAJ* **1983**, *33*, 1096–1098. [CrossRef]

10. Fleming, Z.L.; Monks, P.S.; Manning, A.J. Review: Untangling the influence of air-mass history in interpreting observed atmospheric composition. *Atmos. Res.* **2012**, *104–105*, 1–39. [CrossRef]

11. Kulshrestha, U.; Kumar, B. Airmass Trajectories and Long Range Transport of Pollutants: Review of Wet Deposition Scenario in South Asia. *Adv. Meteorol.* **2014**. [CrossRef]

12. Stein, A.F.; Draxler, R.R.; Rolph, G.D.; Stunder, B.J.B.; Cohen, M.D.; Ngan, F. NOAA's HYSPLIT atmospheric transport and dispersion modeling system. *Bull. Amer. Meteor. Soc.* **2015**, *96*, 2059–2077. [CrossRef]

13. Meteorological Archive Data from NCEP's GDAS. Available online: https://www.ready.noaa.gov/gdas1.php (accessed on 20 January 2019).

14. Draxler, R.; Hess, G. *Description of the HYSPLIT_4 Modeling System*; National Oceanic and Atmospheric Administration: Silver Spring, MD, USA, 1997.

15. Carlos, H.A.; Shi, X.; Sargent, J.; Tanski, S.; Berke, E.M. Density estimation and adaptive bandwidths: A primer for public health practitioners. *Int. J. Health Geogr.* **2010**, *9*, 39. [CrossRef]

16. Smith, M.-J.; Goodchild, M.-F.; Longley, P.-A. *Geospatial Analysis: A Comprehensive Guide to Principles*; Techniques and Software Tools; The Winchelsea Press: Winchelsea, UK, 2015.

17. Bartolini, S.; Cappello, A.; Martí, J.; Del Negro, C. QVAST: A new Quantum GIS plugin for estimating volcanic susceptibility. *Nat. Hazards Earth Syst. Sci.* **2013**, *13*, 3031–3042. [CrossRef]

18. Krisp, J.M.; Peters, S.; Murphy, C.E.; Fan, H. Visual Bandwidth Selection for Kernel Density Maps. *Photogramm. Fernerkund. Geoinf.* **2009**, *5*, 441–450. [CrossRef]

19. Shi, X. Selection of bandwidth type and adjustment side in kernel density estimation over inhomogeneous backgrounds. *J. Int. J. Geogr. Inf. Sci.* **2010**, *24*, 643–660. [CrossRef]

20. Martin, A.J.; Umeda, K.; Connor, C.B.; Weller, J.N.; Zhao, D.; Takahashi, M. Modeling long-term volcanic hazards through Bayesian inference: An example from the Tohuku volcanic arc Japan. *J. Geophys. Res.* **2004**, *109*. [CrossRef]

21. Lutz, T.M.; Gutmann, J.T. An improved method for determining and characterizing alignments of point-like features and its implications for the Pinacate volcanic field, Sonoran, Mexico. *J. Geophys. Res.* **1995**, *100*, 17659–17670. [CrossRef]

22. Connor, C.B.; Hill, B.E. Three nonhomogenous Poisson models for the probability of basaltic volcanism: Application to the Yucca Mountain region, Nevada. *J. Geophys. Res.* **1995**, *100*, 10107–10125. [CrossRef]

23. Wand, M.P.; Jones, M.C. *Kernel Smoothing Monographs on Statistics and Applied Probability*; Chapman & Hall/CRC. Taylor &Francis Group: Boca Raton, FL, USA, 1995.

24. Wolf, M.; Asche, H. Exploring Crime Hotspots: Geospatial Analysis and 3D Mapping. In Proceedings of the 14th International Conference on Urban Planning, Regional Development and Information Society, Schwechat, Austria, 22–25 April 2009.

25. QGIS Development Team. QGIS Geographic Information System. *Open Source Geospatial Foundation.* Available online: http://qgis.org (accessed on 20 September 2018).

26. Institute De Radioprotection et de Surete Nucleaire (ISRN), Chernobyl 25 years on. 2011. Available online: https://www.irsn.fr/EN/publications/thematicsafety/chernobyl/Documents/irsn_booklet_chernobyl_2011.pdf (accessed on 10 January 2019).

27. Bartnicki, J.; Salbu, B.; Saltbones, J.; Foss, A.; Lind, O.L. *Gravitational Settling of Particles in Dispersion Model Simulations Using the Chernobyl Accident as a Test Case*; Research Report; Norwegian Meteorological Institute: Oslo, Norway, 2001; ISSN 0332-9879.

28. Kaviani, F.; Memarian, M.H.; Eslami-Kalantari, M. Simulation of atmospheric dispersion, transport and deposition of nuclear pollutants released from a hypothetical accident at Bushehr Power Plant. *J. Earth Space Phys.* **2017**, *43*, 635–650. [CrossRef]

29. Langner, J.; Robertson, L.; Persson, C.; Ullerstig, A. Validation of the operational emergency response model at the Swedish meteorological and hydrological institute using data from ETEX and the Chernobyl accident. *Atmos. Environ.* **1998**, *32*, 4325–4333. [CrossRef]

30. Talerko, N. Mesoscale modelling of radioactive contamination formation in the Ukraine caused by the Chernobyl accident. *J. Environ. Radioact.* **2005**, *78*, 311–329. [CrossRef] [PubMed]

31. Simsek, V.; Pozzoli, L.; Unal, A.; Kindap, T.; Karaca, M. Simulation of 137Cs transport and deposition after Chernobyl Nuclear Power Plant accident and radiological doses over the Anatolian Peninsula. *Sci. Total Environ.* **2014**, *499*, 74–88. [CrossRef] [PubMed]

32. De Cort, M.; Dubois, G.; Fridman, S.D.; Germenchuk, M.G.; Izrael, Y.A.; Janssens, A.; Jones, A.R.; Kelly, G.N.; Kvasnikova, E.V.; Matveenko, I.I.; et al. *Atlas of Caesium Deposition on Europe after the Chernobyl Accident*; Office of the European Union: Luxembourg, 1998; p. 176.

33. Dorman, C.E.; Beardsley, R.C.; Limeburner, R. Winds in the strait of Gibraltar. *Quaterly J. R. Meteorol. Soc.* **1995**, *121*, 1093–1921. [CrossRef]

34. Vogt, P.J.; Pobanz, B.M.; Aluzzi, F.J.; Baskett, R.L.; Sullivan, T.J. ARAC Modeling of the Algeciras, Spain Steel Mill CS-137 Release. Available online: https://www.osti.gov/servlets/purl/15013432 (accessed on 20 January 2019).

15

Spatiotemporal Change of Plum Rains in the Yangtze River Delta and its Relation with EASM, ENSO and PDO during the Period of 1960–2012

Nina Zhu [1,2,3], **Jianhua Xu** [1,2,3,*], **Kaiming Li** [4], **Yang Luo** [5], **Dongyang Yang** [1,2,3] **and Cheng Zhou** [6]

[1] Key Laboratory of Geographic Information Science (Ministry of Education), East China Normal University, Shanghai 200241, China; ninaecnu@126.com (N.Z.); yangdy@lreis.ac.cn (D.Y.)

[2] Research Center for East-West Cooperation in China, East China Normal University, Shanghai 200241, China

[3] School of Geographic Sciences, East China Normal University, Shanghai 200241, China

[4] School of Urban and Geography, Lanzhou City College, Lanzhou 730070, China; lkm_wd@126.com

[5] Jianhu Data Technology (Shanghai) Co., Ltd, Shanghai 201700, China; luoyang@jianhushuju.com

[6] Faculty of Tourism Management, Shanxi University of Finance & Economics, Taiyuan 030006, China; zhoutravel@163.com

* Correspondence: jhxu@geo.ecnu.edu.cn.

Abstract: The Plum Rains process is a complex process, and its spatiotemporal variations and influencing factors on different time scales still need further study. Based on a dataset on the Plum Rains in the Yangtze River Delta, from 33 meteorological stations during the period of 1960 to 2012, we investigated the spatiotemporal variations of Plum Rains and their relation with the East Asian Summer Monsoon (EASM), the El Niño-Southern Oscillation (ENSO), and the Pacific Decadal Oscillation (PDO) using an integrated approach that combines ensemble empirical mode decomposition (EEMD), empirical orthogonal function (EOF), and correlation analysis. The main conclusions were as follows: (1) the plum rainfall (i.e., the rainfall during the period of Plum Rains) showed a trend of increasing first and then decreasing, and it had a three-year and six-year cycle on the inter-annual scale and a 13-year and 33-year cycle on the inter-decadal scale. The effect of the onset and termination of Plum Rains and the daily intensity of plum rainfall on plum rainfall on the inter-annual scale was greater than the inter-decadal scale, (2) the EOF analysis of plum rainfall revealed a dominant basin-wide in-phase pattern (EOF1) and a north-south out-of-phase pattern (EOF2), and (3) ENSO and EASM were the main influencing factors in the three-year and six-year periods, respectively.

Keywords: plum rains; ensemble empirical mode decomposition (EEMD); empirical orthogonal function (EOF); East Asian summer monsoon; El Niño-southern oscillation; pacific decadal oscillation; the Yangtze River Delta; multi-time scales

1. Introduction

The Plum Rains are one of the typical climatic phenomena in the subtropical monsoon region. They refer to persistent rain in June and July every year in the Yangtze-Huaihe River region, Southern Taiwan, Liaodong Peninsula of China, and Southern Japan [1]. Every year in the spring, warm and humid air of the tropical ocean will enter the mainland from the sea, and the forces of the warm and humid air will gradually strengthen. Especially at the low altitude of two or three kilometers, there is often a very humid and strong southerly airflow from the ocean [1]. When it enters mainland China, it will meet the cold air from the north to the south [2,3]. When the warm and cold air meets, the junction will form a frontal surface, and precipitation will occur near the frontal surface. This forms the Plum Rains [4]. The Plum Rains are also called "Meiyu" in China, "Baiu" in Japan, and "changma" in Korea [2,5,6]. The temperature and humidity are very high during this period. In the 1930s, scholars began to study Plum Rains from different perspectives [7–12]. In a previous study, scholars have mainly concentrated on some typical large or medium regions (such as the middle and lower reaches of the Yangtze River Delta, located at the longitude of 110–122° E and the latitude of 26°–34° N) [13–15]. As for the research methods applied in the previous studies, the classical statistical regression, wavelet analysis, neural network, and mechanism models have been used [16–18]. As for the research content, meteorological scholars in the previous studies have mainly focused on the internal dynamic process and external forcing factors of the atmosphere (such as sea temperature) influencing the degree and change mechanisms of the Plum Rains [19–23].

Yu et al. [24] built a new index of Plum Rains' intensity (IPRI) to analyze the spatiotemporal variation characteristics of Plum Rains in the areas along the Huaihe River in the Anhui province. The results were of great help in analyzing, assessing, and identifying flooding disasters. Chen et al. [25] simulated a heavy rainfall event during the period of Plum Rains using the Penn State–NCAR Mesoscale Model Version 5 (MM5). They found that there were strong interaction and positive feedback between the convective rainstorms embedded within the Mei-Yu front and the Mei-Yu front itself. Zhang et al. [26] developed indicators for the onset and retreat dates, duration, and Meiyu precipitation and analyzed the variations of Meiyu in the Yangtze-Huaihe River valley, which points out that Meiyu rainfall showed an increased trend during the period from 1954 to 2003. Zhu et al. [27] analyzed the Meiyu onset dates (MODs) in the middle and lower reaches of the Yangtze River valley and found that the beginning of June displays the average MOD in this region. However, due to the complexity of the Plum Rains process, there are still many issues that require further research, especially the relationship between the atmospheric circulation and the Plum Rains along with the spatiotemporal variations of the Plum Rains.

In previous studies, when analyzing the relationship between atmospheric circulation and Plum Rains, most meteorologists studied them from an overall perspective and rarely analyzed them on different time scales [28–30]. The relationship between atmospheric circulation and Plum Rains may be different on multi-time scales. Therefore, it is necessary to specifically analyze the relationship between them from the perspective of multi-time scales. In addition, due to regional differences, how to determine the characteristic of local Plum Rains and analyze its climate change, as well as study the consistency and particularity of local Plum Rains and regional Plum Rains climate change, is important. Industry and agriculture of the Yangtze River Delta play major roles in China's economy [31,32]. Frequent floods and droughts in the summer are harmful to agricultural production, transportation, and life, which will all cause huge economic losses [33,34]. Damage caused by Plum Rains is one of the main factors responsible for flood and drought disasters [35]. Therefore, the study of Plum Rains is not only conducive to industrial and agricultural production, but is also advantageous for improving the governments' awareness of disaster prevention. Thus, it can reduce loss of life and property during the period of floods and droughts and promote economic and social development.

When facing global climate change and frequent extreme climate events [36], how to understand the particularity and generality of the local Plum Rains and the relationship between the Plum Rains and atmospheric circulation on different time scales is an important scientific issue. For the above reasons, we select Shanghai, the Jiangsu province, and the Zhejiang province as a typical region. Based on a dataset from 33 meteorological stations during the period of 1960 to 2012, this study applies selected methods, including ensemble empirical mode decomposition (EEMD), empirical orthogonal function (EOF), and correlation analysis to investigate the spatiotemporal changes of plum rainfall and the relationship between plum rainfall and the East Asian Summer Monsoon (EASM), the El Niño-Southern Oscillation (ENSO), and the Pacific Decadal Oscillation (PDO).

2. Materials and Methods

2.1. Study Area and Data

The study area includes most of the Jiangsu province, the Zhejiang province, and Shanghai, which are part of the middle and lower regions of the Yangtze River Delta. It is situated at a longitude of 116°18′ E–123°00′ E and latitude of 27°12′ N–35°20′ N. Its climate belongs to the sub-tropical monsoon climate, and the average temperature ranges from 2 to 4 °C in January and is above 28 °C in July. Annual precipitation is above 800 mm and the precipitation is mainly concentrated in the period of Plum Rains. The precipitation during the period of Plum Rains accounts for approximately 22.5% of the total annual precipitation. In addition, there are numerous lakes, and the recharge of the water mainly comes from atmospheric precipitation. It is well-known that this study area is one of the most developed areas in China with high levels of urbanization and industrialization. According to the Bulletin of the Sixth National Census, in 2010, the resident population had reached approximately 156 million [37].

Figure 1 shows the distribution of 33 meteorological stations, which are fairly uniformly distributed. The daily precipitation of 33 meteorological stations in the Yangtze River Delta from 1960 to 2012 comes from the China meteorological data network (http://data.cma.com). A uniform standard concerning the onset and termination of the Plum Rains was published on 12 May 2017 and implemented on 1 December 2017 (GB/T 33671-2017 Mei-yu Monitoring Indicators) [38]. From the first rainy day, the second day, the third day, etc., to the tenth day, if the ratio of the number of rainy days to the total number of days in the corresponding period is greater than or equal to 50%, the first rainy day is the beginning of Plum Rains. There is no new rainy beginning date after July 20. From the last rainy day of the rainy season, 2 days ago, 3 days ago, etc., to 10 days ago, if the number of rainy-day accounts for more than or equal to 50% of the number of rainy days in the corresponding period, the last rainy day is the termination of Plum Rains. If the number of rainy days accounts for less than 50% of the number of rainy days in the corresponding period, the last rainy day cannot be the termination of Plum Rains. It is necessary to push one day forward from the last rainy day of the rainy season, and then calculate whether the day is the termination of the Plum Rains, according to the conditions of the termination of the Plum Rains. If it is still not, this method is followed until the termination of the Plum Rains is found. For the abnormal rain season, the termination of Plum Rains occurs the day before the first non-rainy day after the rain period enters August.

The termination of Plum Rains should appear before the start of the fall. For the sake of simplicity, it was assumed that the onset and termination of the Plum Rains were the same for each station, and the plum rainfall of each station for every year accumulated, according to the daily rainfall and the onset and termination of the Plum Rains, while the annual plum rainfall in the whole area was the average of the 33 stations. According to previous studies [25–27], we selected the PDO index, ENSO index, and EASM index to investigate the relationship between them and plum rainfall. The PDO index was obtained from http://research.jisao.washington.edu/pdo/PDO.latest. The ENSO data were gathered from http://research.jisao.washington.edu/datasets/globalsstenso. EASM index was defined as an area-averaged seasonally dynamic normalized seasonality index at 850 hPa within the East Asian monsoon domain (10°–40° N, 110°–140° E). The EASM index data were from the literature [39,40].

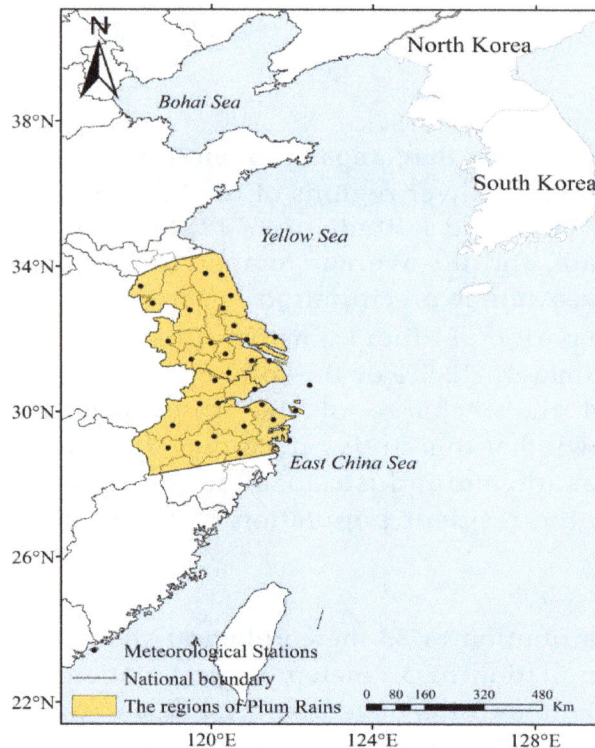

Figure 1. The spatial distribution of 33 meteorological stations and the study area.

2.2. Methodology

To investigate the spatiotemporal variation of Plum Rains and its relation to EASM, ENSO, and PDO, we used an integrated approach combining the EEMD, EOF, and correlation analysis (Figure 2). We first showed the features of plum rainfall. Then, we used the EEMD method to decompose the plum rainfall into four intrinsic mode functions (i.e., IMF1, IMF2, IMF3, and IMF4) and a trend (RES). The influencing factors of the plum rainfall were also studied. We revealed the spatial pattern of the plum rainfall over the Yangtze River Delta during the period of 1960 to 2012. Lastly, we assessed the relationship between the EASM, ENSO, PDO, and plum rainfall on different time scales.

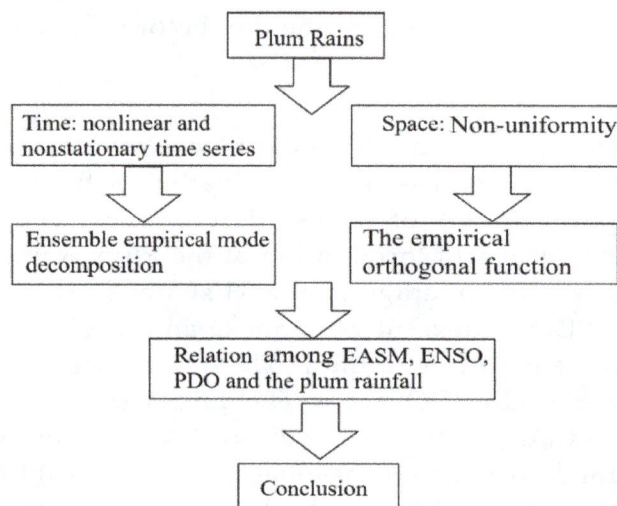

Figure 2. The framework of this study.

2.2.1. Ensemble Empirical Mode Decomposition

Ensemble empirical mode decomposition (EEMD) is utilized to analyze the nonlinear and periodic characteristics of the plum rainfall. It is based on empirical mode decomposition (EMD), and improvement of the EMD [41,42]. EMD is an effective method for dealing with the nonlinear and nonstationary time series problem, and it also has some advantages, such as self-adaptability, orthogonality, and completeness, etc. [43]. However, some defects still exist. One of the biggest defects is mode mixing, which makes a signal have different scales or frequencies in the same component or be decomposed into different components [44]. In order to solve this problem, Wu and Huang [45] proposed a new method, known as the EEMD, to better solve nonlinear problems. It can adaptively decompose the time-frequency, according to the local time variation features, and is completely free from the constraints of the Fourier transform so that it can obtain a high time-frequency resolution [41]. Therefore, we use the EEMD, which is the best decomposition method to extract the changes of various scales in the plum rainfall signal from the plum rainfall time series. The steps of the EEMD are as follows.

1. White noise with specified amplitude is added to the sequence of the original signal:

$$x_i(t) = x(t) + n_i(t) \tag{1}$$

where $x_i(t)$ is the new signal after adding the white noise, $x(t)$ is the original signal, and $n_i(t)$ is white noise.

2. By decomposing the signal, to which white noise has been added, using the EMD can result in IMF1 (the first Intrinsic Mode Function).

3. Adding the same white noise in the sequence that has separated out the IMF1 can result in IMF2 (the Second Intrinsic Mode Function) by using the EMD.

4. Repeating the above steps can result in different IMFs (the Intrinsic Mode Functions). In general, the standard deviation (SD) (generally be set as 0.2–0.3) between two consecutive results is used as a criterion for stopping the generation of IMFs [45]. When the SD reaches a certain threshold, the generation of IMFs is stopped. In this study, we set the threshold to 0.2 according to a previous study [18] and obtained four IMFs.

5. The IMFs obtained by each decomposition are collectively averaged so that the added white noises cancel each other out and can be used as the final decomposition result.

$$C_j(t) = \frac{1}{N} \sum_{i=1}^{N} C_{ij}(t) \tag{2}$$

where $C_j(t)$ is the final jth IMF component, N is the number of white noise series, and $C_{ij}(t)$ denotes the jth IMF from the added white noise trial. From the above decomposition, the IMFs and trend term can be obtained at different scales. Next, the original signal is reconstructed.

$$x(t) = \sum_{j=1}^{n} C_j(t) + r_n(t) \tag{3}$$

After decomposing the original signal, each component should be tested for its significance. They can be tested by means of a set of white noise ensemble disturbance to obtain each IMF's credibility [33]. The energy spectral density of the kth IMF is assumed to be the following.

$$E_k = \frac{1}{N} \sum_{j=1}^{N} |I_k(j)|^2 \tag{4}$$

where N is the length of the IMF component and $I_k(j)$ denotes the kth IMF component. The white noise sequence is tested by the Monte Carlo method [44]. Then, a simple equation that relates the averaged energy density $\overline{E}_k$ to the averaged period $\overline{T}_k$ is obtained.

$$\overline{E}_k + \ln\{\overline{T}_k\} = 0 \tag{5}$$

In a figure of $\ln\left(\overline{T}_k\right)$ as the x-axis and $\ln(E_k)$ as the y-axis, the relations between them can be expressed by a straight line whose slope is -1. The IMF component of the white noise series should, in theory, be distributed in a line. However, a little actual deviation is produced, so the confidence interval for the energy spectrum distribution of the white noise is presented as follows.

$$\ln \overline{E}_k = -\ln\{\overline{T}_k\}_a \pm a \sqrt{\frac{2}{Ne}}^{-\ln\left(\frac{\{T_k\}a}{2}\right)} \tag{6}$$

where a is the significance level. If the energy of the IMF is located above the confidence curve at a given significance level (e.g., $a = 0.05$), periodic oscillation has passed the significance test, and it can be assumed that the information at the selected confidence level contains physical meaning. Its corresponding oscillation period is the main oscillation period of the original sequence. However, if the energy of the IMFs falls outside the confidence curve at a given significance, it is considered to have not passed the significance test [42,46].

2.2.2. The Empirical Orthogonal Function

The spatiotemporal variation of plum rainfall has always been a hot issue. Different methods have been used to analyze it. The empirical orthogonal function (EOF), also known as the eigenvector analysis or principal component analysis (PCA), can describe the original variable field with fewer spatial distribution modalities and can cover most of the information in the original variable field [47,48]. It is helpful to analyze the spatiotemporal variations of the climate element field, and has become an important method for analyzing the characteristics of the variable field in climate research. The method was first proposed by the statistician Pearson in 1902, and then introduced into the analysis of climate problems by Lorenz (1956) [49]. Its analysis principle is as follows.

A climate variable field can be seen as a function of time and space. Suppose a climate variable field has m elements, and the time length is n, then the following is true.

$$X = \begin{bmatrix} x_{11} & x_{12} & \cdots & x_{1n} \\ x_{21} & x_{22} & \cdots & x_{2n} \\ \vdots & \vdots & \vdots & \vdots \\ x_{m1} & x_{m2} & \cdots & x_{mn} \end{bmatrix} (i = 1, 2, \ldots, m; j = 1, 2, \ldots, n) \tag{7}$$

where m is the spatial point, which can be the grid point or the observation point, n is the point in time, which is the number of samples, and x_{ij} represents the observation value of the point i at the point j. EOF expansion is employed to decompose the above equation into the sum of the product of the space function and the time function, which is shown below.

$$X = VT = \sum_{k=1}^{m} v_{kj}t_{kj}, \ i = 1, 2, \ldots, m; j = 1, 2, \ldots, n \tag{8}$$

where V represents the eigenvector matrix, and T represents the time coefficient matrix. According to the orthogonality, V and T satisfy the following conditions.

$$\begin{cases} \sum_{i=1}^{m} v_{ik}v_{il} = 1 \quad k = l \\ \sum_{i=1}^{m} v_{ik}v_{il} = 0 \quad k \neq l \end{cases} \qquad (9)$$

Next, according to Equation (10), the eigenvector, the eigenvalue, and the time coefficient matrix T can be obtained. The eigenvalues are arranged in descending order ($\lambda_1 \geq \lambda_2 \geq \ldots \geq \lambda_m \geq 0$), and the variance contribution rate of each eigenvector (R_k) can be obtained by the following formula.

$$R_k = \lambda_k / \sum_{i=1}^{m} \lambda_i, k = 1, 2, \ldots, p(p < m) \qquad (10)$$

3. Results

3.1. The Time Variation of Plum Rainfall

As seen in Figure 3, the plum rainfall in the Yangtze River Delta over the period of 1960 to 2012 presented an increasing and then decreasing trend. A turning point appeared in the 1990s. Before that, the plum rainfall showed an increasing trend, and after that, the plum rainfall showed a decreasing trend. The plum rainfall was higher in the years of 1969, 1982, 1996, and 1998, and lower in the years of 1965, 1971, 1978, and 2006. In addition, it can be seen that the plum rainfall was not linear, but shows a nonlinear variation trend (see Figure 3). Therefore, a nonlinear method should be used to analyze the nonlinear and non-stationary variations of plum rainfall.

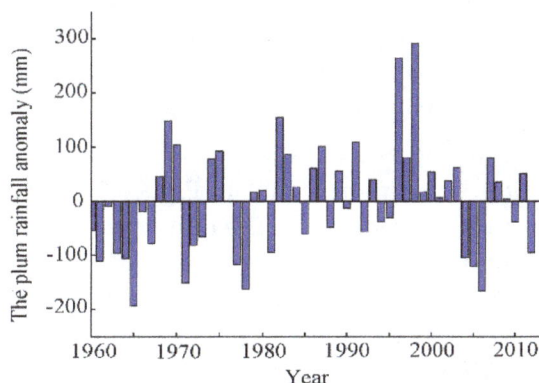

Figure 3. Anomaly in plum rainfall during the period of 1960 to 2012 (The axis of abscissas represents the years, and the axis of ordinates represents a plum rainfall anomaly, which is the result of subtracting the mean from the original value).

As we know, the formation process of Plum Rains is a complex nonlinear system, which is influenced by natural factors and human activities [50], and has many uncertainties, such as ambiguity and randomness. Simple statistical analysis, such as linear regression analysis and correlation analysis, can only describe linear time series, and cannot reveal the regularity of nonlinear and non-stationary time series. Studies have shown that EEMD is an improvement of EMD [41] and exhibits a stronger local performance than wavelet analysis [51,52]. Therefore, we used the EEMD to extract the variations of various scales in terms of the plum rainfall, the onset of Plum Rains, the termination of Plum Rains, and the daily intensity of the plum rainfall signal from their original time series to reveal the oscillating mode structure characteristics at different time scales and explore evolution characteristics of these different scale oscillations.

Figure 4 shows the EEMD results of the four-time series. There were four IMFs (IMF1-4) and one trend component (RES) of plum rainfall, the onset of Plum Rains, the termination of Plum Rains, and the daily intensity of plum rainfall, respectively. The components represent the oscillation characteristics from high to low frequency of the original sequence on different scales and all have a specific physical meaning, while the tendency term represents the variation characteristic as a whole [42]. As shown in Figure 4a, plum rainfall displayed three-year (IMF1) and six-year (IMF 2) periodic fluctuation at the inter-annual scale and 14-year (IMF3) and 33-year (IMF4) periodic variation at the inter-decadal scale. In other words, there may be some dry Plum Rains seasons after a few wet Plum Rains seasons. The result is useful for local governments and farmers to predict and mitigate the effects of droughts and floods. The trend term reflects the variations of plum rainfall, which increased gradually before the 1990s and then decreased gradually. However, previous studies showed that the plum rainfall in the YRD presented an increasing trend in past decades [28,53]. The difference is that they analyzed the variations of plum rainfall from the perspective of the linear trend, but we used the nonlinear method. It can be seen that previous studies did not reflect details of the variations of plum rainfall, which illustrates that the EEMD is an optimal method for dealing with nonlinear problems. In addition, we produced EEMD results for the onset and termination of Plum Rains and the daily intensity of plum rainfall (Figure 4b–d). The onset of Plum Rains exhibited three-year (IMF1) and six-year (IMF2) periodic fluctuation at the inter-annual scale and 14-year (IMF3) and 49-year (IMF4) periodic variation at the inter-decadal scale. The termination of Plum Rains displayed three-year (IMF1), six-year (IMF2), and nine-year (IMF3) periodic fluctuation at the inter-annual scale and 24-year (IMF4) periodic variation at the inter-decadal scale. The daily intensity of plum rainfall showed three-year (IMF1) and five-year (IMF2) periodic fluctuation at the inter-annual scale and 27-year (IMF3) and 51-year (IMF4) periodic variation at the inter-decadal scale. It can be seen that the plum rainfall had similar cycles with the onset and termination of Plum Rains and the daily intensity of plum rainfall on the inter-annual scale. Their trend terms were the same as those for plum rainfall, which means that the variation of plum rainfall is mainly affected by the onset and termination of Plum Rains and the daily intensity of plum rainfall, and the result was consistent with previous studies [13,15]. However, unlike previous studies, the effect of the onset and termination of Plum Rains and the daily intensity of plum rainfall on the inter-annual scale is greater than on the inter-decadal scale.

It is necessary to conduct significant tests in order to determine whether each component is a result of simple noise, or a signal that has physical meaning. In Figure 5, the horizontal coordinate represents periodic data and the vertical coordinate represents the energy spectrum density with four confidence levels, which are 80%, 90%, 95%, and 99%. It indicates that the IMF passes the significance test and has many physical meanings if it falls above the confidence level. Otherwise, IMF contains pure white noise. In other words, more or less physical meaning actually means strong or weak oscillations within a certain range. IMF1, IMF2, and IMF3 of plum rainfall fell above the confidence level of 90% and IMF4 of plum rainfall fell between the confidence level of 80% and 90%. IMF1, IMF2, and IMF3 comprised a more definite physical meaning but not white noise, and IMF4 contained less physical meaning. IMF1 and IMF4 of the onset of Plum Rains fell above the confidence level of 95%, while IMF2 and IMF3 of the onset of Plum Rains fell below the confidence level of 90%. IMF1, IMF2, and IMF3 of the termination of Plum Rains fell above the confidence level of 90% while IMF4 fell below the confidence level of 90%. IMF1, IMF2, and IMF3 of the intensity of plum rainfall fell above the confidence level of 90%, while IMF4 fell below the confidence level of 80%.

Each component's variation contribution rate was calculated to compare each component and the inherent oscillation characteristics in the original sequence. Table 1 presents the results of the variance contribution rates of the EEMD for the plum rainfall, the onset and termination of Plum Rains, and the intensity of plum rainfall. Significant IMFs are shown in bold in Table 1. The changes of each component are the common consequences of the internal dynamic process and the external force factors of the atmosphere. We can note that the largest variance contribution rate was IMF1, followed by IMF2, RES, IMF3, and IMF4 for plum rainfall and the intensity of plum rainfall. The largest variance contribution rate was IMF1, followed by IMF2, RES, IMF4, and IMF3 for the onset of plum rainfall. Additionality, the largest variance contribution rate was IMF1, followed by IMF2, IMF3, IMF4, and RES for the termination of Plum Rains. Although the variance contribution rate of each component is different for plum rainfall, the onset and termination of Plum Rains, and the intensity of plum rainfall, their common point is that the variance contribution rate on the inter-annual scale is greater than the inter-decadal scale.

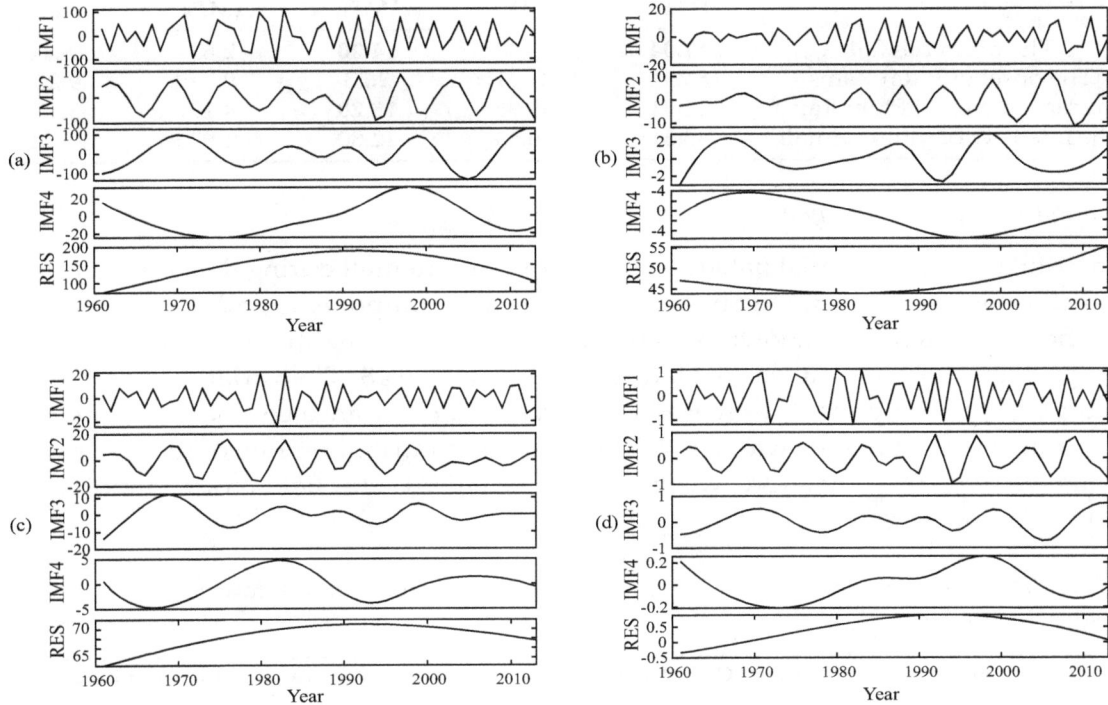

Figure 4. IMFs and the trend component of (**a**) The plum rainfall. (**b**) The onset of Plum Rains. (**c**) The termination of Plum Rains. (**d**) The daily intensity of plum rainfall during the period of 1960 to 2012.

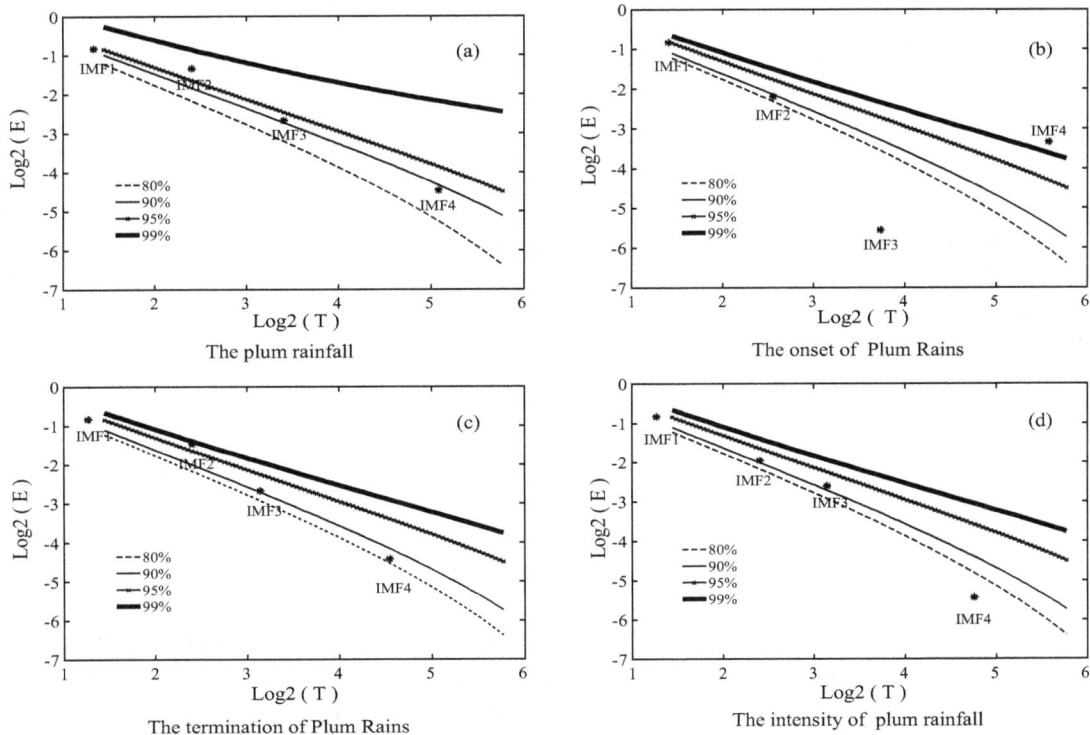

Figure 5. Significance test for the IMFs of (**a**) the plum rainfall, (**b**) the onset of Plum Rains, (**c**) the termination of Plum Rains, and (**d**) the intensity of plum rainfall during the period of 1960 to 2012.

Table 1. The variance contribution rates of EEMD for plum rainfall, the onset and termination of Plum Rains, and the intensity of plum rainfall.

	Variance Contribution Rate (%)				
	IMF1	**IMF2**	**IMF3**	**IMF4**	**RES**
The plum rainfall	**41.04**	**28.33**	**14.20**	2.63	16.14
The onset of Plum Rains	**55.91**	21.21	2.23	**9.17**	9.22
The termination of Plum Rains	**51.91**	**33.85**	**15.20**	4.19	3.12
The intensity of plum rainfall	**43.61**	**20.43**	**12.80**	1.93	14.31

3.2. Spatial Pattern of Plum Rainfall

In order to explore the spatial pattern variation of plum rainfall during the period of 1960 to 2012, the EOF method is used. First, we obtained six principal components and they passed the North test [54]. The main results are shown in Table 2. It can be seen that the variance contribution rate of the first principal component was the largest, which was 50.48. The variance contribution rate of the second principal component decreased rapidly, and the variance contribution rate of the third and other principal components was small. The variance contribution rate of the first two principal components reached 70%. Therefore, we only gave the eigenvector fields corresponding to the first two principal components.

Table 2. The variance contribution rates and the cumulative variance contribution rates of six principle components.

Principal Component	Variance Contribution Rate (%)	Cumulative Variance Contribution Rate (%)
1	50.48	50.48
2	18.57	69.05
3	8.05	77.10
4	5.23	82.33
5	3.72	86.05
6	1.53	87.58

Figure 6a shows the first eigenvector field and all values are positive. This means that the increase or decrease of plum rainfall is consistent in the entire region. That is, if the value is positive, the plum rainfall in the entire study region will increase, and, if the value is negative, the plum rainfall in the entire study region will decrease, so it can be said that it is affected by the large-scale atmospheric circulation. The high value was mainly distributed in the northern part of the region, which indicates that it was the region with the highest variability of plum rainfall and it was the most sensitive region of drought and flood. The low value was mainly distributed in the southwest part of this region, which indicates that it was the region with a low variability of plum rainfall. These results illustrate that the plum rainfall is also affected by local factors. The related temporal coefficient curve (Figure 6c) mostly depicts variations on 4-year to 6-year time scales, which are similar to the EEMD result. Results of EOF and EEMD can be verified with each other. The results of EOF can prove the rationality of EEMD, and the results of EEMD can also prove the rationality of EOF. In addition, the time coefficient in 1996 was the largest and positive, which indicates that the plum rainfall in this year was the highest. This was a typical rainy year, followed by 1982 and 1991. The time coefficients in 1965, 1971, 1978, and 2006 were the smallest and negative, which indicates that they were the typical years of little plum rainfall.

It can be seen from Figure 6b that there was a large difference between the first eigenvector field and the second eigenvector field. From Figure 6b, we can see that the positive values were distributed in the north of the region, while the negative values were distributed in the south of the region. This indicates that the plum rainfall showed an out-of-phase pattern of about 30° N between the north and south area. This means that there is less plum rainfall in the south when there is more plum rainfall in the north and there is more plum rainfall in the south when there is less plum rainfall in the north. The reason for this phenomenon is mainly due to the different location of the Plum Rains' belt every year. When the Plum Rains' belt stays in the south of the Yangtze River for a long time, there is a lot of plum rainfall in the south. When the Plum Rains' belt stays in the Jiang-Huai area for a long time, there is more plum rainfall in the north. Figure 6d shows that the time coefficient in 1991 was the largest and positive, which meant that it was the typical year with more plum rainfall in the north and less plum rainfall in the south, which was followed by 1996 and 2003. The time coefficient in 1992 was the smallest and negative, which meant that it was the typical year with more plum rainfall in the south and less plum rainfall in the north.

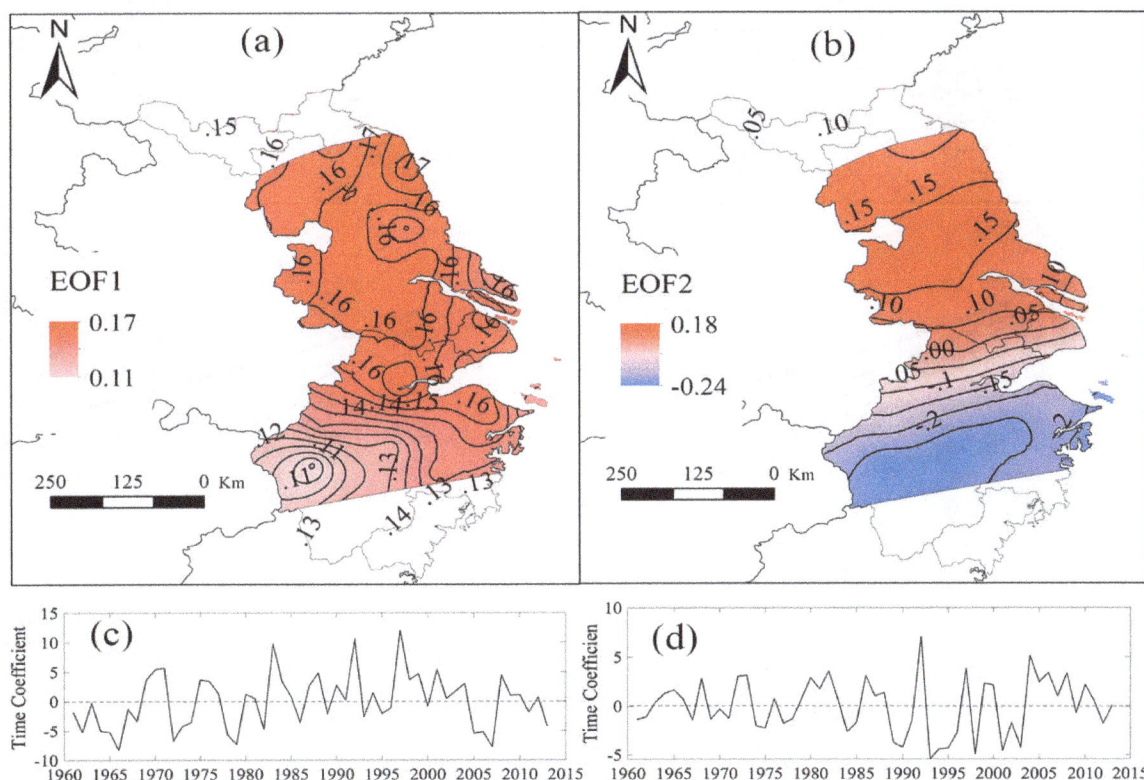

Figure 6. The spatial patterns of the two leading EOF modes of plum rainfall: (**a**) EOF 1, (**b**) EOF 2, and the corresponding (**c,d**) normalized temporal coefficients.

3.3. The Relation among EASM, ENSO, PDO, and Plum Rainfall

The variations of plum rainfall are complex and have a direct relationship with atmospheric circulation. Previous studies have analyzed the influencing factors of plum rainfall on the whole [55–57], while ignoring the relationship between them in detail. Therefore, we have tried to explore the impact of EASM, ENSO, and PDO on plum rainfall from multi-time scales. Table 3 shows the EEMD results for plum rainfall, EASM, ENSO, and PDO. There were four IMF components for each index and they all had similar periods on both the inter-annual scale and the inter-decadal scale. This means that there is a relationship between EASM, ENSO, PDO, and plum rainfall, but the specific relationship between them at different time scales still needs to be discussed. However, the fourth IMF component of plum rainfall was different from EASM and ENSO, and the period of EASM and ENSO was longer than the period of plum rainfall. The period of PDO was shorter than the period of plum rainfall.

We performed significance tests to determine whether each component of EASM, ENSO, and PDO is a physically meaningful signal, or the result of simple noise. In Figure 7, IMF2 and IMF4 of EASM fell above the confidence level of 90%, while IMF1 and IMF3 fell below the confidence level of 90%, which meant that IMF2 and IMF4 contained more physical meaning and IMF1 and IMF3 contained more white noise. IMF1 and IMF4 of ENSO fell above the confidence level of 90%, while IMF2 and IMF3 of ENSO fell below the confidence level of 80%, which meant that IMF1 and IMF4 contained more physical meaning while IMF2 and IMF3 contained less physical meaning. IMF1, IMF2, and IMF4 of PDO fell above the confidence level of 90% and IMF3 of PDO fell between the confidence level of 80% and 90%. This meant that IMF1, IMF2, and IMF4 comprised a more definite physical meaning, and IMF3 contained a less physical meaning. Significant IMFs are shown in bold in Table 3.

Table 3. The EEMD results of the plum rainfall, EASM, ENSO, and PDO (The numbers in the table represent the cycle of the plum rainfall, EASM, ENSO, and PDO on the four components (IMF1-4), respectively).

	Cycle (Year)			
	IMF1	**IMF2**	**IMF3**	**IMF4**
Plum rainfall	**3**	**6**	**14**	33
EASM	3	**6**	14	**48**
ENSO	**4**	6	11	**49**
PDO	**4**	**7**	14	**28**

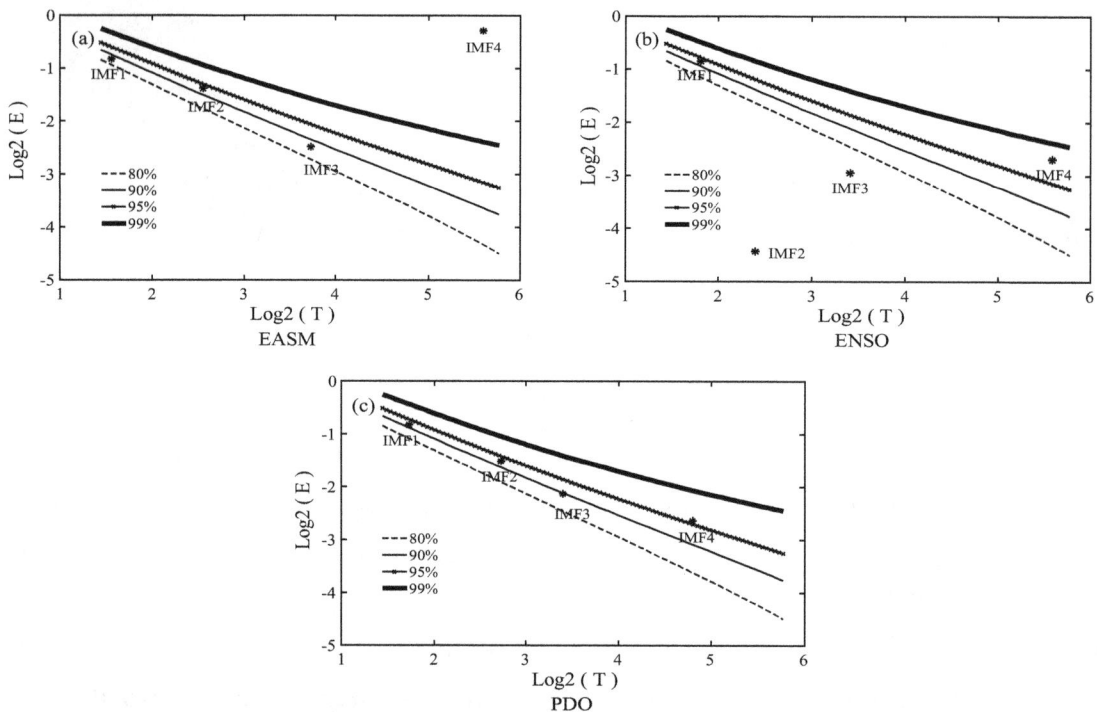

Figure 7. Significance test for the IMFs of (**a**) EASM, (**b**) ENSO, and (**c**) PDO during the period of 1960 to 2012.

Then, the correlation analysis was applied in order to understand the relation between plum rainfall and EASM, ENSO, and PDO more clearly. We first obtained the components of plum rainfall, EASM, ENSO, and PDO on four-time scales, and then calculated the correlation of plum rainfall and EASM, ENSO, and PDO on each time scale, respectively. The results are shown in Table 4. Significant correlations are shown in bold. On the inter-annual scale, IMF2 of plum rainfall had a negative and significant correlation with IMF2 of EASM. The result was similar to the results of some previous studies [58–60], but the difference was that they did not discuss the relation between plum rainfall and EASM on multi-time scales. Our result indicated that EASM had a significant impact on plum rainfall on an interannual scale rather than on the entire time scale. The correlation between IMF1 of plum rainfall and IMF1 of ENSO was positive and significant, which indicated that ENSO had a significant influence on plum rainfall in a three-year periodic fluctuation. On the inter-decadal scale, although all components of plum rainfall had some association with the components of EASM and

ENSO, neither of them passed the significance test. It can be said that EASM and ENSO were not the main influencing factors of plum rainfall on the inter-decadal scale. The correlation between IMF4 of plum rainfall and IMF4 of PDO was positive, but the correlation between IMF3 of plum rainfall and IMF3 of PDO was negative. From the above results, we know that IMF4 of plum rainfall and IMF3 of PDO were not significant. Therefore, we cannot say that PDO was the main influencing factor of plum rainfall on the inter-decadal scale during the period of 1960 to 2012, and the relationship between plum rainfall and PDO needs to be further confirmed. This study gives a simple correlation between them. In summary, ENSO and EASM were the main influencing factors in the three-year and six-year periods, respectively, and the main influencing factors on the inter-decadal scale remains to be further studied. This means that on the time scale of three years, the plum rainfall is mainly affected by ENSO. On the time scale of six years, the plum rainfall is mainly affected by EASM. Therefore, it is necessary to separate the diverse time scales when forecast and climate changes are studied.

Table 4. The Pearson correlation coefficients between plum rainfall and EASM, ENSO, and PDO (1960 to 2012).

	IMF1	IMF2	IMF3	IMF4
EASM	0.05 (0.37)	**−0.18** **(0.09)**	−0.18 (0.10)	0.11 (0.21)
ENSO	**0.24** **(0.09)**	0.04 (0.78)	0.16 (0.26)	−0.00 (0.98)
PDO	0.07 (0.60)	−0.01 (0.96)	**−0.24** **(0.08)**	**0.24** **(0.08)**

Notes: The value in parentheses indicates the p value.

4. Discussion

This study first investigated the multi-time scale temporal variation of plum rainfall in the YRD by using the EEMD method and then analyzed the spatial pattern of plum rainfall by using the EOF method. Lastly, we showed the relationship between plum rainfall and EASM, ENSO, and PDO. We found that, in the past 53 years, plum rainfall has three-year and six-year cycles on the inter-annual scale and 14-year and 33-year cycles on the inter-decadal scale. The plum rainfall variability was dominated by the inter-annual scale. In addition, the first EOF showed in-phase changes of plum rainfall across the whole study region and the second EOF showed an out-of-phase plum rainfall pattern between the north and south areas of about 30° N. The changes of plum rainfall were affected by ENSO and EASM in the three-year and six-year periods, respectively. Whether PDO was the main influencing factor on the inter-decadal scale remains to be further studied.

Few studies have analyzed variations of the Plum Rains from multi-time scales, and most of them have used the wavelet analysis method and rarely used the EEMD method [17,26]. It has been proven that the EEMD method is superior to the wavelet analysis method [51,52]. Our study used the EEMD method to analyze the multi-time-scale variations of plum rainfall, the multi-time-scale variations of the onset of Plum Rains, the termination of Plum Rains, and the intensity of plum rainfall, which can, more accurately, identify their variation periods. We found that plum rainfall had three-year and six-year cycles on the inter-annual scale and 14-year and 33-year cycles on the inter-decadal scale. IMF1, IMF2, and IMF3 all passed the significance test. However, IMF4 of plum rainfall was not significant, even though it accounted for a large fraction of the variance, which meant that it was not possible to say whether there was a 33-year cycle in plum rainfall. Wei and Xie [17] used the wavelet analysis method to analyze the multi-time scales variations of plum rainfall, and they found that the plum rainfall had cycles of 2 to 3 years, 6 to 7 years, 23 to 24 years, and 36 to 37 years in the entire Yangtze-Huaihe River region. Zhang et al. [26] pointed out that plum rainfall existed for about a three-year to six-year short cycle and 11-year and 20-year-long cycles by using the wavelet analysis method in the entire Yangtze-Huaihe River region. Bai et al. [18] showed that the plum rainfall had three-year, six-year,

13-year, and 24-year cycles in the Middle-Lower Reaches of the Yangtze River. Compared with previous studies, we found that plum rainfall has similar variations on the interannual scale, but the variation on the interdecadal scale is very large in different regions. This is mainly due to the fact that, on the interdecadal scale, the influence of various factors on plum rainfall in different regions is more complicated, which illustrates the importance of regional differences. In addition, we also found that the onset of Plum Rains had three-year, six-year, 14-year, and 49-year cycles, but six-year and 14-year cycles were not significant, even though they accounted for a large fraction of the variance, which meant that the onset of Plum Rains only had significant three-year and 49-year cycles. Similarly, the termination of Plum Rains had significant three-year, six-year, and nine-year cycles (IMF4 is not significant), and the intensity of plum rainfall had significant three-year, five-year, and 27-year cycles (IMF4 is not significant). It can be seen that the onset of Plum Rains, the termination of Plum Rains, and the intensity of plum rainfall had similar cycles with plum rainfall, which reveals their relationship with plum rainfall on multi-time scales. Previous studies have analyzed the relationship between the onset of Plum Rains, the termination of Plum Rains, and the intensity of plum rainfall as well as plum rainfall from the perspective of the linear trend [61,62]. Prior studies did not discuss it from multi-time scales. We can see from the above results that variations of the onset of Plum Rains, the termination of Plum Rains, and the intensity of plum rainfall are nonlinear, so it is necessary to explore their relationship at multi-time scales.

As far as the spatial pattern of plum rainfall is concerned, Chen and Li [63] analyzed the spatial pattern in the Yangtze-Huaihe River region during the period of 1960 to 2012 using the EOF method. They found that there were four modes, the first of which was characterized by the simultaneous increase or decrease of the plum rainfall in the whole region, and the second of which was characterized by more plum rainfall in the north (south) and less plum rainfall in the south (north). Our result was consistent with this. However, the difference was that there were two other local spatial patterns in their study. It may be because our study region is small, which cannot recognize the smaller spatial pattern. Li et al. [64] found a different spatial pattern of plum rainfall. They pointed out that the first mode of plum rainfall showed a spatial pattern of more (less) plum rainfall in the south and less (more) plum rainfall in the north, and the second mode showed a spatial pattern of more (less) plum rainfall in the middle and less (more) plum rainfall on both sides, which was different from our results. This is mainly due to regional differences and differences in research times. Therefore, it can be seen that, although the Plum Rains season is formed by some common factors, it does not mean that the plum rainfall has completely consistent variation. Therefore, it is necessary to study the spatiotemporal variations of plum rainfall in different sub-regions. The sub-regional analysis of the spatiotemporal variations of the plum rainfall is meaningful to formulate agricultural and industrial development policies, according to local conditions. In addition, it is worth mentioning that our study only analyzes an average state for a period of 53 years, and does not discuss the spatial pattern of plum rainfall in different time periods, which needs to be further studied in the future.

In addition, we also assessed the relationship between ENSO, EASM, PDO, and plum rainfall on multi-time scales. Many studies have explored the effects of atmospheric circulation on Plum Rains from statistical methods or physics-based methods [58–60]. The difference between our study and previous studies was that previous studies studied the relationship between plum rainfall and EASM, ENSO, and PDO from the average state, and we explored the relationship between ENSO, EASM, PDO, and plum rainfall from multi-time scales. It was also found that ENSO and EASM are the main influencing factors in the three-year and six-year periods, respectively. In addition, the correlation between IMF3 of plum rainfall and IMF3 of PDO and the correlation between IMF4 of plum rainfall and IMF4 of PDO were significant. However, the IMF4 of plum rainfall and IMF3 of PDO were not significant. Therefore, we cannot say that PDO was the main influencing factor of plum rainfall during the period of 1960 to 2012. Jiang and Gao [61] pointed out that the amount of plum rainfall would increase when PDO was in a positive phase compared to a normal year. In this study, we did not get the same results, and the difference is that Jiang and Gao did not study the relationship between plum rainfall and PDO on multi-time scales, as we did. Due to the short observational record in this study, we will study the relationship between plum rainfall and PDO at multi-time scales for longer observational records. Whether PDO is the main influencing factor of plum rainfall on the inter-decadal scale remains to be further studied. This study builds on previous ones by exploring different aspects of the Plum Rains onset and withdrawal variability. However, this study has only analyzed the relationship between ENSO, EASM, PDO, and plum rainfall at multi-time scales by using statistical methods and has not explained the reasons for the different relationships at different time scales in terms of the physical mechanism, which requires further research.

5. Conclusions

The spatiotemporal variation of plum rainfall and its relation to EASM, ENSO, and PDO was investigated using a dataset from 33 meteorological stations on the onset and termination of Plum Rains as well as daily precipitation during the period of 1960 to 2012. An integrated approach combining the EEMD, EOF, and correlation analysis was employed. The main conclusions are reproduced below.

(1)   By analyzing the time series at the onset and termination of Plum Rains, the daily intensity of plum rainfall, and plum rainfall, we found that, in the past 53 years, the plum rainfall had three-year and six-year cycles on the inter-annual scale and 14-year and 33-year cycles on the inter-decadal scale. Additionally, it showed a trend of increasing first and then decreasing. In addition, the onset of Plum Rains, the termination of Plum Rains, and the daily intensity of plum rainfall had similar cycles and trends to plum rainfall. Furthermore, the effect of the onset and termination of Plum Rains and the daily intensity of plum rainfall on plum rainfall on the inter-annual scale was greater than on the inter-decadal scale.

(2)   There are two obvious characteristics of the spatial pattern of plum rainfall. The first EOF was characterized by in-phase changes across the whole study region. The second EOF showed an out-of-phase plum rainfall pattern of about 30° N between the areas north and south. It is necessary to analyze the spatiotemporal variations of plum rainfall in different sub-regions.

(3)   The changes in plum rainfall were affected by different factors on different time scales. The difference was that ENSO and EASM were the main influencing factors in the three-year and six-year periods, respectively, and whether PDO was the main influencing factor on the inter-decadal scale remains to be further studied. Studying the influence of atmospheric circulation on plum rainfall at multi-time scales can give us a deeper understanding of it.

Author Contributions: N.Z. designed, carried out the analysis, and wrote the manuscript. J.X. revised the paper and refined the results, conclusion, and abstract. K.L. and C.Z. discussed the results. Y.L. edited the figures. All authors approved the manuscript.

Acknowledgments: The authors are grateful to the Resource and Environmental Science Data Center (http://www. resdc.cn) of the Chinese Academy of Sciences and the China Meteorological Data Sharing Service System (http://cdc.cma.gov.cn/) for providing data. The authors appreciate the insightful comments of anonymous reviewers.

References

1.   Ninomiya, K. Characteristics of Baiu front as a predominant subtropical front in the summer Northern Hemisphere. *J. Meteor. Soc. Jpn.* **1984**, *62*, 880–894. [CrossRef]

2.   Akiyama, T. The large-scale aspects of the characteristic features of the Baiu front. *Meteor. Geophys.* **1973**, *24*, 157–188. [CrossRef]

3.   Kodama, Y.-M. Large-scale common features of subtropical precipitation zones (the Baiu frontal zone, the SPCZ, and the SACZ). Part I: Characteristics of subtropical frontal zones. *J. Meteorol. Soc. Jpn.* **1992**, *70*, 813–836. [CrossRef]

4.   Sampe, T.; Xie, S.P. Large-scale dynamics of the Meiyu-Baiu rainband: Environmental forcing by the westerly jet. *J. Climatol.* **2010**, *23*, 113–134. [CrossRef]

5.   Huang, Z.; Xu, H.; Hu, J. Review and Discussion on the Plum Rain Research in China. *Meteorol. Environ. Res.* **2011**, *2*, 25–29. [CrossRef]

6.   You, C.H.; Lee, D.I.; Jang, S.M.; Jang, M.; Uyeda, H.; Shinoda, T.; Kobayashi, F. Characteristics of rainfall systems accompanied with Changma front at Chujado in Korea. *Asia-Pac. J. Atmos. Sci.* **2010**, *46*, 41–51. [CrossRef]

7.   Zhu, K.Z. The southeast summer monsoon and precipitation of China. *J. Geogr. Sci.* **1934**, *1*, 1–27.

8.   Tu, C.W. Chinese air mass properties. *Q. J. R. Meteorol. Soc.* **1939**, *65*, 33–51. [CrossRef]

9.   Lu, A. Precipitation in the South Chinese-Tibetan Borderland. *Geogr. Rev.* **1947**, *37*, 88–93. [CrossRef]

10.   Kang, I.S.; Ho, C.H.; Lim, Y.K. Principle modes of climatological seasonal and intraseasonal variations of the Asian summer monsoon. *Mon. Weather Rev.* **1999**, *127*, 322–340. [CrossRef]

11.   Zhou, H. Study on the mesoscale structure of the heavy rainfall on Meiyu front with dual-Doppler RADAR. *Atmos. Res.* **2009**, *93*, 335–357. [CrossRef]

12.   Huang, D.Q.; Qian, Y.F.; Zhu, J. The heterogeneity of Meiyu rainfall over Yangtze-Huaihe River vally and its relationship with oceanic surface heating and intraseasonal variability. *Theor. Appl. Climatol.* **2012**, *108*, 601–611. [CrossRef]

13.   Gao, Q.; Sun, Y.; You, Q. The northward shift of Meiyu rain belt and its possible association with rainfall intensity changes and the Pacific-Japan pattern. *Dyn. Atmos. Ocean.* **2016**, *76*, 52–62. [CrossRef]

14.   Matsumoto, J. Heavy rainfall over East Asia. *Int. J. Climatol.* **1988**, *9*, 407–423. [CrossRef]

15.   Lee, D.K.; Kim, Y.A. Variability of the East Asian summer monsoon during the period of 1980–1989. *J. Korean Meteorol. Soc.* **1992**, *28*, 315–331.

16.   Chen, C. Investigation of a heavy rainfall event over southwestern Taiwan associated with a sub-synoptic cyclone during the 2003 Mei-Yu season. *Atmos. Res.* **2010**, *95*, 235–254. [CrossRef]

17.   Wei, F.Y.; Xie, Y. Interannual and interdecadal oscillations of Meiyu over the middle-lower reaches of the Changjiang River for 1885–2000. *J. Appl. Meteorol. Climtol.* **2005**, *16*, 492–499. [CrossRef]

18.   Bai, L.; Chen, Z.S.; Zhao, B.F. Application of Ensemble Empirical Mode Decomposition Method in Multiscale analysis of Meiyu in Middle-Lower Reaches of Yangtze River. *Resour. Environ. Yangtze Basin* **2015**, *24*, 482–488. [CrossRef]

19.   Jiang, J.Y.; Ni, Y.Q. Diagnostic study on the structural characteristics of a typical Meiyu front system and its maintenance mechanism. *Adv. Atmos. Sci.* **2004**, *21*, 802–813. [CrossRef]

20.   Shinoda, T.; Uyeda, H.; Yoshimura, K. Structure of moist layer and sources of water over the southern region far from the Meiyu/Baiu front. *J. Meteorol. Soc. Jpn.* **2005**, *83*, 137–152. [CrossRef]

21.   Zhou, X.J.; Zha, P.; Liu, G.; Zhou, T.J. Characteristics of decadal-centennial-scale changes in East Asian summer monsoon circulation and precipitation during the medieval warm period and little ice age and in the present day. *Chin. Sci. Bull.* **2011**, *56*, 3003–3011. [CrossRef]

22.   Liu, D.N.; He, J.H.; Yao, Y.H.; Qi, L. Characteristics and Evolution of Atmospheric Circulation Patterns during Meiyu over the Jianghuai valley. *Asia-Pac. J. Atmos. Sci.* **2012**, *48*, 145–152. [CrossRef]

23.   Wu, Y.T.; Shaw, T.A. The Impact of the Asian Summer Monsoon Circulation on the Tropopause. *J. Climatol.* **2016**, *8*, 8689–8701. [CrossRef]

24.   Yu, J.; Huang, X.; Yu, Y.; Guo, X.; Dong, B.; Lu, T.; Wang, H. Analysis on the new index of plum rains intensity and its spatio-temporal characteristics: A case study on the reaches of the region along Huaihe River in Anhui Province. *Sci. Agric. Sin.* **2009**, *42*, 1325–1330.

25.   Chen, S.J.; Kuo, Y.W.; Wang, W.; Tao, Z.Y.; Cui, B. A Modeling Case Study of Heavy Rainstorms along the Mei-Yu Front. *Mon. Weather Rev.* **1998**, *126*, 2330. [CrossRef]

26.   Zhang, Y.; Zhai, P.; Qian, Y. Variations of Meiyu indicators in the Yangtze-Huaihe River basin during 1954–2003. *Acta Meteorol. Sin.* **2005**, *19*, 479–484.

27.   Zhu, X.; Wu, Z.; He, J. Anomalous Meiyu onset averaged over the Yangtze River valley. *Theor. Appl. Climatol.* **2008**, *94*, 81–95. [CrossRef]

28.   Zhu, J.; Huang, D.Q.; Zhang, Y.C.; Huang, A.N.; Kuang, X.Y.; Huang, Y. Decadal changes of Meiyu rainfall around 1991 and its relationship with two types of ENSO. *J. Geophys Res. Atmos.* **2013**, *118*, 9766–9777. [CrossRef]

29. Xue, F.; Liu, C.Z. The influence of moderate ENSO on summer rainfall in eastern China and its comparison with strong ENSO. *Chin. Sci. Bull.* **2008**, *53*, 791–800. [CrossRef]

30. Wang, H.; Yao, J.Q.; Shi, C.H.; Chen, B.M. The relationship between Meiyu and the intensity of East Asian Summer Monsoon. *Plateau Meteorol.* **2008**, *27*, 109–117.

31. Zhang, N.; Gao, Z.; Wang, X.; Chen, Y. Modeling the impact of urbanization on the local and regional climate in Yangtze River Delta, China. *Theor. Appl. Climatol.* **2010**, *102*, 331–342. [CrossRef]

32. Feng, Z.; Jin, M.; Zhang, F.; Huang, Y. Effects of ground-level ozone (O3) pollution on the yields of rice and winter wheat in the Yangtze River Delta. *J. Environ. Sci. China* **2003**, *15*, 360–362. [CrossRef]

33. Zhang, Q.; Gemmer, M.; Chen, J. Climate changes and flood/drought risk in the Yangtze Delta, China, during the past millennium. *Quat. Int.* **2008**, *176*, 62–69. [CrossRef]

34. Wang, Y.; Xu, Y.; Lei, C.; Li, G.; Han, L.; Song, S.; Yang, L.; Deng, X. Spatio-temporal characteristics of precipitation and dryness/wetness in Yangtze River Delta, eastern China, during 1960–2012. *Atmos. Res.* **2016**, *172*, 196–205. [CrossRef]

35. Chen, W.; Cutter, S.L.; Emrivh, C.T.; Shi, P. Measuring social vulnerability to natural hazards in the Yangtze River Delta region, China. *Int. J. Disaster Risk Sci.* **2013**, *4*, 169–181. [CrossRef]

36. Zhu, N.; Xu, J.; Li, W.; Li, K.; Zhou, C. A Comprehensive Approach to Assess the Hydrological Drought of Inland River Basin in Northwest China. *Atmosphere* **2018**, *9*, 370. [CrossRef]

37. The Sixth Census Commission. *The Main Data Bulletin for the Sixth National Census in 2010 (No. 1)*; Sixth Census Commission: Beijing, China, 2010.

38. General Administration of Quality Supervision, Inspection and Quarantine of the People's Republic of China and China National Standardization Management Committee. *Mei-Yu Monitoring Indicators*; Standards Press of China: Beijing, China, 2017.

39. Li, J.P.; Zeng, Q.C. A new monsoon index and the geographical distribution of the global monsoons. *Adv. Atmos. Sci.* **2003**, *20*, 299–302. [CrossRef]

40. Li, J.P.; Zeng, Q.C. A unified monsoon index. *Geophys. Res. Lett.* **2002**, *29*, 1274. [CrossRef]

41. Chu, W.; Qiu, S.; Xu, J. Temperature change of Shanghai and its response to global warming and urbanization. *Atmosphere* **2016**, *7*, 114. [CrossRef]

42. Xu, J.; Chen, Y.; Bai, L.; Xu, Y. A hybrid model to simulate the annual runoff of the Kaidu River in northwest China. *Hydrol. Earth Syst. Sci.* **2016**, *20*, 1447–1457. [CrossRef]

43. Bai, L.; Xu, J.; Chen, Z.; Li, W.; Liu, Z.; Zhao, B.; Wang, Z. The regional features of temperature variation trends over Xinjiang in China by the ensemble empirical mode decomposition method. *Int. J. Climatol.* **2015**, *35*, 3229–3237. [CrossRef]

44. Wu, Z.H.; Huang, N.E. Ensemble empirical mode decomposition: A noise-assisted data analysis method. *Adv. Adap. Data Anal.* **2009**, *1*, 1–41. [CrossRef]

45. Huang, N.E.; Shen, Z.; Long, S.R.; Wu, M.C.; Shih, H.H.; Zheng, Q.; Yen, N.C.; Tung, C.C.; Liu, H.H. The empirical mode decomposition and the Hilbert spectrum for nonlinear and non-stationary time series analysis. Proceedings of the Royal Society of London. Series A: Mathematical. *Phys. Eng. Sci.* **1998**, *454*, 903–995. [CrossRef]

46. Zhu, N.; Xu, J.; Wang, C.; Chen, Z.; Luo, Y. Modeling the multiple time scale response of hydrological drought to climate change in the data-scarce inland river basin of Northwest China. *Arab. J. Geosci.* **2019**, *12*, 225. [CrossRef]

47. Xoplaki, E.; Luterbacher, J.; Burkard, R.; Patrikas, I.; Maheras, P. Connection between the large-scale 500 hPa geopotential height fields and precipitation over Greece during wintertime. *Climatol. Res.* **2000**, *14*, 129–146. [CrossRef]

48. Kikuchi, K.; Wang, B. Diurnal precipitation regimes in the global tropics. *J. Climate* **2008**, *21*, 2680–2696. [CrossRef]

49. Lorenz, E.N. Empirical orthogonal functions and statistical weather prediction. In *Technical Report*; Statistical Forecast Project Report 1; Dept. of Meteor. MIT: Cambridge, MA, USA, 1956; p. 49.

50. Fu, C. Potential impacts of human-induced land cover change on East Asia monsoon. *Glob. Planet. Chang.* **2003**, *37*, 219–229. [CrossRef]

51. Ayenu-Prah, A.Y.; Attoh-Okine, N.O. Comparative study of Hilbert-Huang transform, Fourier transform and wavelet transform in pavement profile analysis. *Veh. Syst. Dyn.* **2009**, *47*, 437–456. [CrossRef]

52. Alexandrov, T. A method of trend extraction using singular spectrum analysis. *Rev. Stat. J.* **2009**, *7*, 1–22.

53. He, J.H.; Wu, Z.W.; Jiang, Z.H.; Miao, C.S.; Han, G.R. "Climate effect" of the northeast cold vortex and its influences on Meiyu. *Chin. Sci. Bull.* **2007**, *52*, 671–679. [CrossRef]

54. North, G.R.; Bell, T.L.; Cahalan, R.F. Sampling errors in the estimation of the empirical orthogonal functions. *Mon. Weather Rev.* **1982**, *110*, 699–706. [CrossRef]

55. Hao, Z.; Zheng, J.; Ge, Q. Variations in the summer monsoon rainbands across eastern China over the past 300 years. *Adv. Atmos. Sci.* **2009**, *26*, 614–620. [CrossRef]

56. Gao, H.; Jiang, W.; Li, W.J. Changed Relationships Between the East Asian Summer Monsoon Circulations and the Summer Rainfall in Eastern China. *Acta Meteorol. Sin. (Engl. Ed.)* **2014**, *28*, 1075–1084. [CrossRef]

57. Gong, D.; Wang, S. Impacts of ENSO on rainfall of global land and China. *Chin. Sci. Bull.* **1999**, *44*, 852–857. [CrossRef]

58. Liu, C.Z.; Wang, H.J.; Jiang, D.B. The Configurable Relationships between Summer Monsoon and Precipitation over East Asia. *Chin. J. Atmos. Sci.* **2004**, *28*, 700–712.

59. Ge, Q.S.; Guo, X.F.; Zheng, J.Y.; He, Z.X. Meiyu in the middle and lower reaches of the Yangtze River since 1736. *Chin. Sci. Bull.* **2007**, *52*, 2792–2797. [CrossRef]

60. Hu, C.S.; Xu, Y.P.; Han, L.F.; Yang, L. Long-term trends in daily precipitation over the Yangtze River Delta region during 1960–2012, Eastern China. *Theor. Appl. Climatol.* **2016**, *125*, 131–147. [CrossRef]

61. Jiang, W.; Gao, H. New features of Meiyu over middle-lower reaches of Yangtze River in the 21st century and the possible causes. *Meteorol. Mon.* **2013**, *39*, 1139–1144.

62. Zhao, J.; Chen, L.; Xiong, K. Climate characteristics and influential systems of Meiyu to the south of the Yangtze River based on the new monitoring rules. *Acta Meteorol. Sin.* **2018**, *76*, 680–698. [CrossRef]

63. Chen, X.; Li, D. The features of Meiyu under the new standard. *J. Meteorol. Sci.* **2016**, *36*, 165–175. [CrossRef]

64. Li, K.; Yu, J.; Wang, Y.; Song, J.; Zhuang, Y. Abnormal distribution of Meiyu precipitation over Jianghuai region and its relation with East Asia subtropical westerly jet. *J. Meteorol. Sci.* **2018**, *38*, 302–309. [CrossRef]

PERMISSIONS

LIST OF CONTRIBUTORS

Natalie J. Harvey and Helen F. Dacre
Department of Meteorology, University of Reading, Reading RG6 6BB, UK

Michael C. Cooke
Met Office, FitzRoy Road, Exeter EX1 3PB, UK

Helen N. Webster
Met Office, FitzRoy Road, Exeter EX1 3PB, UK
College of Engineering, Mathematics and Physical Sciences, University of Exeter, Exeter EX4 4QF, UK

Isabelle A. Taylor and Roy G. Grainger
COMET, Sub-Department of Atmospheric, Oceanic and Planetary Physics, University of Oxford, Oxford OX1 3PU, UK

Sujan Khanal
Sub-Department of Atmospheric, Oceanic and Planetary Physics, University of Oxford, Oxford OX1 3PU, UK

Quan-Liang Chen
Plateau Atmosphere and Environment Key Laboratory of Sichuan Province, College of Atmospheric Science, Chengdu University of Information Technology, Chengdu 610225, China

Yan Wang
Plateau Atmosphere and Environment Key Laboratory of Sichuan Province, College of Atmospheric Science, Chengdu University of Information Technology, Chengdu 610225, China
State Key Laboratory of Severe Weather, Chinese Academy of Meteorological Sciences, Beijing 100081, China
Laboratory for Climate Studies & CMA-NJU Joint Laboratory for Climate Prediction Studies, National Climate Center, China Meteorological Administration, Beijing 100081, China

Hong-Li Ren
State Key Laboratory of Severe Weather, Chinese Academy of Meteorological Sciences, Beijing 100081, China
Laboratory for Climate Studies & CMA-NJU Joint Laboratory for Climate Prediction Studies, National Climate Center, China Meteorological Administration, Beijing 100081, China

Jie Wu, Wei-Hua Jie and Pei-Qun Zhang
Laboratory for Climate Studies & CMA-NJU Joint Laboratory for Climate Prediction Studies, National Climate Center, China Meteorological Administration, Beijing 100081, China

Fang Zhou
Climate Change Research Center, Institute of Atmospheric Physics, and Nansen–Zhu International Research Centre, Chinese Academy of Sciences, Beijing 100029, China

Joshua-Xiouhua Fu
Department of Atmospheric and Oceanic Sciences & Institute of Atmospheric Sciences, Fudan University, Shanghai 200433, China

Trang Thi Kieu Tran and Taesam Lee
Department of Civil Engineering, ERI, Gyeongsang National University, 501 Jinju-daero, Jinju 660-701, Korea

Jong-Suk Kim
State Key Laboratory of Water Resources and Hydropower Engineering Science, Wuhan University, Wuhan 430072, China

Jose Manuel Jiménez-Gutiérrez and Francisco Valero
Departamento de Astrofísica y Ciencias de la Atmósfera, Universidad Complutense de Madrid, 28040 Madrid, Spain

Sonia Jerez and Juan Pedro Montávez
Departamento de Física, Universidad de Murcia, 30100 Murcia, Spain

Fan Song, Xiaohua Yang and Feifei Wu
State Key Laboratory of Water Environment Simulation, School of Environment, Beijing Normal University, Beijing 100875, China

Ventsislav Danchovski
Department of Meteorology and Geophysics, Faculty of Physics, Sofia University, Sofia 1164, Bulgaria

Zeeshan Javed, Chengzhi Xing and Haoran Liu
School of Earth and Space Sciences, University of Science and Technology of China, Hefei 230026, China

Cheng Liu
School of Earth and Space Sciences, University of Science and Technology of China, Hefei 230026, China
Key Lab of Environmental Optics & Technology, Anhui Institute of Optics and Fine Mechanics, Chinese Academy of Sciences, Hefei 230031, China
Center for Excellence in Regional Atmospheric Environment, Institute of Urban Environment, Chinese Academy of Sciences, Xiamen 361021, China
Anhui Province Key Laboratory of Polar Environment and Global Change, USTC, Hefei 230026, China

Wei Tan
Center for Excellence in Regional Atmospheric Environment, Institute of Urban Environment, Chinese Academy of Sciences, Xiamen 361021, China

Kalim Ullah
Department of Meteorology, COMSATS University Islamabad, Islamabad 44000, Pakistan

Mi Zhang, Wei Wang, Yongbo Hu and Shoudong Liu
Yale-NUIST Center on Atmospheric Environment, Nanjing University of Information, Science and Technology, Nanjing 210044, China

Jingzheng Xu
Radio Science Research Institute Inc., Wuxi 214073, China

Wenjing Huang and Wei Xiao
Yale-NUIST Center on Atmospheric Environment, Nanjing University of Information, Science and Technology, Nanjing 210044, China
NUIST-Wuxi Research Institute, Wuxi 214073, China

Cheng Hu
Yale-NUIST Center on Atmospheric Environment, Nanjing University of Information, Science and Technology, Nanjing 210044, China
Department of Soil, Water, and Climate, University of Minnesota-Twin Cities, St. Paul, MN 55108, USA

Xuhui Lee
Yale-NUIST Center on Atmospheric Environment, Nanjing University of Information, Science and Technology, Nanjing 210044, China
NUIST-Wuxi Research Institute, Wuxi 214073, China
School of Forestry and Environmental Studies, Yale University, New Haven, CT 06511, USA

Ying Zhang, Zhengqiang Li, Hua Xu, Kaitao Li, Donghui Li and Yisong Xie
State Environmental Protection Key Laboratory of Satellite Remote Sensing, Institute of Remote Sensing and Digital Earth, Chinese Academy of Sciences, Beijing 100101, China

Chi Zhang
State Environmental Protection Key Laboratory of Satellite Remote Sensing, Institute of Remote Sensing and Digital Earth, Chinese Academy of Sciences, Beijing 100101, China
University of Chinese Academy of Sciences, Beijing 100049, China

Yongqian Wang and Yang Zhang
College of Resources and Environment, Chengdu University of Information Technology, Chengdu 610103, China

Yifang Ren
The Center of Jiangsu Meteorological Service, Nanjing 21008, China

Jun A. Zhang
Cooperative Institute for Marine and Atmospheric Studies, University of Miami, Miami, FL 33149, USA
NOAA/AOML/Hurricane Research Division, Miami, FL 33149, USA

Stephen R. Guimond
Joint Center for Earth Systems Technology, University of Maryland Baltimore County, Baltimore, MD 21250, USA
NASA Goddard Space Flight Center (GSFC), Greenbelt, MD 20771, USA

Xiang Wang
Centre of Data Assimilation for Research and Application, Nanjing University of Information Science & Technology, Nanjing 210044, China

Qian Wang, Haiming Xu and Jiechun Deng
Key Laboratory of Meteorological Disaster, Ministry of Education (KLME)/Joint International Research Laboratory of Climate and Environment Change (ILCEC)/Collaborative Innovation Center on Forecast and Evaluation of Meteorological Disasters (CIC-FEMD), Nanjing University of Information Science & Technology, Nanjing 210044, China
School of Atmospheric Sciences, Nanjing University of Information Science & Technology, Nanjing 210044, China

Leying Zhang
Joint Innovation Center for Modern Forestry Studies, College of Biology and Environment, Nanjing Forestry University, Nanjing 210037, China

Yunying Li
College of Meteorology and Oceanography, National University of Defense Technology, Nanjing 211101, China

Cuicui Gao
College of Meteorology and Oceanography, National University of Defense Technology, Nanjing 211101, China
Shaoguan Meteorological Office, Shaoguan, Guangdong 512000, China

Haowei Chen
Shaoguan Meteorological Office, Shaoguan, Guangdong 512000, China

Yufei Zhao, Qiang Zhang, Xiaowei Jiang and Aixia Feng
National Meteorological Information Centre, China Meteorological Administration, Beijing 100081, China

Jianping Li
College of Global Change and Earth System Science, Beijing Normal University, Beijing 100875, China
Laboratory for Regional Oceanography and Numerical Modeling, Qingdao National Laboratory for Marine Science and Technology, Qingdao 266237, China

Miguel Ángel Hernández-Ceballos and Luca De Felice
European Commission, Joint Research Centre, 21027 Ispra, Italy

Nina Zhu, Jianhua Xu and Dongyang Yang
Key Laboratory of Geographic Information Science (Ministry of Education), East China Normal University, Shanghai 200241, China
Research Center for East-West Cooperation in China, East China Normal University, Shanghai 200241, China
School of Geographic Sciences, East China Normal University, Shanghai 200241, China

Kaiming Li
School of Urban and Geography, Lanzhou City College, Lanzhou 730070, China

Yang Luo
Jianhu Data Technology (Shanghai) Co, Ltd, Shanghai 201700, China

Cheng Zhou
Faculty of Tourism Management, Shanxi University of Finance & Economics, Taiyuan 030006, China

Index